D0205893

Nutrition is a major environmental factor in plant production, and is therefore of significant practical concern to ecologists and agriculturalists worldwide. In this book, the role of nutrition in regulating plant growth is explored. Case studies are used to illustrate the practical implications of the interaction between plant and environment for crop and resource management.

This book will be of interest to graduate students and researchers of agriculture, horticulture, forestry and ecology who are concerned with the complex ways in which plants interact with their environments.

SOCIETY FOR EXPERIMENTAL BIOLOGY
SEMINAR SERIES: 43

PLANT GROWTH: *interactions with nutrition and environment*

SOCIETY FOR EXPERIMENTAL BIOLOGY SEMINAR SERIES

A series of multi-author volumes developed from seminars held by the Society for Experimental Biology. Each volume serves not only as an introductory review of a specific topic, but also introduces the reader to experimental evidence to support the theories and principles discussed, and points the way to new research.

1. Effects of air pollution on plants. *Edited by T.A. Manfield*
2. Effects of pollutants on aquatic organisms. *Edited by A.P.M. Lockwood*
3. Analytical and quantitative methods. *Edited by J.A. Meek and H.Y. Elder*
4. Isolation of plant growth substances. *Edited by J.R. Hillman*
5. Aspects of animal movement. *Edited by H.Y. Elder and E.R. Trueman*
6. Neurones without impulses: their significance for vertebrate and invertebrate systems. *Edited by A. Roberts and B.M.H. Bush*
7. Development and specialisation of skeletal muscle. *Edited by D.F. Goldspink*
8. Stomatal physiology. *Edited by P.G. Jarvis and T.A. Mansfield*
9. Brain mechanisms of behaviour in lower vertebrates. *Edited by P.R. Laming*
10. The cell cycle. *Edited by P.C.L. John*
11. Effects of disease on the physiology of the growing plant. *Edited by P.G. Ayres*
12. Biology of the chemotactic response. *Edited by J.M. Lackie and P.C. Williamson*
13. Animal migration. *Edited by D.J. Aidley*
14. Biological timekeeping. *Edited by J. Brady*
15. The nucleolus. *Edited by E.G. Jordan and C.A. Cullis*
16. Gills. *Edited by D.F. Houlihan, J.C. Rankin and T.J. Shuttleworth*
17. Cellular acclimatisation to environmental change. *Edited by A.R. Cossins and P. Sheterline*
18. Plant biotechnology. *Edited by S.H. Mantell and H. Smith*
19. Storage carbohydrates in vascular plants. Edited by *D.H. Lewis*
20. The physiology and biochemistry of plant respiration. *Edited by J.M. Palmer*
21. Chloroplast biogenesis. *Edited by R.J. Ellis*
22. Instrumentation for environmental physiology. *Edited by B. Marshall and F.I. Woodward*
23. The biosynthesis and metabolism of plant hormones. *Edited by A. Crozier and J.R. Hillman*
24. Coordination of motor behaviour. *Edited by B.M.H. Bush and F. Clarac*
25. Cell ageing and cell death. *Edited by I. Davies and D.C. Sigee*
26 The cell division cycle in plants. Edited by *J.A. Bryant and D. Francis*
27. Control of leaf growth. *Edited by N.R. Baker, W.J. Davies and C. Ong*
28. Biochemistry of plant cell walls. *Edited by C.T. Brett and J.R. Hillman*
29. Immunology in plant science. *Edited by T.L. Wang*
30. Root development and function. *Edited by P.J. Gregory, J.V. Lake and D.A. Rose*
31. Plant canopies: their growth, form and function. *Edited by G. Russell, B. Marshall and P.G. Jarvis*
32. Developmental mutants in higher plants. *Edited by H. Thomas and D. Grierson*
33. Neurohormones in invertebrates. *Edited by M. Thorndyke and G. Goldsworthy*
34. Acid toxicity and aquatic animals. *Edited by R. Morris, E.W. Taylor, D.J.A. Brown and J.A. Brown*
35. Division and segregation of organelles. *Edited by S.A. Boffey and D. Lloyd*
36. Biomechanics in evolution. *Edited by J.M.V. Rayner*
37. Techniques in comparative respiratory physiology: An experimental approach. *Edited by C.R. Bridges and P.J. Butler*
38. Herbicides and plant metabolism. *Edited by A.D. Dodge*
39. Plants under stress. *Edited by H.G. Jones, T.J. Flowers and M.B. Jones*
40. *In situ* hybridisation: application to developmental biology and medicine. *Edited by N. Harris and D.G. Wilkinson*
41. Physiological strategies for gas exchange and metabolism. *Edited by A.J. Woakes, M.K. Grieshaber and C.R. Bridges*
42. Compartmentation of plant metabolism in non-photosynthetic tissues. *Edited by M.J. Emes*
43. Plant growth: interactions with nutrition and environment. *Edited by J.R. Porter and D.W. Lawlor*

PLANT GROWTH: *interactions with nutrition and environment*

Edited by

J.R. Porter

Research Fellow in Plant Physiology, University of Bristol

and

D.W. Lawlor

AFRC Institute of Arable Crops,
Rothamsted Experimental Station, Harpenden

CAMBRIDGE UNIVERSITY PRESS

Cambridge

New York   Port Chester   Melbourne   Sydney

Published by the Press Syndicate of the University of Cambridge
The Pitt Building, Trumpington Street, Cambridge CB2 1RP
40 West 20th Street, New York, NY 10011-4211, USA
10 Stamford Road, Oakleigh, Melbourne 3166, Australia

First published 1991

Printed in Great Britain at the University Press, Cambridge

British Library cataloguing in publication data

Plant growth.
1. Plants. Growth
I. Porter, J. R. (John R.) II. Lawlor, D. W. (David W.)
III. Series
581.31

Library of Congress cataloguing in publication data

Plant growth: interactions with nutrition and environment/edited by J. R. Porter and D. W. Lawlor.
p. cm. — (Seminar series/Society for Experimental Biology; 43)
Includes index.
ISBN 0–521–36133–8
1. Growth (Plants) 2. Plants—Nutrition. 3. Botany—Ecology.
I. Porter, J. R. (John R.) II. Lawlor, D. W. (David W.)
III. Series: Seminar series (Society for Experimental Biology (Great Britain)); 43.
QK731.P587 1991
581.3′1—dc20 90–15024 CIP

ISBN 0 521 36133 8 hardback

UP

CONTENTS

CONTRIBUTORS

FINK, S.
Institute of Biology, Eberhard-Karls University of Tuebingen, Auf der Morgenstelle 1, 7400 Tuebingen, Federal German Republic.
GRIME, J.P.
Unit of Comparative Plant Ecology (NERC), Dept. of Animal and Plant Sciences, The University, Sheffield S10 2TN, UK.
GROOT, J.J.R.
Institute for Soil Fertility, P.O. Box 30003, 9750 RA Haren, The Netherlands.
HUETTL, R.F.
Forestry Dept., Kali und Salz AG, P.O. Box 102029, 3500 Kassel, Federal German Republic.
LAWLOR, D.W.
AFRC Institute of Arable Crops Research, Rothamsted Experimental Station, Harpenden, Hertfordshire AL5 2JQ, UK.
LEIGH, R.A.
AFRC Institute of Arable Crops Research, Rothamsted Experimental Station, Harpenden, Hertfordshire AL5 2JQ, UK.
MARSCHNER, H.
Institute of Plant Nutrition, University of Hohenheim, P.O. Box 700562, 700 Stuttgart 70, Federal German Republic.
MARSHALL, B.
Scottish Crop Research Institute, Invergowrie, Dundee DD2 5DA, UK.
PORTER, J.R.
Dept. of Agricultural Sciences, University of Bristol, Long Ashton Research Station, Bristol BS18 9AF, UK.
RAO, I.M.
Dept. of Plant and Soil Biology, University of California, Berkeley, CA 94720, USA.

ROBINSON, D.
Scottish Crop Research Institute, Invergowrie, Dundee DD2 5DA, UK.
RORISON, I.H.
Unit of Comparative Plant Ecology (NERC), Dept. of Animal and Plant Sciences, The University, Sheffield S10 2TN, UK.
SPIERTZ, J.H.J.
Centre for Agrobiological Research, P.O Box 14, 6700 AA Wageningen, The Netherlands.
STEWART, G.R.
Dept. of Biology, University College London, Gower Street, London WC1E 6BT, UK.
STOREY, R.
CSIRO Division of Horticulture, Merbein, Victoria 3505, Australia.
TERRY, N.
Dept. of Plant and Soil Biology, University of California, Berkeley, CA 94720, USA.

PREFACE

I do not believe in things, I believe only in their relationships

Georges Braque, painter (1882–1963).

Interest in whole-plant physiology has ebbed in the last few years with the current intellectual tide of biology running with studies of processes at deeper, *circum*-molecular, levels of organisation. However, the Seminar Series, organised by the Environmental Physiology Group of the Society for Experimental Biology, has done much to illustrate that there are many exciting and relevant scientific problems at each level of biological organisation. It was with such an integrative approach in mind that the present volume of papers and the conference that preceded it were conceived. The conference was held during the April 1989 meeting of the Society for Experimental Biology at Edinburgh. All the invited speakers to the meeting have contributed chapters to this volume, some in collaboration with colleagues.

The fundamental question asked by the organisers was the extent to which we can understand the nature and scope of the tri-partite interaction between plant growth, nutrition and the aerial environment. Traditionally, whole-plant physiologists and ecologists have looked at the influence of above-ground conditions *or* nutrition on the growth and distribution of plants; it was our intention to replace '*or*' with '*and*' and to explore some of the complexity of the plant–nutrient–environment system.

In attempting this we have turned convention on its head by having papers on the conceptual models of processes not as epilogues but before those devoted to experimental analysis. We have also tried to be broad in including work from managed and natural eco-systems and processes with short-and-long response times.

We would like to record the organisers' thanks for financial and other support from the Society for Experimental Biology, the British Ecological Society, the Association of Applied Biologists and the Potash Development Association.

Finally, we thank the other members of the Organising Committee for their time in planning the meeting and help with the publication of this volume.

John Porter
Bristol

David Lawlor
Harpenden

D.W. LAWLOR

Concepts of nutrition in relation to cellular processes and environment

Introduction

Plants increase in mass and energy as they grow; carbon, mineral nutrients, absorbed in suitable form from the environment (Epstein, 1972; Marschner, 1986) and water provide the mass whilst energy is derived from the sun (Lawlor, 1987). As part of the growth process, cells with complex structures such as membranes and organelles are formed (Schnepf, 1983). This review aims to provide a conceptual framework of the relation between nutrition and cellular processes, of how the supply of chemical elements and energy are intimately related to the needs of the plants' cellular metabolism and thence to growth. One central theme is that the overall nutritional requirements of plants are determined by the biochemical and chemical components of cells and that these, in turn, are needed for biochemical function. Another theme is that the potential size of a plant and its organs is genetically determined (Strickberger, 1976) and thus the absolute amounts of nutrients required for growth are defined.

The growth process requires the synthesis of proteins, both structural and enzymatic, and it is a central tenet of biology that the production of proteins is genetically controlled by the information stored in DNA (Gerloff & Gabelman, 1983; Darnell, Lodish & Baltimore, 1986). This genetic control is strong, as shown by the fidelity with which the phenotypes of plant varieties and species are reproduced through generations (Strickberger, 1976). However, plant populations are genetically heterogeneous (Bradshaw, 1983; 1984) so variation occurs in phenotype, in structure and cellular and sub-cellular composition and in metabolic processes. The rates of growth and the size and composition of plants, as seen from this viewpoint, depend on the interaction between the genetic constitution, the needs of cellular metabolism and on the environmental supply of nutrients, energy and the conditions, such as temperature which determine growth. Genes regulate the rate of production of cellular components and therefore the potential rate of growth in different species

(Stebbins, 1971; Grime & Hunt, 1975; Bradshaw, 1983). Hence, the forms, potential amounts and potential rates of demand for the nutrients and energy required for growth are also genetically determined (Fig. 1). This *potential* growth rate (Grime & Hunt, 1975) will be attained when the supply of materials necessary for full expression of the genotype under particular conditions is equal to the genetically determined demand. The *actual* growth rate will fall below the potential rate when the above conditions are inadequate (Hunt & Lloyd, 1987) or exceed the ability of the plants' metabolism and physiological systems to cope so that damage ensues.

These concepts are not simply theoretical, they provide a framework for analysing how plants of different genotypes respond to their environments (both aerial and edaphic) and for understanding how nutrition interacts with other features of the environment in determining the growth of particular species. Any conceptual framework to improve analysis of the links between processes in the plant and its environment must combine with a quantitative understanding of the biological and physiological mechanisms by which growth occurs (Grime, 1991, this volume). Better understanding of such mechanisms and their genetic control will aid prediction of the form, amounts, timing and rates of supply of nutrients and also the other resources and conditions required by plants for growth in natural and managed ecosystems (Ågren, 1988). Such understanding is needed if the maximum efficiency of plant growth is to be achieved either by altering the environment or by changing the plant by genetic engineering (Gasser & Fraley, 1989). Similarly, minimising the damage to ecosystems caused by human activities (Huettl & Fink, 1991, this volume) depends on understanding how plants respond to the coincidental impacts of nutrition and their aerial and edaphic environment.

Nutrients and biochemical composition

Autotrophic plants absorb their chemical constituents from the environment either as simple molecules or ions of molecular mass less than 100 (Epstein, 1972); energy is derived from photons of solar radiation (Lawlor, 1987). Water is also essential but its absorption follows rather than determines growth, unless the supply is inadequate and water stress restricts and slows metabolism (Kramer, 1969). Therefore, the role of water is not further considered. Fig. 1 illustrates the main pathways for the synthesis of cellular components and structures, which perform the metabolic functions and determine the links to the environment. Plants are composed predominantly of H, C and O which are derived from CO_2 and H_2O incorporated via photosynthesis. Additionally, plants contain 13 mineral elements (Table 1) which are essential for metabolism, growth and

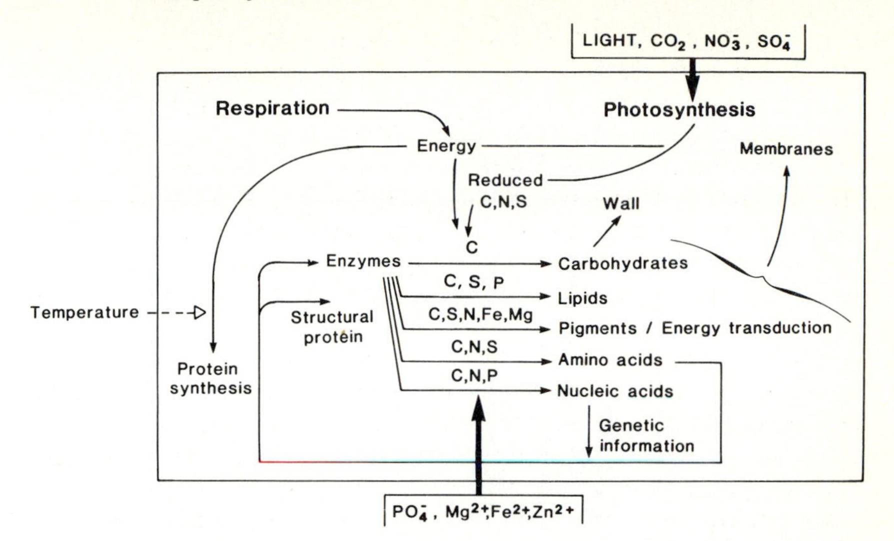

Fig. 1. Scheme of the interactions between the environment and plant metabolism. The synthesis of cell components and the role of the major environmental factors is illustrated to emphasise the cyclic nature of metabolic processes and how it is they generate structures which serve to capture nutrients and energy.

development and, ultimately for successful reproduction (Epstein, 1972; Rains, 1976; Marschner, 1986). These elements form either an integral part of the cellular structure (covalently or strongly bound) or are essential for providing the conditions needed for metabolism. An example of the former is nitrogen, which is a constituent of proteins; an example of the latter is potassium which is not covalently linked into organic molecules (Rains, 1976) but is essential for ionic and osmotic regulation in the cell (Leigh & Storey, 1991, this volume). Magnesium functions both as a constituent of molecules (e.g. chlorophyll) and in ionic regulation of enzyme activities.

Elements other than those in Table 1 may stimulate growth of particular species; Na is beneficial for sugar beet and Si for grasses but neither appear essential for growth. Substitution of one element is possible; for example, Na and K are partially interchangeable especially in halophytes. Also K may be replaced by Rb in regulation of stomatal activity (Rains, 1976). However, essential elements cannot be completely replaced by others, so growth depends on a supply of the required amounts of all essential components from the environment.

Mineral elements in plants are classified as macro- or micro- based on their proportion in tissues. There is a 10-fold difference in molar concentration per unit of dry matter between sulphur, the least abundant

Table 1. *Chemical elements essential for growth and their average content in material from cultivated higher plants (from Epstein, 1972), and approximate concentration in the environment.*

Element	Mass conc. (g dry matter^{-1})	Molar conc. (mmol (kg dry matter)$^{-1}$)	No. of atoms relative to Mo	Conc. in environment (mol m^{-3})	Examples of function in cell
Macro-					
H	60	60000	60×10^6	$1–111 \times 10^3$	Oxidation/reduction structure in organic molecules
C	450	35000	35×10^6	14×10^{-3}	Structure of cell organic molecules, oxidation/reduction
O	450	30000	30×10^6	14×10^{-3}	Structure of cell organic molecules, oxidation/reduction
N	15	1000	1×10^6	1–3	Constituent proteins, nucleic acids, regulation
K	10	250	2.5×10^5	1–2	Ionic and osmotic regulation, medium for synthetic processes, e.g. protein synthesis
Ca	5	125	1.25×10^5	0.5–1.5	Cell walls, pectinates, regulation of cell metabolism, hormones, membrane stability
Mg	2	80	8×10^4	2–4	Constituent of chlorophyll, enzyme co-factor, control cell metabolism/enzymes

P	2	60	6×10^4	0.0005–0.002	Constituent of nucleic acids, lipids, essential cellular energetics, co-enzyme regulation
S	1	30	3×10^4	0.3–0.7 0.001–10	Constituent of proteins, sulpholipids, energy transfer
Micro-					
Cl	0.1	3	3×10^3	0.001	Chloroplast photosystem II, metabolism, growth
B	0.02	2	2×10^3	0.001	
Fe	0.01	2	2×10^3	0.001	Energy transfer proteins, co-enzyme factor prosthetic groups
Mn	0.05	1	1×10^3	0.001	Co-factor in water splitting enzyme, aminopeptidase etc.
Zn	0.02	0.3	3×10^2	7×10^{-4}	Enzyme co-factor, carbonic anhydrase, alkaline phosphatase, enzyme regulation
Cu	0.06	0.1	1×10^2	3×10^{-4}	Constituent of plastocyanin, ascorbic acid oxidase etc.
Mo	0.0001	0.001	1	5×10^{-4}	Constituent of nitrate reductase

macro-element, and chlorine the most abundant micro-element (Rains, 1976). However, there is little physiological justification for the distinction. Macro-elements occur in large molar ratios in the more abundant molecules (e.g. N in proteins) and at high concentrations in the cell solution (e.g. K in vacuoles). Micro-elements often form prosthetic groups in enzymes (Table 1) and have specific, vital functions in metabolism (Marschner, 1983; 1986). For example, zinc is a component of the active catalytic centre of the enzyme carbonic anhydrase (6 Zn atoms per molecule) which increases the rate at which an equilibrium is achieved between CO_2 and bicarbonate ions in solution. The reaction is very fast (a turnover time of 10^{-6} s) and therefore the concentration of the enzyme and thus of Zn of this particular component of leaf tissue is very small (Poincelot, 1979). Another example is iron which is a component of electron transport molecules such as haem groups in iron sulphur centres. Terry & Rao (this volume) discuss the role of nutrition in the photosynthetic processes. In contrast, nitrogen occurs in all proteins at about 16% by mass (Bray, 1983; Rains, 1976).

Although the functions of nutrients are generally understood, details of the mechanisms of action and how they are incorporated into plant structures are often lacking. Manganese, for example, is essential for the water splitting and oxygen evolving processes in photosynthesis (Lawlor, 1987) but its location and mode of action in the photosystem II complex of chloroplast are uncertain (Rutherford, 1989).

Nutrient absorption

Plants obtain mineral nutrients mainly from the soil and the atmosphere is the source of CO_2. However, ammonia, oxides of nitrogen and sulphur dioxide as well as particulate materials, often pollutants produced by human activity, may contribute to the plant's mineral balance and also cause damage (Huettl & Fink, 1991, this volume). The forms, amounts and proportions of different nutrients in the environment vary greatly as does the timing and the rates of availability to plants. These depend on the type of soil and the biological activity of microorganisms and of the plants themselves (Marschner, 1986; 1991, this volume). Many plant strategies for exploitation of nutrient resources depend on matching the supply with plant demand (Grime & Hunt, 1975). The mineral nutrients exploited by plants are largely in ionic form, such as K^+, NO_3^- and SO_4^{2-}, and plants may exploit forms depending on the conditions, e.g. NO_3^-, NH_3 and NH_4^+ are sources of N for higher plants, ultimately and largely metabolised to amino acids (Bray, 1983) and used in protein synthesis. The total amount of a nutrient available to plants depends on the concentrations of usable forms of the nutrient in the environment, the efficiency of the plant in extracting the element, and the total volume of the medium exploited. This is very

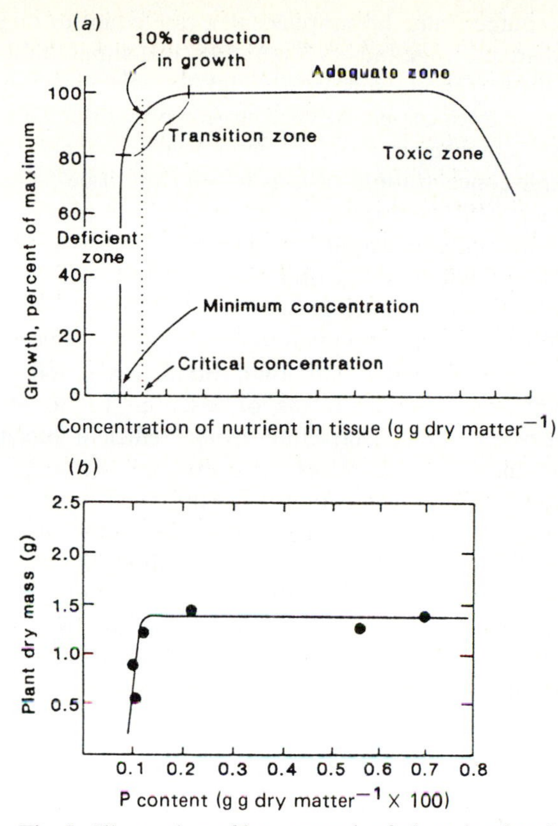

Fig. 2. Illustration of how growth of plants is related to the concentration of nutrients. (*a*) General response curve showing dependence of growth on external nutrient concentration and showing the 'critical' concentrations for a 10% reduction in growth, the minimum concentration for growth to occur, the zone of adequate nutrient supply and the concentration at which toxic symptoms appear. (*b*) Growth in mass of maize in relation to the phosphorus content (g g dry matter^{-1}) of the tissue (after Epstein, 1972).

dependent on the extent and efficiency of the root system (Epstein, 1972) which are genetically determined but depend also on the environment for expression.

The response of plants to nutrient concentrations in the environment and also to light, CO_2 and temperature generally takes the form shown in Fig. 2*a* which summarises experimental information of the type shown in Fig. 2*b*. Such a 'diminished returns' type of relationship is characteristic of plant responses to all essential resources and conditions. The response shows that to maintain a genetically determined growth rate (Grime & Hunt, 1975) at its maximum, all resources must be supplied at a rate equal to or greater than the potential demand (Ingestad, 1970*a*,*b*; 1982). If supply falls below demand, photosynthetic processes decrease and growth slows. At very small concentrations, growth ceases. At very large supply or concentration of a resource a plant's biochemical and physiological mechanism may be damaged, e.g. by large concentrations of ions which cannot be excluded by the roots (Marschner, 1986). Species have different requirements for nutrients and different abilities to acquire resources and to tolerate or avoid damage by particular conditions, so the quantitative details of Figs 2*a* or 2*b* are very species specific and therefore genetic.

In order to transport selected nutrients into cells against a concentration gradient 100 to 1000 times greater than that found in the soil solution (Table 1) and to maintain similar ratios of essential to non-essential elements in cells (Epstein, 1972; Marschner, 1986) efficient biochemical transport and accumulation mechanisms have evolved (Fromter, 1983). These are coupled to a supply of energy and to the intermediary metabolism which maintains cellular homeostasis and, at the same time, ensures that fluxes of nutrients are available for growth when required. However, these mechanisms may not completely regulate the entry of elements which may damage plant metabolism; sodium, for example, may be absorbed from saline soils. Heavy metals (Cd, Pb, Zn) from mine and industrial wastes may accumulate in plants and damage them and pose a threat to the animals, including man, that consume the plants (Bollard, 1983). There are great differences in the adaptation of species of genotypes of plants to the chemical composition of the environment. Some, such as *Agrostis tenuis* have populations that differ in susceptibility to heavy metals (Bradshaw, 1975, 1984). Similarly there are large differences between species and sub-populations in their tolerance in calcareous soils (Jefferies & Willis, 1964). *Caluna vulgaris* (an extreme calcifuge) cannot grow in such soils whereas *Scabiosa columbaria* (an extreme calcicole) is adapted to them (Grime, Hodgson & Hunt, 1988).

Nutrients and cell composition

Mineral nutrients absorbed by cells accumulate in particular cell compartments, where they may regulate metabolism or are stored before entering into metabolism. They are incorporated into about 10^5 types of organic molecules, synthesised in very different amounts (Lehninger, 1975). Half are of molecular mass less than 500 and are often metabolic intermediates in the synthesis of structural, large molecular mass components such as proteins (10^4–10^5 kD), nucleic acids (10^9 kD) and carbohydrate polymers such as cellulose. The main classes of components are given in Table 2. They are carbohydrates (cell walls, membrane glycolipids, starch) con-

Table 2. *Elemental composition (percent by mass) of the main classes of biochemical components and of a whole plant of maize (Zea mays) (after Epstein, 1972).*

Element	Carbo-hydrates	Fats	Protein	Nucleic acid	Plant
O	51.4	11.3	24	30	44.4
C	42.1	76.5	52	39	43.6
H	6.5	12.2	7	3.5	6.2
N	—	—	16	17	1.5
S	—	—	0.1	—	1.2
K	—	—	—	—	0.9
P	—	—	—	10	0.2
Mg	—	—	—	—	0.2
Fe	—	—	—	—	0.1
Other	—	—	0.9	0.5	1.7
	100	100	100	100	100

taining only C, H and O; proteins (enzymes, cytoskeleton and membrane proteins) and nucleic acids containing C, H, O, N, P and S as well as many metallic mineral elements in prosthetic groups and as electron carriers (Lehninger, 1975; Marschner, 1986); and lipids (phospholipids, oils) with C, H, O, P and S. The amounts of N, S and P that occur in the major groups are known; although proteins have, on average, 16% by mass of N, particular proteins differ in their proportions of amino acids and therefore in elemental composition. However, the major differences in composition are *between* classes of biochemical components rather than *within* them. Of great significance in composition of tissues is the proportion of biochemical components which depends on the type of cell and its functions and thus differ in the proportion of constituents. For example, leaf mesophyll cells contain much protein, chlorophyll and lipid associated with the photosynthetic apparatus, whereas storage parenchyma cells contain much carbohydrate but little protein.

The important general principle, stated earlier and worthy of repetition, is that it is the biochemical components of tissues that determine the amounts of mineral and other elements in cells. Further, it is by biochemical processes that plant nutrition, composition and the influence of the environment are linked.

Tissue and plant composition

Different tissues and organs of plants contain different amounts of nutrients (Epstein, 1972); for example, leaves contain more N than roots or stems per unit of dry matter. Differences in the gross nutrient composition of plants are often due to different amounts and proportions of organs and their age (Lawlor *et al.*, 1981). Young plants have proportionally more material in leaves than in stems and storage organs than old plants. The cellular proteins of older leaves may be broken down and used for growth of new leaves or of reproductive organs (Lawlor *et al.*, 1981; Lawlor, Kontturi & Young, 1989; Groot & Spiertz, 1991, this volume), so the C/N ratio of young plants is smaller than that of old plants. Nutrient requirements of plants may be estimated if the biochemical composition of the tissues and the proportions are known. Penning de Vries (1974, 1975) used this approach to estimate the costs of synthesising plant tissues. Stated very simply, the rate of dry matter production, D (g s^{-1}) is a function of photosynthetic rate, Pn (g m^{-2} s^{-1}), total leaf area, L (m^2), respiration per unit of dry matter, Rd (g^{-1} s^{-1}) and the mass of the respiring crop, W (g), all expressed as a mass of carbon. Thus:

$$D = (Pn \times L) - (Rd \times W) \tag{1}$$

To estimate D over a period, Pn must be summed over L and Rd summed over W; both L and W change with time. Pn is a well defined function of the photon flux, of CO_2 concentration in the atmosphere, and of temperature; it is also a function of nutrient supply. From such information the capacity of the leaf for photosynthesis may be estimated. Similarly the respiration of crops may be calculated if the relation between respiration and temperature is known in addition to the mass of the standing crop.

It is instructive to consider how assimilation rates and nutrient uptake rates are related for a barley crop growing under field conditions (Lawlor *et al.*, 1981). The aerial parts of the plant grew at a rate of 50 mg dry matter day^{-1}; as the photosynthetically produced dry matter accumulating in the roots is *c.* 15% of the shoot mass and the equivalent of 30% of the total dry matter is lost as respiration (Mitchell & Lawlor, unpublished), the total daily demand for assimilates is equivalent to 74.8 mg. It is established that 1 g of dry matter contains an average of 0.5 g carbon, corresponding to 1.83 g of CO_2 (Penning de Vries, 1974, 1975) so the daily assimilation of CO_2 is 3.11 mmol. A typical photosynthetic rate is 10 μmol m^{-2} s^{-1} and an average daily radiation flux of 600 μmol photons m^{-2} s^{-1}. An average barley plant has a leaf area of 0.01 m^2, so in a 10 hour day total photosynthesis would be 3.6 mmol CO_2 day^{-1}, compared to the 3.11 mmol of dry matter accumulated. Experimental measurements show that a soluble protein content of *c.* 10 g m^{-2} is necessary to maintain such rates of photosynthesis. About half of this large amount of soluble protein is in the photosynthetic enzyme ribulose bisphosphate carboxylase–oxygenase (Rubisco), a large protein of 550 kD mass. This enzyme accumulates CO_2 slowly with the catalytic sites turning over in *c*. 2 s. Even with eight active sites per molecule the rate of photosynthesis needed to maintain the growth rate of crops is large, and correspondingly a large amount of Rubisco is also required (Mooney & Chiarello, 1984; Lawlor, 1987; Lawlor *et al.*, 1989; Groot & Spiertz, 1991, this volume). In addition to the CO_2 assimilating enzymes, chloroplasts, light harvesting components and systems to produce ATP and NADPH are needed (Lawlor, 1987). Hence, actively photosynthesising leaves need large amounts of protein and have correspondingly large amounts of N per unit of dry matter. With 16% of N in proteins, and 40 g of crop dry matter per plant containing *c*. 30% of protein, this is equivalent to 4.4% of N in dry matter, a figure often observed (Lawlor *et al.*, 1981, 1989). Most of the N is in the three youngest expanded leaves of a cereal crop. As one leaf grows at a time and forms its protein in about 6 days, it is possible to calculate the rate of demand for N; in the case of the crops mentioned above the accumulation of N is 0.4 $g\ m^{-2}\ d^{-1}$ during leaf growth.

Such calculations indicate how growth and nutrient requirements at the whole plant level are linked by the biochemical composition and emphasise the considerable degree of interaction between nutrition, largely determined by the soil, the aerial environment and cellular processes. To model such a process requires empirical knowledge of the growth rate of the plant and its biochemical composition; improved understanding of the mechanisms might allow simulations from more basic information.

Cell metabolism, link between nutrition and environment

Interaction between nutrient supply, other resources such as light energy, and environmental conditions e.g. temperature which determine plant growth, occurs at the biochemical level of cellular metabolism. Cells function as complex metabolic machines which convert the materials and energy of the environment into structures (i.e. growth) which ultimately allows replication of the genome (i.e. reproduction) (Darnell *et al.*, 1986). Within the average mature plant cell (Fig. 3; see Schnepf, 1983; Hall, 1983) the cytosol and organelles are the sites of metabolism and stable regulated conditions in them are crucial to acquisition of resources and to their assimilation into intermediary metabolites and synthesis of structures. To maintain homeostasis in cells, the concentrations of solutes, substrates, ions and regulatory metabolites at the sites of enzymatic reactions must be controlled (Lehninger, 1975; Huber & Bennett, 1983). Despite relatively poor understanding of integrated multi-enzyme systems *in vivo*, *in vitro*

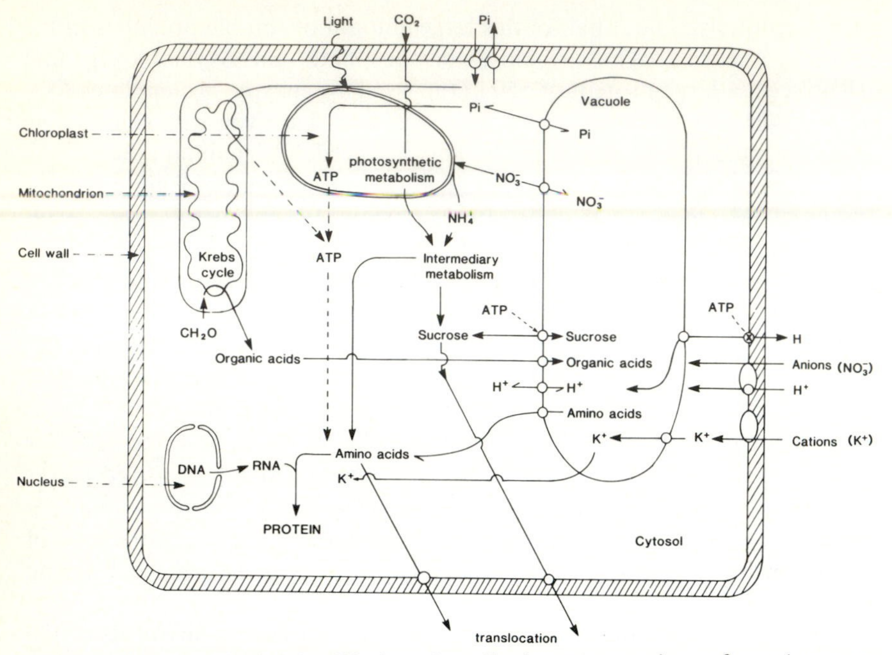

Fig. 3. Much simplified and stylised representation of an 'average photosynthetic cell' to emphasise the interactions between energy transduction in chloroplasts and mitochondria, storage of ions, sucrose, amino acids and organic acids in the vacuole and metabolism in the cytosol. The links between sucrose synthesis, protein synthesis and intermediary metabolism are indicated.

studies suggest the conditions that are needed for stable and active rates of some metabolic sequences. Each enzyme reaction in a complex sequence has particular requirements of both the types and concentrations of substrate and control molecules (Fig. 4) (Wyn Jones & Pollard, 1983) and on the behaviour of other enzymes in the system. Systems analysis suggest that such control mechanisms are essential for maintenance of the stability and efficiency of metabolic systems (Kacser, 1987). An example is given by CO_2 assimilation which can only proceed, at a given rate, if the ATP and NADPH supply to the photosynthetic carbon reduction cycle is adequate; this depends on both the light energy and on the supply of inorganic phosphate, amongst other factors. If any single substrate is deficient, the CO_2 assimilation rate will decrease. In many systems metabolism is regulated to ensure that deficiency in one metabolic process does not disrupt the whole organism but allows at least some growth and reproduction to proceed.

To illustrate the importance of cellular homeostasis in determining the

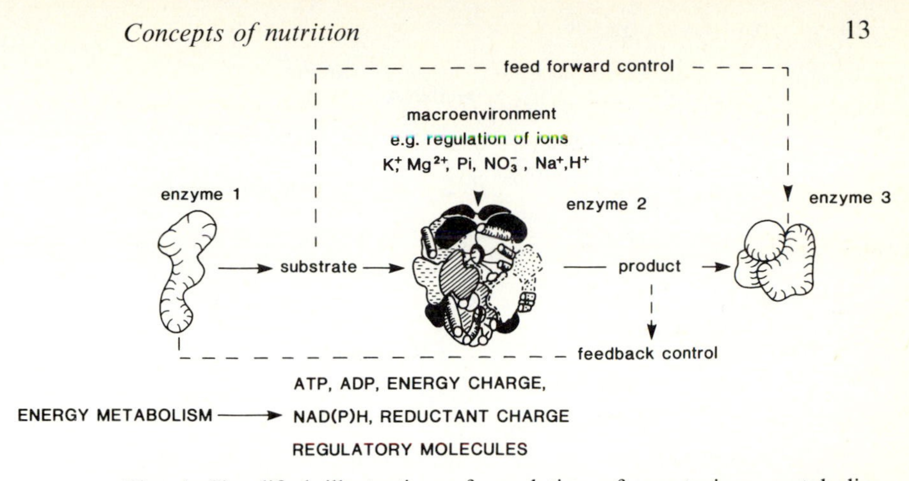

Fig. 4. Simplified illustration of regulation of events in a metabolic sequence of three enzymes, with a complex enzyme, depending on its environment (concentration of ions, metabolic products etc.) to maintain its structure and efficiency. The products of the reactions regulate other steps in the sequence by feedback and feed forward control; the way that some control steps depend on the energy status of the cell is indicated.

uptake of nutrients and linking it to environmental conditions thus ensuring the growth of plants, protein synthesis will be considered and then the role of energy metabolism and the associated phosphorus metabolism.

Protein synthesis is central to all cellular functions (Lehninger, 1975; Darnell *et al.*, 1986). In its absence, the growth and maintenance of organs stops (Gerloff & Gabelman, 1983) and thus it represents the rate-limiting step for potential growth rate. The rate of protein synthesis depends on both the rate of transcription and of translation of the genetic information and also on the amino acid supply for the growth of the polypeptide chain (Fig. 5). In the nucleus and other organelles where replication of DNA and synthesis of mRNA and tRNA occurs (often against a background of very active metabolism) the micro-environment of very large macro-molecular complexes might affect hydrogen bonding and water structure which are important for the structural integrity and fidelity of translation (Darnell *et al.*, 1986), particularly during cell replication (Tschesche, 1983). The synthesis of proteins also requires amino acids demanding a large supply of energy and nutrients, especially of nitrogen. The supply of N depends on the amount of nitrate or ammonium ions from the environment (Fig. 1) and on the rate that they are metabolised to amino acids. However, the link between the rate of protein synthesis and N supply is not direct in higher plants for they store NO_3^- and amino acids (made in excess of demand when substrates are available) in vacuoles; these provide a buffer between the supply of N and demands for growth and often result in complex changes

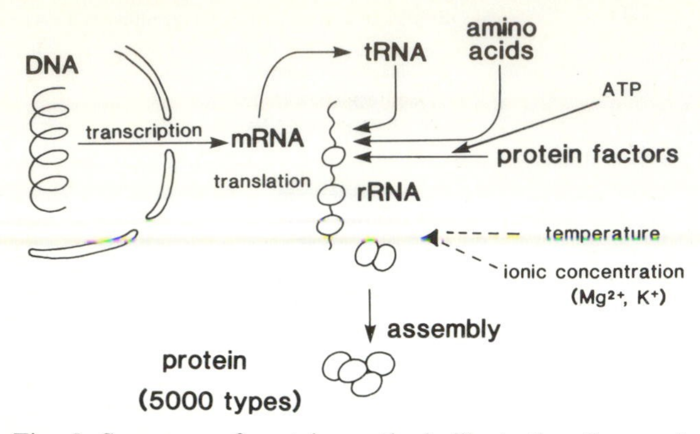

Fig. 5. Summary of protein synthesis illustrating the requirements for substrates of amino acids, energy and regulatory molecules. The link to the genetic information and the role of RNA species in the synthetic machinery are illustrated.

in the content of N-compounds in plants depending on the conditions, e.g. temperature (Lawlor *et al.*, 1988). These processes of N metabolism are also linked closely to the energy supply. For example, photosynthetic cells use light energy to generate the reductant required to convert NO_3^- to NH_3 and the ATP necessary for the assimilation of NH_3 to glutamic acid by the glutamine synthetase/glutamate synthetase cycle (Stewart, this volume).

The ionic environment of protein synthesis is also very important, for Mg^{2+} and K^+ (the latter at a concentration of between 100 and 150 mol m^{-3}, Fig. 6) provide the necessary conditions in the aqueous environment (Wyn Jones & Pollard, 1983; Leigh & Storey, 1991, this volume). Other ions (e.g. Na^+ and H^+) which may inhibit activity must be excluded or carefully controlled. Maintenance of the cytosolic concentrations of these ions is achieved in conjunction with that of other cellular compartments such as the cell vacuoles. Vacuoles play an important role in the storage of many mineral nutrient ions, organic acids, sugars and, as mentioned, amino acids. The vacuole is involved in the dynamic regulation of H^+ and therefore of the pH of the cytosol and organelles (Smith & Raven, 1979), and it has functions in osmotic regulation (Wyn Jones, Brady & Speirs, 1979). The change in vacuolar composition is a major uncertainty in establishing the 'critical' concentrations of nutrients required for particular cell functions (Epstein, 1972; but see Leigh & Storey, 1991, this volume) and contributes to the changes in nutrient composition of plants. It is important to realise that the acquisition of nutrients and the maintenance of homeostasis in cells in the face of greatly varying concentrations in the environment and changes in other regulating factors, is achieved by close regulation of

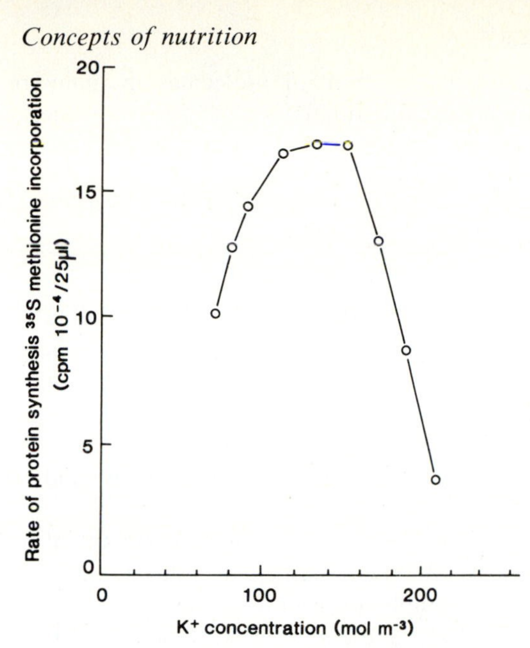

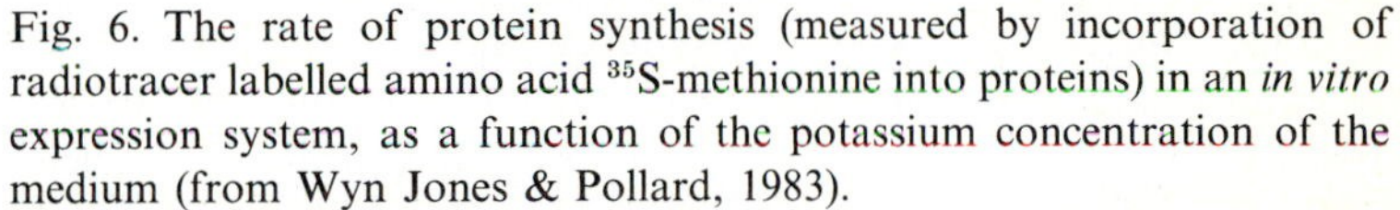

Fig. 6. The rate of protein synthesis (measured by incorporation of radiotracer labelled amino acid ^{35}S-methionine into proteins) in an *in vitro* expression system, as a function of the potassium concentration of the medium (from Wyn Jones & Pollard, 1983).

mechanisms of metabolism in organelles and transport between cell compartments, often involving the control of fluxes at compartmental membranes (Hall & Baker, 1977).

Other metabolic processes of great importance in cells are linked to the provision of energy for synthetic metabolism, such as pyrimidine synthesis and for maintenance processes such as ion transport and ionic regulation. Associated very closely with cellular energy regulation is phosphorus metabolism (Bieleski & Ferguson, 1983) and some comments showing how these processes interact and link the supply of energy to nutrition and plant growth are in order.

Energy metabolism is, of course, central to all cellular functions. Photosynthesis transduces solar energy into useful metabolic forms such as ATP and mitochondrial oxidative respiration provides energy by exploiting the supply of assimilates of photosynthesis. A continuous supply of energy is required for all synthetic and maintenance processes and is dependent on the regulation of ions and substrates and exclusion of mineral elements (e.g. heavy metals) which might interfere with the electron flow, energy transduction and ATP synthesis (Bollard, 1983). Metabolism is regulated, largely by the provision of ATP as a substrate and by the amounts of ATP,

ADP and AMP which serve as control molecules in many reactions. Enzymes are, for example, often regulated by the adenylate energy charge (EC);

$$EC = \frac{[ATP] + \frac{1}{2}[ADP]}{[AMP] + [ADP] + [ATP]} \qquad (2)$$

Similarly the pyridine nucleotides and reductant energy charge regulate many enzyme functions (Pradet & Raymond, 1983). These authors have discussed the role of adenylate and pyridine nucleotides and energy charge in regulation of cellular processes; clearly the very intricate and delicate balance between the supply of resources (e.g. energy and nutrients) in relation to the demands by growth require mechanisms to overcome or reduce the effects of fluctuating supply of resources and changing conditions.

The phosphorus supply to plants is a major determinant of growth in many environments (Fox, 1979) and efficient uptake mechanisms and accumulation of P, for example in vacuoles, serve to maintain plant function with deficient and variable P supply. Metabolism of P dominates cellular functions linking the energy supply with fluxes of cellular metabolites (e.g. carbon fluxes in photosynthesis and ion fluxes into cells). Phosphorus is found in slowly turned-over structures (e.g. nucleic acids), in more rapidly transformed compounds (P-lipids) and also in rapidly metabolised materials (e.g. ATP) and in very different amounts (Table 3; Bieleski & Ferguson, 1983). However, the concentration of inorganic P in the cytosol is close to 10 mol m^{-3} and the molar ratio of ATP:ADP:AMP is held at approximately 10:3:1 by the adenylate kinase system so that the energy charge lies between 0.8 and 0.9 (Pradet & Raymond, 1983). When ATP is consumed by processes in the cytosol and organelles, and the phosphorus is bound into metabolites, the inorganic P (P_i) content falls; a flux of P_i from the vacuole maintains the concentration and, providing energy is available, ATP is synthesised and homeostasis is maintained. However, with deficient P but continued demand by cell processes for it, the P_i from the vacuole is depleted and the P_i and ATP concentrations decrease in cell compartments thus slowing cellular metabolism. Growth is impaired in the long-term. Adjustments in the tissue to P-deficiency involve fast regulation of intermediary metabolism to maintain homeostasis, slower changes in cell structure and decreased growth rates and, in the very long-term, genetic changes in plant and cell composition, growth rates and regulatory mechanisms evolve.

Table 3. *Phosphate in leaf tissue (after Bieleski & Ferguson, 1983): (a) phosphate fractions in young leaf tissue (μg atoms P (g fresh mass^{-1})) and (b) average mol per cent of components of the P lipid fraction.*

(*a*)		(*b*)	
Component	μg atom P (g f wt)$^{-1}$	Component	%
P_i	10	P choline	45
RNA	2	P ethanolamine	25
DNA	2	P glycerol	15
P lipid	1.5	P inositol	5
P ester	1.0	P serine	5
		Other	5

Protein synthesis and the effects of temperature

Temperature determines the rate of cellular processes and of organ and plant growth; there are considerable differences between species in response to temperature (Warren-Wilson, 1966; Berry & Raison, 1981). Protein synthesis is dependent on the temperature (Freifelder, 1987) and this constitutes a major regulator of whole plant growth (Lawlor *et al.*, 1989) if other factors such as N supply are not limiting. In mature plant systems, processes other than protein synthesis may regulate plant response to temperature (Berry & Raison, 1981), but protein synthesis is known to be the major rate determining process in many biological systems. The bacterium *Escherichia coli*, for example, when grown at 37 °C incorporates 17 amino acids per second into a protein molecule but at 30 °C the rate is 35% slower (Freifelder, 1987). A specific example of the effects of temperature on the mechanisms of protein synthesis is given by the amino acid: tRNA ligase reaction:

$$\text{amino acid} + \text{ATP} \rightarrow \text{amino acyl–tRNA} + \text{AMP} + P_i \qquad (3)$$

This reaction 'primes' amino acids for incorporation into the growing polypeptide chain. The rate of reaction of the wheat enzyme is slow at 5 °C and increases up to a maximum at 35 °C and decreases thereafter (Fig. 7; Weidner, Mattee & Schmitz, 1982). Other metabolic activities and the growth of the whole plant have similar responses to temperature and it is therefore reasonable to assume that plant growth rate is determined by protein synthesis the rate of which is controlled by genetic factors (which

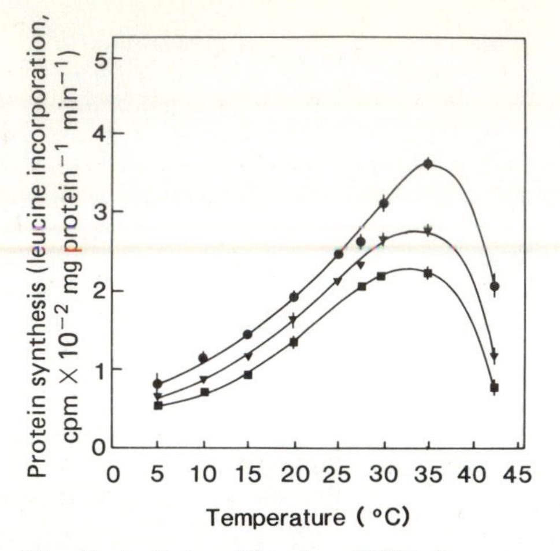

Fig. 7. Activity of leucine:tRNA ligase, an enzyme essential for protein synthesis, depends on the temperature of the reaction. The enzyme was extracted from wheat plants grown at 4 °C (●), 20 °C (▼), and 30 °C (■). The temperature optimum for the reaction was between 30 °C and 35 °C and decreased slightly with temperature at which the plants were grown, as did the specific activity per unit of protein (after Weidner *et al.*, 1982).

define the nature of the protein synthesised) and by temperature. This may be regarded as the process which regulates the potential growth rate as discussed in the Introduction. In addition, the amino acid:tRNA ligase reaction clearly illustrates the points made previously, namely the dependence of plant growth on interactions between nutrition (N supply for amino acid synthesis and P for ATP synthesis) and energy (for ATP synthesis) and temperature (as determined by the rates of the system reactions) all acting at the biochemical level of organisation.

Cellular regulation, genes and species differences in nutrition

As the composition of the biochemical constituents of cellular structures and of tissues and organs formed from them are genetically determined it follows that the nutritional requirements are also genetically defined. The potential rate of growth and the potential size of a plant are also defined by the genotype. At the simplest level it is known that the genome determines both the amount and rates of synthesis of cell constituents and also their proportions and therefore lays down the potential for resource capture and growth. However, there is little information on specific differences in regulatory mechanisms, but it can be predicted that these will be found at the level of protein production and involve selective translation and transduction of genes and gene families. The rates of protein synthesis may be varied by gene amplification, by production of multiple copies of the mRNA, by extending the life time of mRNA or by altering the rate of translation of mRNA (Fig. 8). Such mechanisms enable proteins to be made in the required amounts and to respond to specific environmental signals. For example the supply of nitrate ions induces synthesis of the enzyme nitrate reductase and the increased N supply allows formation of more proteins and of larger plants (Lawlor *et al.*, 1989). Changes in the type, as well as the amounts of proteins are regulated by such signals as the light intensity or day length. Key developmental processes are often tied closely to such stable features of the environment. Light, acting through the phytochrome system, affects the expression of genes for some 60 chloroplast components, photosynthetic enzymes, pigments and nucleic acids involved in their synthesis (Tobin & Silverthorne, 1985; Thompson, 1988) and there is rapid progress in understanding the mechanisms of this system (Kuhlemeier, Green & Chua, 1987).

Understanding the genetic regulation of nutritional requirements and the mechanisms involved is also progressing. Genetic differences in response to essential nutrients such as P and N are well established. However, because macro-nutrients are involved in most, if not all, plant activities the regulatory mechanisms are many and multigenic and their interactions complex. The role of genetic regulation has been better defined in relation to micro-elements, which are involved in fewer, more specific reactions and controlled, therefore, by fewer genes (Bollard, 1983). The role of molybdenum is well established and may be used as an example; it is a co-factor in nitrate reductase which reduces nitrate ions to nitrite. A Mo-phosphorylated pterin–protein complex is essential for activity and the genes for its synthesis have been identified. Mutants lacking the capacity to synthesise the pterin nucleus, or the ability to transport or oxidise or reduce Mo, or without the capacity to ligate Mo to the sulphur residues of the pterin all prevent formation of the active complex (Wray, 1986). Similarly, understanding of the genetics of the plant adaptation to heavy metal toxicity has been advanced. Copper tolerant grasses, for example, synthesise metallothienein proteins with much cysteine which binds the metal, in the presence of Cu (Rauser & Curvetto, 1980; Woolhouse, 1983). Current evidence is, however, that metallothieneins are not the primary protective agents against heavy metals in plants, rather peptides called phytochelatins play a more important role (Grill *et al.*, 1989). Single gene differences are responsible for the susceptibility of soya bean and maize to Fe deficiency, of celery to Mg^{2+} and B deficiency and of beans to K^+ (Marschner, 1986; 1991, this volume). Aluminium tolerance in wheat is qualitatively inherited with perhaps three genes, one on the 4D chromosome (Rhue, 1979). In this

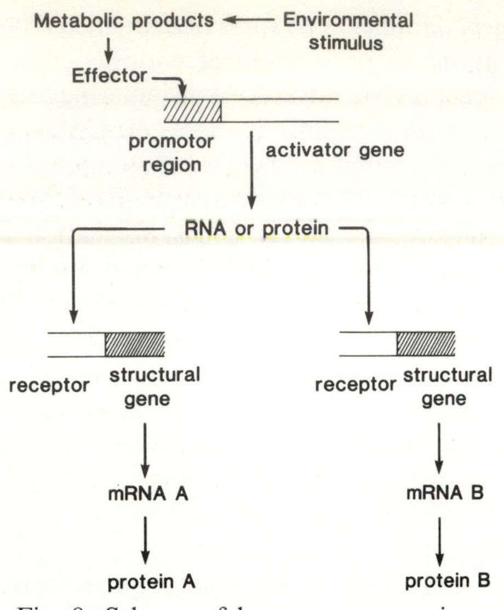

Fig. 8. Scheme of how gene expression may be regulated in plants to produce different proteins in the required proportions. An environmental stimulus is detected in the plant, for example light is detected by the phytochrome system which causes the production of an effector. This stimulates the production of large quantities of a particular protein. Regulation may reside in other processes, e.g. at the level of turn-over of the mRNA (after Freifelder, 1987).

context, it is perhaps significant that salt tolerance in wheat is also associated with the 4D chromosome (Gorham *et al.*, 1987). In all these cases adaptation to the elements probably involves changes in some amino acids of proteins which produce tolerant metabolic systems and often the production of more gene products, i.e. more copies of genes to overcome the presence of high concentrations of metals.

Species differences in nutrition

The mechanisms conferring adaptation to mineral elements in different species of plants exhibit great ecological diversity, from slow growing large perennial woody species to rapidly growing, small herbaceous ephemerals. There are large differences in biochemical composition and hence the amounts, rates and times at which nutrients and other resources are required (Chapin, 1980). A classification (Grime, 1979; Grime *et al.*, 1988) into stress tolerant species, competitor species, and ruderal (disturbance tolerant) species is based on competition, growth rates and resource requirements. Stress tolerant species have low growth rates, slow rates of photosynthesis and structural adaptations which minimise loss of essential resources. For example, many plants of arid, nutrient-poor soils have small, thick-walled, spiny and often perennial leaves which protect against herbivores thus reducing water and nutrient losses.

Many biochemical processes contribute to ecological strategies; slow turn-over of cell components and tolerance of adverse conditions require the presence of adequate mechanisms to withstand the effects of the adverse conditions. For example, chloroplasts must contain adequate energy dissipating systems to reduce the excess light energy and thus to avoid photoinhibition (Powles, 1984) under drought or cold conditions when photosynthesis is slowed but light energy is captured by the chloroplasts. Another example of a major biochemical feature is the use of nitrogen containing metabolites such as glycinebetaine for osmotic regulation and as a protective compound of cellular processes in halophytic plants (Stewart, 1991, this volume).

The theme of this review is the demand for resources determined by cell and plant composition and it is important to consider the strong link between growth rates and nutrient requirements (amongst other ecological adaptations) of species (Grime & Hunt, 1975). Grime *et al.* (1988) considered that species show 'sensitive adjustment of general levels of potential maximum growth rate' to soil fertility. From earlier discussion it is qualitatively apparent why this is so. Given the requirements for mineral elements (including N in Rubisco) in photosynthesis, the correlation is explicable and if these resources can only be acquired slowly in a nutrient-limited environment then growth rate must necessarily be slow; perenniality of organs and protection from herbivores is thus also explicable. As an example, the nettle (*Urtica dioica*) has a fast potential growth rate and grows under fertile conditions and this contrasts to the slow growing grass *Deschampsia flexuosa* which is found in poor soils. Nettle maintains a high P content and does not grow at P concentrations below 10^{-5} M whereas *Deschampsia* has a low P content and grows in solutions 100-fold more dilute in P (Rorison, 1966).

In contrast, competitive species have rapid rates of growth to maximise resource capture in undisturbed habitats, for example by producing leaves with high potential photosynthesis and displaying them for effective light capture in the canopy. This places large demands on the metabolic systems and, of course, on the need for resources (Chapin, 1980). Ruderal species, particularly, rely on a strategy of rapid growth and maximising resource capture in disturbed habitats (Grime & Hunt, 1975; Grime *et al.*, 1988). Differences in response to nutrition and to other environmental factors are seen between closely related species. *Plantago maritima* grows under saline conditions, whereas *P. major* subspecies *major* does not. However, *P. major* subspecies *intermedia* is tolerant of salinity (Grime *et al.*, 1988); this is perhaps an example of speciation in relation to selection pressure (Stebbins, 1971; Bradshaw, 1983, 1984).

There are interesting, if poorly understood, correlations between ecological behaviour and cell size and the amount of nuclear DNA which may have implications for understanding regulation of metabolism in relation to nutrient and other resource capture (Grime, 1983; Grime, Shacklock & Band, 1985). Generally, large cells are associated with large complements of DNA. Plants growing at low temperatures or plants from mediterranean types of environment have more DNA per cell than those from warmer and less droughted environments and species with much DNA per cell tend to grow more slowly than those with little. Ruderal species (e.g. *Aribidopsis thaliana*) have often very small genomes, possibly allowing very rapid replication and hence fast growth. The significance of genome size in determining metabolism, nutrient requirements and ecological behaviour is not understood, although large amounts of DNA per cell are associated with 'stored growth' that is the formation of cells under previously good conditions and the rapid expansion of leaves when new growth occurs under adverse conditions such as cool early springs. It is possible that the extra DNA increases the number of gene copies allowing more rapid protein synthesis per cell and therefore increasing the growth rate under adverse conditions. Thus, polyploid genotypes of wheat have large cells containing more constituents than the equivalent diploids (Dean & Leech, 1982) but much of the DNA in species with large genomes is not translated but is in the form of repeated base sequences (Grime *et al.*, 1988). As the genome regulates formation of cells and their constituents and also the potential rates of growth, it is important to understand the relation between genome size, its expression, metabolic function and the relation to plant growth strategies.

Growth rate, mineral nutrition and temperature

Having examined some of the mechanisms by which plant growth is determined at the genetic, metabolic, cellular and environmental levels of organisation, I return to the central theme and consider how the growth rate of the whole plant is related to resources and to temperature. By using the relative growth rate (R_w = g dry matter (g plant dry matter)$^{-1}$ d^{-1}) rather than the absolute growth rate (G = g dry matter d^{-1}) differences in plant size (which affects the absolute growth rate – large plants grow more than small ones) are taken into account. The potential growth rate may be equated with the maximum R_w ($R_{w,max}$) obtained when the concentrations of any essential nutrients at the site of action $C_{n,opt}$ is not limiting (Fig. 9). Clearly, $C_{n,opt}$ may differ from the nutrient concentration in the medium,

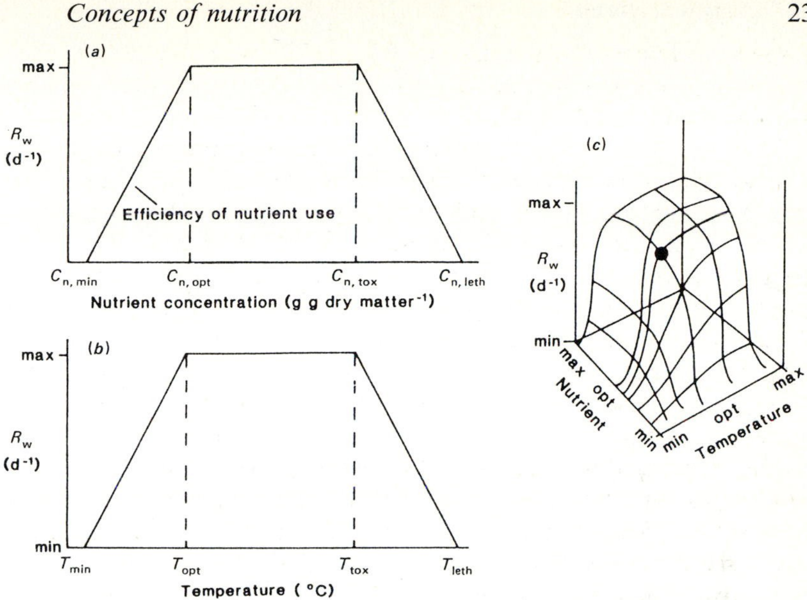

Fig. 9. Growth rate of a plant as a function of nutrition and temperature. (*a*) Relative growth rate (R_w) of a plant as a function of the internal concentrations, C_n, of a nutrient. The point at which the nutrient concentration is optimal ($C_{n,opt}$) for growth ($R_{w,max}$) and the minimum concentration for growth ($C_{n,min}$ and $R_{w,min}$ respectively) are shown. The situation of nutrient toxicity is illustrated, where at $C_{n,tox}$ R_w is decreased and at $C_{n,leth}$ R_w ceases and the plant is killed (after Ågren, 1988). (*b*) The response of R_w to temperature showing $R_{w,max}$ at an optimum temperature, decreasing at temperatures above T_{tox} until growth ceases at a temperature T_{leth} and the plant dies. (*c*) Illustration of how the two factors may combine in a response surface to give the maximum growth rate at the optimum for both conditions.

depending on the uptake mechanisms. Other resources, such as light, may be treated in the same way. R_w also depends on temperature (Warren-Wilson, 1966), for reasons considered previously and the potential maximum rate of R_w will occur (other factors not limiting) at a particular temperature (T_{opt}). R_w is zero ($R_{w,min}$) when the nutrient concentration is at or below a minimum ($C_{n,min}$); by analogy growth ceases below a given minimum temperature (T_{min}). As discussed earlier, the nutrient concentration may increase until it is toxic (Fig. 9) but this is not further discussed (see Ågren, 1988). By analogy, higher temperatures also slow growth and eventually prevent it.

To relate growth of the plant to nutrition and temperature it is assumed that R_w increases linearly between $R_{w,min}$ and $R_{w,max}$, i.e. between $C_{n,min}$ and

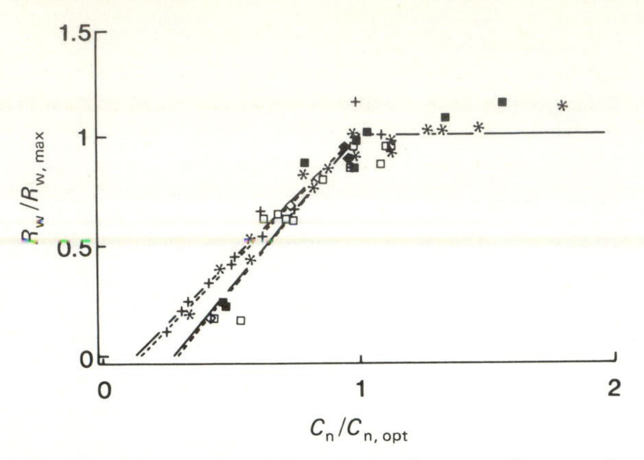

Fig. 10. An example of the relation between the growth rate of plants and nutrient concentration. The relative growth rate, scaled to the maximum relative growth rate is shown for *Betula pendula* (birch), in relation to internal P (∗) and N (+) concentrations (C_n), scaled according to the concentration giving the maximum growth rate ($C_{n,opt}$). Also the growth of a flagellate algae *Monochrysis lutheri* with P (■, large excess; □, lower concentration in growing medium) is shown, expressed in the same manner. The response of two different organisms to nutrients is very similar, with the slope indicating greater efficiency of birch and tolerance to lower nutrient concentrations (modified from Ågren, 1988).

$C_{n,opt}$, and between T_{min} and T_{opt}. Ågren (1988), following the analysis of Droop (1973, 1974), has shown that:

$$R_w = PR_{n,min} \cdot C_{n,opt} - C_{n,min} \tag{4}$$

where $_{min}$ is the minimum concentration of the nutrient, taken over all nutrients and PR_n is a productivity factor relating R_w to C, the nutrient concentration. Temperature may be treated in the same way with a response factor PR_t relating R_w to T between T_{min} and T_{max}. Ågren's (1988) analysis of the growth rates of birch (*Betula pendula*) and a flagellate alga (*Monochrysis lutheri*) in relation to nutrient concentrations (Fig. 10) shows how very different organisms respond in similar manner to the nutrient content although the growth rate of birch is greater than that of the flagellate alga at low relative nutrient concentrations. This quantitative analysis, a form of Blackman's Law of Limiting Factors, expresses the relation of growth rate to the supply of external nutrients and other resources and shows how deficiency of one nutrient decreases R_w below the potential (Ingestad, 1982). Ågren (1988) has extended the analysis to co-limitation of R_w by several nutrients, with R_w determined by the minimum concentration of any one nutrient. R_w is independent of nutrition when all nutrients exceed $C_{n,opt}$ but is dependent on temperature when below T_{max}. If T_{max} and $C_{n,opt}$ are exceeded (but are below the toxic levels) then the genetic factors determine the growth rate. This quantitative analysis is a development of the concepts that the genetically determined potential growth rate, regulated by temperature, sets the requirements for resources including nutrition. A limited supply of any one essential nutrient or other resource from the environment prevents full expression of potential and R_w decreases. The proportionality factors, PR_n and PR_t are a measure of the efficiency of the complete metabolic system and of the so-called rate-limiting processes (Kacser, 1987) and ultimately of the multi-dimensional response surfaces which are required to describe the interactions and environment.

Conclusions

This chapter analyses the relation of nutrition, temperature and the genetically determined cellular metabolic processes. A central theme is that the production of cellular components, the size of cells and the plant and the growth rates are largely determined by genotype which, in turn, determines the nutrient and energy requirements of plants. The cellular machinery provides a stable, regulated resource and substrate supply within a homeostatic environment which is essential for the maintenance of cellular processes, particularly protein synthesis. The pivotal role of protein synthesis in cell functions and growth requires that it is closely linked to both energy and P metabolism to ensure that balance is achieved between processes within a particular range of resources. The role of phosphate in intermediary energy and synthetic metabolism and as a component of cell structure is such that it forms a mechanism for manipulating both energy transduction and growth in plants and is, itself, a major determinant of productivity. The genetically determined rates of protein synthesis are regulated by temperature which determines the actual growth rate of plants. This is regulated in relation to the ecological strategies and to the plant requirements for material resources and environmental conditions. Rapid growth, as in species with competitor and ruderal strategies, demands rapid and abundant supply of resources, whereas slow growth, as exhibited by plants with stress tolerance strategies, is an evolutionary adaptation to the amount of resources and the times at which they are available and is therefore genetically determined. The potential growth rate of plants is achieved when the resources and conditions available exceed the demand. If any resource is deficient then growth slows. The cellular mechanisms which maintain homeostasis with reduced resources allow cellular functions to be maintained at relatively high efficiency but with reduced plant size.

With much decreased resources the regulated mechanisms become impaired, efficiency decreases and metabolism is damaged. The ecological performance of species is the result of the interaction of the genome with the environmental resources, such as nutrients, and conditions, such as temperature.

References

Ågren, G.I. (1988). Ideal nutrient productivities and nutrient proportions in plant growth. *Plant, Cell and Environment*, **11**, 613–20.

Berry, J. & Raison, J.K. (1981). Response of macrophytes to temperature. In *Encyclopedia of Plant Physiology, New Series, Vol. 12A, Physiological Plant Ecology. I. Responses to the Physical Environment*, eds. O.L. Lange, P.S. Nobel & H. Ziegler, pp. 277–338. Berlin: Springer-Verlag.

Bieleski, R.L. & Ferguson, I.B. (1983). Physiology and metabolism of phosphate and its compounds. In *Encyclopedia of Plant Physiology, Vol. 15A, Inorganic Plant Nutrition*, eds. A. Lauchli & R.L. Bieleski, pp. 422–49. Berlin: Springer-Verlag.

Bollard, E. G. (1983). Involvement of unusual elements in plant growth and nutrition. In *Encyclopedia of Plant Physiology, New Series, Vol. 15B, Inorganic Plant Nutrition*, eds. A. Lauchli & R.L. Bieleski, pp. 695–744. Berlin: Springer-Verlag.

Bradshaw, A.D. (1975). The evolution of heavy metal tolerance and its significance for vegetation establishment on metal contaminated sites. In *Heavy Metals in the Environment*, ed. T.C. Hutchinson, pp. 599–622. Toronto: University of Toronto Press.

Bradshaw, A.D. (1983). The importance of evolutionary ideas in ecology – and vice versa. In *Evolutionary Ecology*, ed. B. Sharrocks, pp. 1–25. Oxford: Blackwell.

Bradshaw, A.D. (1984). Ecological significance of genetic variation between populations. In *Perspectives of Plant Population Ecology*, eds. R. Dirzo & J. Sarukhan, pp. 213–28. Sunderland, Massachusetts: Sinauer Associates Inc.

Bray, C.M. (1983). *Nitrogen Metabolism in Plants*. London: Longman.

Chapin, F.S. (1980). The mineral nutrition of wild plants. *Annual Review of Ecology and Systematics*, **11**, 233–60.

Darnell, J., Lodish, H. & Baltimore, D. (1986). *Molecular Cell Biology*. New York: Scientific American Books Inc., Freeman & Co.

Dean, C. & Leech, R.M. (1982). Genome expression during normal leaf development. Direct correlation between ribulose bis phosphate carboxylase content and nuclear ploidy in a polyploid series of wheat. *Plant Physiology*, **70**, 1605–8.

Droop, M.R. (1973). Some thoughts on nutrient limitation in algae. *Journal of Phycology*, **9**, 265–72.

Droop, M.R. (1974). The nutrient status of algal cells in continuous culture. *Journal of the Marine Biology Association of the United Kingdom*, **54**, 825–55.

Epstein, E., (1972). *Mineral Nutrition of Plants: Principles and Perspectives*. New York: John Wiley & Sons.

Fox, R.L. (1979). Comparative responses of field grown crops to phosphate concentrations in soil solutions. In *Stress Physiology in Crop Plants*, eds. H. Mussell & R.C. Staples, pp. 81–106. New York: John Wiley & Sons.

Freifelder, D. (1987). *Molecular Biology*, 2nd Edn. Boston: Jones & Bartlett, Publishers Inc.

Fromter, E. (1983). Transport of matter through biological membranes. In *Biophysics*, eds. W. Hoppe, W. Lohmann, H. Markl & H. Ziegler, pp. 65–501. Berlin: Springer-Verlag.

Gasser, C.S. & Fraley, R.T. (1989). Genetic engineering of plants for crop improvement. *Science*, **244**, 1293–9.

Gerloff, G.C. & Gabelman, W.H. (1983). Genetic basis of inorganic plant nutrition. In *Encyclopedia of Plant Physiology, New Series, Vol. 15B, Inorganic Plant Nutrition*, eds. A. Lauchli & R.L. Bieleski, pp. 453–80. Berlin: Springer-Verlag.

Gorham, J., Hardy, C., Wyn Jones, R.G., Joppa, L.R. & Law, C.N. (1987). Chromosomal location of a K–Na discrimination character in the D genome of wheat. *Theoretical and Applied Genetics*, **74**, 584–8.

Grill, E., Loffler, F., Winnacker, E. & Zenk, M.H. (1989). Phytochelatins, the heavy-metal-binding peptides of plants, are synthesised from glutathione by a specific γ-glutamylcysteine dipeptidyl transpeptidase (phytochelatin synthase). *Proceedings of the National Academy of Science, USA*, **86**, 6838–42.

Grime, J.P. (1979). *Plant Strategies and Vegetation Processes*. Chichester: John Wiley & Sons.

Grime, J.P. (1983). Prediction of weed and crop response to climate based upon measurement of nuclear DNA content. In *Aspects of Applied Biology. 4. Influence of Environmental Factors on Herbicide Performance and Crop and Weed Biology*. National Vegetable Research Station, Wellesbourne.

Grime, J.P. (1991). Nutrition, environment and plant ecology: an overview. In *Plant Growth: Interactions with Nutrition and Environment*, eds. J.R. Porter & D.W. Lawlor, Society for Experimental Biology, Seminar Series 43, pp. 249–67. Cambridge University Press.

Grime, J.P., Hodgson, J.G. & Hunt, R. (1988). *Comparative Plant Ecology*. London: Unwin Hyman.

Grime, J.P. & Hunt, R. (1975). Relative growth rate; its range and adaptive significance in a local flora. *Journal of Ecology*, **63**, 393–422.

Grime, J.P., Shacklock, J.M.L. & Band, S.R. (1985). Nuclear DNA contents, shoot phenology and species coexistence in a limestone grassland community. *New Phytologist*, **100**, 435–45.

Groot, J.J.R. & Spiertz, J.H.J. (1991). The role of nitrogen in yield formation and achievement of quality standards in cereals. In *Plant Growth: Interactions with Nutrition and Environment*, eds. J.R. Porter & D.W. Lawlor, Society for Experimental Biology, Seminar Series 43, pp. 227–47. Cambridge University Press.

Hall, J.L. (1983). Cells and their organization: current concepts. In *Plant Physiology, Vol. VII, Energy and Carbon Metabolism*, eds. F.C. Steward & R.G.S. Bidwell, pp. 3–156. New York: Academic Press.

Hall, J.L. & Baker, D.A. (1977). *Cell Membranes and Ion Transport*. London: Longman.

Huber, R. & Bennett, W.S., Jr. (1983). Enzymes as biological catalysts. In *Biophysics*. eds. W. Hoppe, W. Lohmann, H. Markl & H. Ziegler, pp. 372–93. Berlin: Springer-Verlag.

Hunt, R. & Lloyd, P.S. (1987). Growth and partitioning. In *Frontiers of Comparative Plant Ecology*, eds. I.H. Rorison, J.P. Grime, G.A.F. Hendry & D.H. Lewis, pp. 235–50. London: Academic Press.

Huettl, R.F. & Fink, S. (1991). Pollution, nutrition and plant function. In *Plant Growth: Interactions with Nutrition and Environment*, eds. J.R. Porter & D.W. Lawlor, Society for Experimental Biology, Seminar Series 43, pp. 207–26. Cambridge University Press.

Ingestad, T. (1970*a*). A definition of optimum nutrient requirements in birch seedlings. I. *Physiologia Plantarum*, **23**, 1127–38.

Ingestad, T. (1970*b*). A definition of optimum nutrient requirements in birch seedlings. II. *Physiologia Plantarum*, **24**, 118–25.

Ingestad, T. (1982). A fertilizer model based on the concepts of nutrient flux density and nutrient productivity. *Scandinavian Journal of Forest Research*, **3**, 157–73.

Jefferies, R.L. & Willis, A.J. (1964). Studies on the calcicole–calcifuge habitat: II. The influence of calcium on the growth and establishment of four species in soil and sand cultures. *Journal of Ecology*, **52**, 691–707.

Kacser, H. (1987). Control of metabolism. In *The Biochemistry of Plants, Vol. II, Biochemistry of Metabolism*, ed. D.D. Davis, pp. 39–68. London: Academic Press.

Kramer, P.J. (1969). *Plant and Soil Water Relations. A Modern Synthesis*. New York: McGraw Hill Book Company.

Kuhlemeier, C., Green, P.J. & Chua, N.-H. (1987). Regulation of gene expression in higher plants. *Annual Review of Plant Physiology*, **38**, 221–57.

Lawlor, D.W. (1987). *Photosynthesis, Metabolism, Control and Physiology*. Harlow: Longman Interscience.

Lawlor, D.W., Boyle, F.A., Keys, A.J., Kendall, A.C. & Young, A.T. (1988). Nitrate nutrition and temperature effects on wheat: a synthesis of plant growth and nitrogen uptake in relation to metabolic and physiological processes. *Journal of Experimental Botany*, **39**, 329–43.

Lawlor, D.W., Day, W., Johnston, A.E., Legg, B.J. & Parkinson, K.T. (1981). Growth of spring barley under drought: crop development, photosynthesis, dry matter accumulation and nutrient content. *Journal of Agricultural Science, Cambridge*, **96**, 167–86.

Lawlor, D.W., Kontturi, M. & Young, A.T. (1989). Photosynthesis by flag leaves of wheat in relation to protein, ribulose bisphosphate carboxylase activity and nitrogen supply. *Journal of Experimental Botany*, **40**, 43–52.

Lehninger, A. L. (1975). *Biochemistry*, 2nd edn. New York: Worth Publishers Inc.

Leigh, R.A. & Storey, R. (1991). Nutrient compartmentation in cells and its relevance to the nutrition of the whole plant. In *Plant Growth: Interactions with Nutrition and Environment*, eds. J.R. Porter & D.W. Lawlor, Society for Experimental Biology, Seminar Series 43, pp. 33–54. Cambridge University Press.

Marschner, H. (1983). General introduction to the mineral nutrition of plants. In *Encyclopedia of Plant Physiology, Vol. 15A, Inorganic Plant Nutrition*, eds. A. Lauchli & R.L. Bieleski, pp. 5–60. Berlin: Springer-Verlag.

Marschner, H. (1986). *Mineral Nutrition in Higher Plants*. London: Academic Press.

Marschner, H. (1991). Plant–soil relationships: acquisition of mineral nutrients by roots from soils. In *Plant Growth: Interactions with Nutrition and Environment*, eds. J.R. Porter & D.W. Lawlor, Society for Experimental Biology, Seminar Series 43, pp. 125–55. Cambridge University Press.

Mooney, H.H. & Chiarello, N.R. (1984). The study of plant function – the plant as a balanced system. In *Perspectives of Plant Population Ecology*, eds. R. Dirzo & J. Sarukhan, pp. 305–23. Sunderland, Massachusetts: Sinauer Associates Inc.

Penning de Vries, F.W.T. (1974). Substrate utilization and respiration in relation to growth and maintenance in higher plants. *Netherlands Journal of Agricultural Science*, **22**, 40–4.

Penning de Vries, F.W.T. (1975). Use of assimilates in higher plants. In *Photosynthesis and Productivity in Different Environments*, ed. J.P. Cooper, pp. 459–80. Cambridge University Press.

Poincelot, R.P. (1979). Carbonic anhydrase. In *Encyclopedia of Plant Physiology, New Series, Vol. 6, Photosynthesis II*, eds. M. Gibbs & E. Latzko, pp. 230–8. Berlin: Springer-Verlag.

Powles, S.B. (1984). Photoinhibition of photosynthesis induced by visible light. *Annual Review of Plant Physiology*, **35**, 15–44.

Pradet, A. & Raymond, P. (1983). Adenine nucleotide ratios and adenylate energy charge in energy metabolism. *Annual Review of Plant Physiology*, **34**, 199–224.

Rains, D.W. (1976). Mineral nutrition. In *Plant Biochemistry*, 3rd edn., eds. J. Bonner & J.E. Varner, pp. 561–97. New York: Academic Press.

Rauser, W.E. & Curvetto, N.R. (1980). Metallothionein occurs in roots of *Agrostis* tolerant to excess copper. *Nature*, **287**, 563–4.

Rhue, R.D. (1979). Differential aluminium tolerance in crop plants. In *Stress Physiology in Crop Plants*, eds. H. Mussell & R.C. Staples, pp. 61–80. New York: John Wiley & Sons.

Rorison, I.H. (1966). The response to phosphorus of some ecologically distinct plant species. I. Growth rates and phosphorus absorption. *New Phytologist*, **67**, 913–23.

Rutherford, A.W. (1989). Photosynthesis II, the water-splitting enzyme. *Trends in Biochemical Science*, **14**, 227–32.

Schnepf, E. (1983). The structure of cells (prokaryotes, eukaryotes). In *Biophysics*, eds. W. Hoppe, W. Lohmann, H. Markl & H. Ziegler, pp. 1–19. Berlin: Springer-Verlag.

Smith, F.A. & Raven, J.A. (1979). Intracellular pH and its regulation. *Annual Review of Plant Physiology*, **30**, 289–311.

Stebbins, G.L. (1971). *Chromosomal Evolution in Higher Plants*. London: Edward Arnold.

Stewart, G.R. (1991). The comparative ecophysiology of plant nitrogen metabolism. In *Plant Growth: Interactions with Nutrition and Environment*, eds. J.R. Porter & D.W. Lawlor, Society for Experimental Biology, Seminar Series 43, pp. 82–97. Cambridge University Press.

Strickberger, M.L. (1976). *Genetics*, 2nd edn. New York: Machills Publishing Co. Inc.

Terry, N. & Rao, I.M. (1991). Nutrients and photosynthesis: iron and phosphorus as case studies. In *Plant Growth: Interactions with Nutrition and Environment*, eds. J.R. Porter & D.W. Lawlor, Society for Experimental Biology, Seminar Series 43, pp. 55–79. Cambridge University Press.

Thompson, W.F. (1988). Photoregulation: diverse gene responses in greening seedlings. *Plant, Cell and Environment*, **11**, 319–28.

Tobin, E.M. & Silverthorne, J. (1985). Light regulation of gene expression in higher plants. *Annual Review of Plant Physiology*, **36**, 569–93.

Tschesche, H. (1983). The chemical structure of biologically important macromolecules. In *Biophysics*, eds. W. Hoppe, W. Lohmann, L. Markl & H. Ziegler, pp. 20–41. Berlin: Springer-Verlag.

Warren-Wilson, J. (1966). Effect of temperature on net assimilation rate. *Annals of Botany*, **30**, 753–61.

Weidner, M., Mattee, C. & Schmitz, F.K. (1982). Phenotypic temperature adaptation of protein synthesis in wheat seedlings. *Plant Physiology*, **69**, 1281–8.

Woolhouse, H.W. (1983). Toxicity and tolerance in response of plants to metals. In *Encyclopedia of Plant Physiology, New Series, Vol. 12C, Physiological Plant Ecology III. Response to the Chemical and Biological Environment*, eds. O.L. Lange, P. S. Nobel, C.B. Osmond & H. Ziegler, pp. 245–300. Berlin: Springer-Verlag.

Wray, J.L. (1986). The molecular genetics of higher plant nitrate assimilation. In *A Genetic Approach to Plant Biochemistry*, eds. A.D. Blonstein & P.J. King, pp. 101–57. Wien: Springer-Verlag.

Wyn Jones, R.G., Brady, C.J. & Speirs, J. (1979). Ionic and osmotic relations in plant cells. In *Recent Advances in the Biochemistry of Cereals*, eds. D.L. Laidman & R.G. Wyn Jones, pp. 63–103. London: Academic Press.

Wyn Jones, R.G. & Pollard, A. (1983). Proteins, enzymes and inorganic ions. In *Encyclopedia of Plant Physiology, Vol. 15B, Inorganic Plant Nutrition*, eds. A. Lauchli & R.A. Bieleski, pp. 528–55. Berlin: Springer-Verlag.

ROGER A. LEIGH AND RICHARD STOREY

Nutrient compartmentation in cells and its relevance to the nutrition of the whole plant

Introduction

Plants, like all higher organisms, are a collection of cells that function in a coordinated manner to undertake all of the processes needed for successful growth and reproduction. While it is unlikely that a detailed understanding of the activities of each cell type in isolation would give a complete picture of the functioning of a whole plant, the recognition of the cell as the basic unit of biological activity nonetheless focuses attention on the mechanisms that operate at this level and demands that we consider their relevance to the whole organism. For many plant cell types, the relevance of their activities to the whole plant are readily understood, e.g. photosynthesis in leaf mesophyll cells, opening and closing of stomatal guard cells, and cell division in meristematic cells. This is because essential processes that affect the whole plant have been localised to these cells. Further study of these processes and their control leads to a better understanding of the operation of the whole organism. Such studies provide a base from which to build towards an integrated picture of the whole plant (see Lawlor, 1991, this volume).

However, it is not possible to consider the activities of cells without taking into account the fact that they are subdivided into compartments, each of which fulfils specific functions. The contribution of each compartment to a particular process needs to be understood if the integration of activities at the cell level is to be appreciated. In some cases this can lead to complicated interactions between several organelles, e.g. the involvement of chloroplasts, mitochondria and peroxisomes in photorespiration (Schnarrenberger & Fock, 1976). Therefore, just as an understanding of whole organisms requires some appreciation of the functions of whole cells, so an understanding of cells requires knowledge of the activities of subcellular compartments. Through this chain of organisation, events within a single organelle may be responsible for symptoms observed at the tissue or whole plant level.

The last ten years have seen a growing awareness of the importance of compartmentation of nutrients at the cellular level for the performance of plants under conditions of nutrient stress (e.g. Flowers, Troke & Yeo, 1977; Wyn Jones, Brady & Spiers, 1979; Leigh & Wyn Jones, 1986). This has led to the appreciation that compartmentation of nutrients, particularly between vacuole and cytoplasm, is fundamental to our understanding of plant nutrition. The aim of this chapter is to describe how an understanding of nutrient compartmentation at the cellular and subcellular levels is relevant to the operation of the whole plant. This will be done mainly with reference to the behaviour of K^+. This nutrient provides a good example because it is not metabolised and therefore different forms of the nutrient do not need to be considered. However, it will be shown that the general principles established for K have relevance to other nutrients.

We shall begin with a general description of the principles which govern the distribution of nutrients at the subcellular level, particularly between cytoplasm and vacuole, and then show how these can be useful in understanding behaviour at the whole plant level. Throughout, the cell is treated as a simple two compartment system consisting of cytoplasm and a vacuole. This is obviously a gross oversimplification but this is the level at which most information is available for nutrient ions and, fortunately, it appears to be sufficient for explaining responses of plants to changes in nutrient availability. Studies of the response of cytoplasmic organelles, such as chloroplasts, suggest that they behave in a manner similar to that of cytoplasm (Robinson & Downton, 1984; Schröppel-Meier & Kaiser, 1988*a*, *b*).

The distribution of nutrients within plant cells

It has been known for many years that solutes in plant cells are distributed between different pools. Early evidence came from experiments with ^{14}C-labelled metabolites which indicated that for many solutes there were two pools which turned over at different rates; a 'metabolic' pool in which solutes were metabolised rapidly and a 'storage' pool which was only slowly utilised (e.g. Lips & Beevers, 1966; Holleman & Key, 1967; Oaks & Bidwell, 1970). The metabolic pool was assumed to be located in the cytoplasm and the storage pool in the vacuole. In giant algal cells, it was possible to isolate these two compartments directly and analyse their contents. This demonstrated clearly that inorganic ions were asymmetrically distributed between vacuole and cytoplasm (Table 1; see MacRobbie, 1970, for further examples). Studies with higher plants using the technique of compartmental flux analysis also established asymmetric distributions of ions between cytoplasm and vacuole (data for *Avena* roots in Table 1; see also Wyn Jones *et al.*, 1979; Leigh & Wyn Jones, 1986). However, the

Table 1. *Ion concentrations in the cytoplasm and vacuole of internodal cells of giant algae and in cortical cells of roots.*

Cell	Ion concentration (mol m^{-3}) K	Na	Cl	Reference
Nitella translucens				1
Flowing cytoplasm	119	14	65	
Vacuole	75	65	160	
Nitella flexilis				1
Flowing cytoplasm	125	5	36	
Vacuole	80	28	136	
Avena sativa root				2
Cytoplasm	178	15	83	
Vacuole	174	27	65	
Hordeum vulgare root (low salt)[a]				3, 4
Cytoplasm	180	—	5	
Vacuole	15	—	4	
Hordeum vulgare root (high salt)[b]				3, 4
Cytoplasm	208	—	16	
Vacuole	168	—	13	

[a] Grown in 0.5 mol m^{-3} $CaSO_4$. [b] Grown in full nutrient solution with low Cl.
References: 1, MacRobbie (1970); 2, Pierce & Higinbotham (1970); 3, Huang & van Steveninck (1988); 4, Huang & van Steveninck (1989*a*).

emphasis of most of these early studies was on transport processes between the compartments and they had little impact on plant nutrition but this began to change when it became apparent that the compartmentation of nutrients between vacuole and cytoplasm was an important determinant of the response of plants to salinity.

Early experiments established that the ability of halophytes to withstand high concentrations of NaCl within their cells was not the result of modification of their enzymes to allow them to operate in the presence of high salt; enzymes from both halophytes and glycophytes show essentially the same sensitivity to NaCl *in vitro* (see Flowers *et al.*, 1977; Wyn Jones & Pollard, 1983). This suggested that the enzymes and salt were separated within the cells and it was proposed that the NaCl was compartmented in the vacuole where it would have no effect on the activity of salt-sensitive enzymes which were located in the cytoplasm where a relatively constant ionic environment dominated by K was maintained (Wyn Jones *et al.*, 1977; 1979). When salt loads in the vacuole were high it was proposed that osmotic balance between the cytoplasm and vacuole was maintained by the accumulation of 'compatible' solutes, such as glycinebetaine and proline, in the cytoplasm (Wyn Jones *et al.*, 1977). There is now a considerable amount of evidence to support this general model. In particular, studies with the technique of X-ray microanalysis have shown that the cytoplasm of meristematic cells is rich in K even in conditions where the vacuole of mature cells is loaded with NaCl (e.g. Storey *et al.*, 1983; Gorham & Wyn Jones, 1983; Hajibagheri *et al.*, 1988; Koyro & Stelzer, 1988). Although some studies have shown that Na can be accumulated to relatively high concentrations in the cytoplasm of mature halophyte cells (Harvey *et al.*, 1981), there is a general acceptance that mechanisms operate to prevent excessive salt loads in the cytoplasm and allow the accumulation in the cytoplasm only of those ions and solutes that will not adversely affect the operation of metabolism.

In mature plant cells, the vacuole limits the cytoplasm to about 5 to 10% of the intracellular volume, usually as a thin layer around the periphery of the cell (Dainty, 1963). Nonetheless, the cytoplasm and its organelles are the site of the majority of metabolic processes and so its proper functioning is essential to the plant. Solutes in this compartment have either a direct role in metabolism as substrates (e.g. sugar phosphates) or cofactors (e.g. Mg), or have an osmotic role (e.g. glycinebetaine). The ionic composition of the cytoplasm seems to be closely controlled. Potassium is the dominant cation and the K/Na ratio is relatively high even when the tissue contains a preponderance of Na (e.g. Gorham & Wyn Jones, 1983; Storey *et al.*, 1983; Hajibagheri *et al.*, 1988). Free Ca concentration is kept very low (Miller & Sanders, 1987) while free Mg^{2+} concentration is probably regulated at about 1 mol m^{-3} (Gibson, Spiers & Brady, 1984). Cytoplasmic P_i concentration is also maintained even though total tissue concentration may vary (e.g. Lee & Ratcliffe, 1983). In general, the plant cell appears to maintain relatively constant concentrations of ions in the cytoplasm and mechanisms exist to achieve this, often at the expense of the vacuole (see below; Leigh & Wyn Jones, 1986).

The vacuole fills the majority of the intracellular volume of a plant cell and its presence allows large cells to be produced without the need for a large investment of nitrogen and phosphorus which are scarce in many natural ecosystems (Wiebe, 1978). Solutes in the vacuole are separated from the sites of metabolic activity and this allows the accumulation of a large range of solutes in this organelle without any adverse effects on metabolism. In effect, the presence of a vacuole extends the range of solutes that can be accumulated in plant cells. If the cells were rich in cytoplasm, as are animal cells, then the range of solutes that could be accumulated would be restricted to those that are compatible with the functioning of the processes in the cytoplasm. The vacuole also allows storage of organic solutes without disturbance of the metabolic pathways involving these compounds. This has been used to advantage by plants with Crassulacean acid metabolism (CAM) which store and retrieve vacuolar malate diurnally (Kenyon, Kringstad & Black, 1978) and by sugarbeet and sugarcane which store sucrose in vacuoles (Leigh *et al.*, 1979; Thom, Maretzki & Komor, 1982).

Solutes are accumulated in the vacuole for a variety of purposes. Two have already been alluded to; the removal of 'incompatible' solutes from the cytoplasm (which has earned the vacuole the soubriquet of 'dustbin of the cell') and storage of compounds for either short (e.g. CAM plants) or long periods (e.g. sugarbeet). However, depending on the concentration to which solutes are accumulated in the vacuole, these functions are intimately linked with the third, and potentially most important, function – the generation of turgor. The vacuole is not necessary for the generation of turgor (this requires the presence of a cell wall) but the presence of the vacuole allows a greater variety of solutes to be used for this purpose than would otherwise be the case (for the reasons outlined above). There is evidence that mechanisms operate to maintain turgor within a certain range of values (e.g. Perry *et al.*, 1987) and thus total osmotic concentrations of solutes in the vacuole must be maintained within the range determined by this requirement. However, the type and mix of vacuolar solutes that may contribute to turgor appears not to be greatly restricted (e.g. Mott & Steward, 1972) although when in plentiful supply, K-salts are the dominant vacuolar solutes and turgor generation is a major physiological function of K in plants (Leigh & Wyn Jones, 1984).

The general principles outlined above for the behaviour of solutes in vacuole and cytoplasm apply also to nutrient ions and the application of these ideas allows certain predictions to be made about the likely behaviour of nutrient ions as their levels in cells change. Nutrients accumulated in the vacuole will contribute to turgor, but have no unique and essential role in this compartment. Thus providing other solutes are available to maintain turgor, the concentration of any nutrient in the vacuole may vary with supply. In contrast, concentrations in the cytoplasm should remain more constant as mechanisms operate to prevent large-scale fluctuations in the concentrations of nutrient ions in this compartment.

The behaviour and functions of nutrients in the vacuole and cytoplasm have implications for the study of nutrition in whole plants. In particular, for the way in which nutrient concentrations are expressed, and for explaining the responses of plants to changes in nutrient supply.

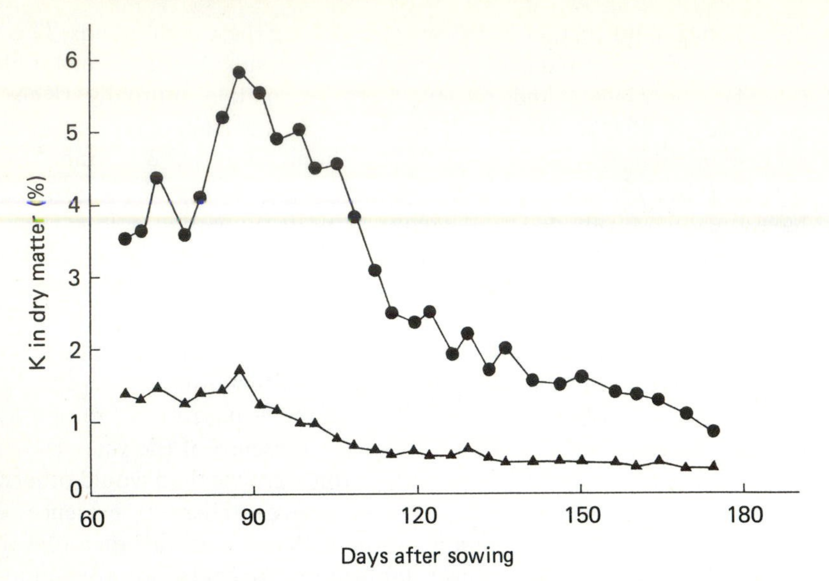

Fig. 1. Time-dependent changes in the concentration of K in shoots of barley when the concentration was expressed as a % in dry matter. The crops were grown in the Hoosfield Experiment at Rothamsted in 1981 on soils containing exchangeable K concentrations of either 382 (●) or 55 (▲) mg kg^{-1} dry soil. Further details can be found in Leigh & Johnston (1983*a*, *b*).

Implications of the basis for expressing whole-plant nutrient concentrations

It is routine in plant nutrition to use dry matter as the basis for expressing the concentrations of nutrients in whole plants or their tissues (e.g. Chapman, 1966; Bates, 1971; Jarrell & Beverly, 1981). There are good practical reasons for this; dry matter is an absolute measure of plant biomass and, unlike fresh weight, will not vary with post-harvest treatments which lead to water loss from the tissue. Thus it provides a good basis for comparison between samples collected at different times or in different ways. However, when the concentrations of nutrients in the dry matter are studied over the life-time of a plant, it is found that they vary with time. The concentrations are generally higher when the plant is young and then decline with time (Fig. 1; see also Jarrell & Beverly, 1981; Leigh & Johnston, 1983*a*; 1985; 1986). This makes it very difficult to define the minimum concentration that is required to support maximum growth without reference to the stage of growth at which the observations are made; young plants will apparently require higher concentrations for maximum growth than older plants. However, for a nutrient like K, which

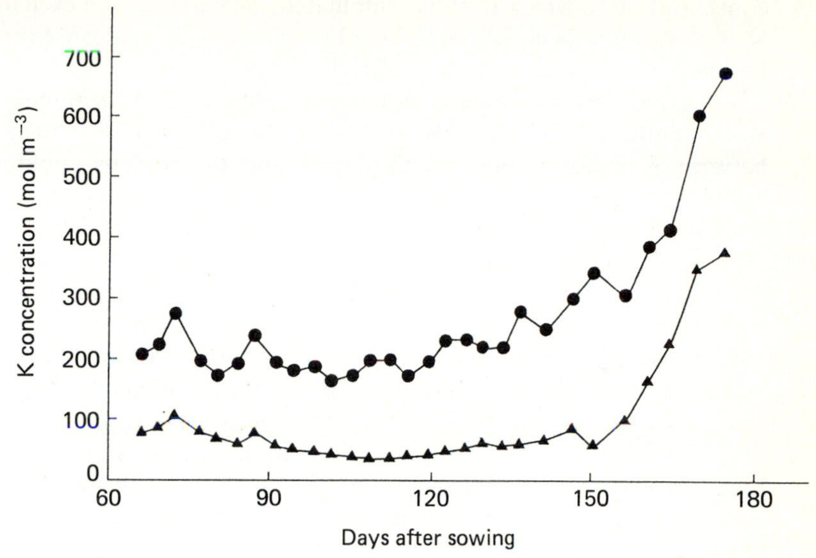

Fig. 2. Time-dependent changes in the concentration of K in shoots of barley when the concentration was expressed on the basis of tissue water. The crops are the same as those in Fig. 1.

makes major contributions to turgor, the convention of relating concentrations to dry matter ignores the fact that its physiological function is related to the concentration in the aqueous phase in the vacuole. In this situation, greater insight into the behaviour of the nutrient in relation to its intracellular function is obtained by expressing concentrations relative to tissue water.

Leigh & Johnston (1983*a*, *b*) made measurements of fresh weight, dry weight and K contents of a range of spring barley crops throughout their growth and used these data to calculate K concentrations in the dry matter and tissue water. They found that the per cent of K in dry matter varied in the way described above (see Fig. 1) and that concentrations of K expressed in this way were affected by a range of treatments e.g. N, P, and K applications, and drought. However, the behaviour of concentrations expressed relative to tissue water was very different. Firstly, the concentrations were more or less constant with time throughout vegetative and early reproductive growth and, in spring barley crops well-supplied with K, the plants maintained a concentration of 200 mol m^{-3} until senescence began (Fig. 2). Behaviour was similar in K-deficient crops except that the concentration was maintained at about 50–70 mol m^{-3} (Fig. 2). Other treatments such as N, P or drought did not affect K concentrations in tissue water and the concentration of 200 mol m^{-3} appeared to be the maximum concentration to which K was accumulated. Increases of soil exchangeable K concentration from 329 to 827 mg kg^{-1} dry soil did not cause this tissue K concentration to be exceeded (Leigh & Johnston, 1983*b*).

In the experiments sampled by Leigh & Johnston (1983*a*, *b*) the range of K concentrations in the soils was relatively restricted so a relationship between K concentrations in the crops and in the soils could not be established. However, similar studies have recently been undertaken on grass growing on the Barnfield Experiment at Rothamsted. This is a long-term field experiment which, as a result of the treatments applied in the past, now has plots with a wide range of soil K levels (Warren & Johnston, 1962). Potassium concentrations in tissue water were somewhat more variable with time than those observed in barley, for reasons that are still not established (Bletcher, Griffiths, Barraclough & Leigh, unpublished results). However, as found in barley, the average concentration maintained over an 11-week period was about 200 mol m^{-3}. This concentration was found in all grass crops growing on soils with exchangeable K concentrations greater than about 140 mg kg^{-1} dry soil; below this concentration the tissue K concentration declined (Fig. 3).

These experiments establish that K concentrations in barley and grass behave in a more predictable manner when they are expressed relative to tissue water. This is consistent with the role of this nutrient in turgor generation and the maintenance of turgor within a relatively confined range of values. A K concentration of about 200 mol m^{-3} has been observed in barley and a number of other plants growing in nutrient solutions with adequate supplies of K (Table 2; see also Ahmad & Wyn Jones, 1982). This consistency between field and laboratory experiments is heartening and suggests that the figure of 200 mol m^{-3} has real physiological significance and can be treated as the maximum value expected for K concentrations in barley and grass plants with adequate K nutrition. This has practical significance. If the concentration of 200 mol m^{-3} is the maximum value for internal K concentration and is independent of growth stage (Fig. 2) then it provides a basis for assessing the adequacy of K-supply to crops at any time during their growth. By definition, any crop which has a K concentration of 200 mol m^{-3} must have a K-supply that is adequate. However, if the concentration is significantly less than 200 mol m^{-3}, then inadequacy of supply may be assumed and steps taken to rectify the situation. The lack of effect of treatments such as N and P supply or drought on K concentrations in tissue water further enhances the usefulness of this approach because it allows concentrations in the plant to be related only to the supply of K; other factors can be ignored.

Although a wide range of plants apparently maintain K concentrations of 200 mol m^{-3} in their shoots when well-supplied with K (Table 2), this is

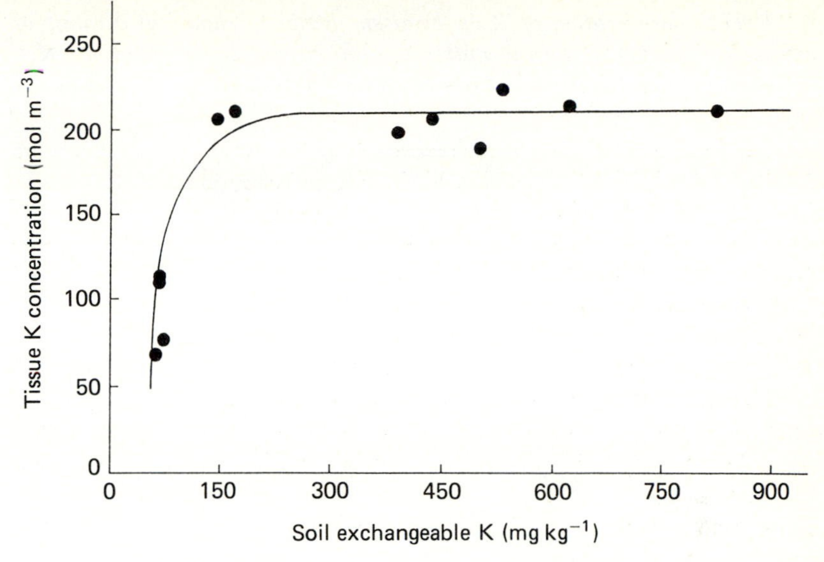

Fig. 3. The relationship between the exchangeable K concentration in soils from the Barnfield Experiment at Rothamsted, and the average concentration of K in the tissue water of grass grown on those soils in 1988. Each point is the average of samples taken over a 75-day period.

not the case for all plants. Measurements made on crops grown in the field at Rothamsted suggest that there can be quite wide variation in the average concentration of K maintained in the shoots of different crops (Table 3). Some, such as oats and wheat had average concentrations of about 200 mol m^{-3} suggesting that they behaved in a manner similar to that for barley, grass and other plants (Figs 2 and 3; Table 2). However, others, particularly field beans (*Vicia faba*) and oilseed rape (*Brassica napus*), maintained concentrations of only about 100 mol m^{-3} which indicates that their K-relations are quite different from those of cereals. It is not known whether the smaller K concentrations in these crops are matched by a concomitantly lower turgor or whether some other solute is accumulated to compensate for the lower K concentration. Oilseed rape contains large amounts of Ca (Barraclough, 1989) but its solubility state is not known so its contribution to turgor has not been estimated.

The use of tissue water as the basis for expressing concentrations can also be extended to nutrients other than K. Although the physiological basis for doing this for other nutrients is not as secure as that for K, there is some justification on the basis that plants are 80–90% water, thus if control of nutrient concentrations occurs, it is more likely that concentrations will be

Table 2. *The concentration of K in the tissue water of shoots of a range of plant species growing in flowing nutrient solution containing 1 mol* m^{-3} *K. Concentrations were calculated from the data given in Asher & Ozanne (1967).*

Plant	K concentration (mol m^{-3})
Pisum arvense	153
Vicia sativa	160
Cryptostemma calendula	176
Erodium botrys	178
Avena sativa	204
Trifolium subterraneum	210
Ornithopus sativus	214
Hordeum vulgare	215
Trifolium hirtum	229
Medicago tribuloides	230
Lolium rigidum	238
Festuca myuros	241
Ehrharta longifolia	267
Bromus rigidus	274
Average for 14 species	214

Table 3. *Average K concentrations maintained in the tissue water of the shoots of various crops sampled at Rothamsted in 1988. The values are the average of a number of samples taken during vegetative growth.*

Crop	Average K concentration (mol m^{-3})
Hordeum vulgare	184
Triticum aestivum	
cv Brimstone	170
cv Square Heads Master	178
Avena sativa	152
Solanum tuberosum	155
Vicia faba	118
Brassica napus	100

regulated relative to the natural (i.e. wet) state of living tissue than to the dessicated state of dead and dried samples. Leigh & Johnston (1985, 1986) explored the usefulness of expressing N and P concentrations in spring barley on the basis of tissue water. The results for N demonstrate that this approach can be useful for nutrients other than K (Leigh & Johnston, 1985). Expressing total N concentrations on the basis of tissue water eliminated differences between crops receiving different N fertiliser applications. Both N-sufficient and N-deficient crops maintained the same concentrations of total N on a tissue water basis, despite very large differences in growth. Leigh & Johnston (1985) proposed that this indicated that barley plants maintain the same average N concentration in cells, irrespective of the total N supply; the latter simply alters the number of cells that can be produced with the required N concentration. Thus with a good supply of N, many cells can be produced with this N concentration while fewer are produced under conditions of N-deficiency. It is these differences in the number of cells produced which lead to N-dependent differences in growth rate (Leigh & Johnston, 1985). The lack of effect of N supply on N concentrations relative to tissue water means that concentrations expressed in this way give no indication of N-deficiency even though the tissues may show visual symptoms of N-deficiency. Thus, unlike K concentrations in tissue water, N concentrations expressed on this basis have physiological but not practical significance.

Implications for the response of crops to changing nutrient supply

The different functions that nutrients play in the vacuole and cytoplasm place constraints on the way the nutrient concentrations in these compartments react to changes in total concentrations in the tissue. Concentrations in the vacuole can be expected to vary widely because nutrients accumulated in this compartment play a mainly osmotic role and can be replaced in this function by any other osmoticum that is available. In contrast, nutrients in the cytoplasm have a biochemical role and thus mechanisms should exist to maintain their concentrations at a more constant value, even in the face of large changes in the total concentration in the tissue. Leigh & Wyn Jones (1984) proposed that this difference in behaviour could be used to explain growth responses to changes in tissue K concentrations, and, in particular, the physiological basis of the 'critical' K concentration.

The relationship between growth and K concentrations in plants is curvilinear with growth showing little response to initial decreases in tissue K concentrations but then declining rapidly below the 'critical' K concentration (see Leigh & Wyn Jones, 1984). This 'critical' concentration is usually defined as the tissue K concentration at which 90% of maximum yield is obtained (Ulrich & Hills, 1967). It generally falls in the range 0.5–2% K in dry matter but the value is affected by the availability of other cations such as Na and Mg.

Potassium is known to be distributed between vacuole and cytoplasm and its functions in each of these compartments are well established (Leigh & Wyn Jones, 1984). In the vacuole, it has a purely osmotic role with K-salts contributing the majority of cell turgor when plants, particularly glycophytes, are grown with adequate supplies of this nutrient. Although K-salts seem to be preferred for turgor generation, they can be replaced by a variety of other solutes including Na- or Ca-salts (Leigh *et al.*, 1986), reducing sugars (Pitman, Mowat & Nair, 1971), or amino acids (Thom *et al.*, 1982). Therefore, as long as these other solutes are available, the concentration of K in the vacuole can be expected to vary with K supply. However, there is evidence that the concentration of K in the vacuole cannot decline to zero. Instead, there seems to be a lower concentration limit of about 10–20 mol m^{-3} below which it will not fall (see Leigh & Wyn Jones, 1984; Huang & van Steveninck, 1989*a*).

In contrast to this biophysical (i.e. osmotic) role in the vacuole, K in the cytoplasm has a biochemical role. It is well-known, for instance, that many enzymes are K-dependent (Evans & Sorger, 1966), of which phosphoenolpyruvate carboxylase is a classic example (Memon, Siddiqi & Glass, 1985). However, most enzymes require only a few mol m^{-3} of K for complete activity yet the K concentration in the cytoplasm is around 100–200 mol m^{-3} (e.g. Table 1; see also Leigh & Wyn Jones, 1984). It is thought that the explanation for this is the high requirement of protein synthesis for K. *In vitro*, this process requires K concentrations in the same range as those found in the cytoplasm (Lubin & Ennis, 1964; Wyn Jones *et al.*, 1979; Gibson *et al.*, 1984) and at concentrations both higher and lower than this range there is an inhibition of the rate of protein synthesis. Sodium is not able to substitute for K and is inhibitory to protein synthesis *in vitro* (Lubin & Ennis, 1964; Gibson *et al.*, 1984). This probably explains the high K:Na ratio that is usually found in cytoplasm (Wyn Jones *et al.*, 1979). Clearly, any major alteration of the cytoplasmic K concentration to values outside the optimal range would have a deleterious effect on the rate of protein synthesis. Therefore, mechanisms should exist to maintain the cytoplasmic K concentration within the range necessary for the operation of efficient protein synthesis. The high K:Na ratio that is found in the cytoplasm may also play a role in buffering cytoplasmic pH and the electrical potential difference across the plasma membrane (Skulachev, 1978). These two parameters are essential to growth because they provide the driving forces for many membrane transport systems (Sanders & Slayman, 1989), and a stable pH near neutrality is needed by many enzymes.

Leigh & Wyn Jones (1984) proposed that the critical K concentration is determined by the point at which cytoplasmic K concentration begins to decline. They suggested that as K supply decreases, the initial decline in tissue K concentration is a reflection of changes in the concentration within the vacuole; concentrations in the cytoplasm remain unchanged at the optimal concentration for protein synthesis. Because no essential processes are affected by the decline in vacuolar K, there is little effect on growth. However, once the vacuolar K concentration reaches its minimum value, then any further decrease is at the expense of the cytoplasm. The small volume of this compartment means that relatively small decreases in tissue K concentrations lead to large changes in cytoplasmic K concentration and this causes large changes in the rate of protein synthesis. Calculations by Leigh & Wyn Jones (1984) suggested that the point at which cytoplasmic K begins to decline does coincide with the 'critical' K concentration in many plants. The importance of maintaining high rates of protein synthesis is also emphasised by Lawlor (1991, this volume).

The model can also explain why the availability of other cations, such as Na, Mg, and Ca, can affect the 'critical' K concentration. In general, the availability of other cations allows growth to be maintained at lower tissue K concentrations, i.e. it decreases the 'critical' K concentration. This can be attributed to the ability of the cations to directly replace K in the vacuole. In their absence, organic solutes have to be accumulated in the vacuole to maintain turgor (e.g. Pitman *et al.*, 1971; Mott & Steward, 1972) thus diverting assimilate from growth processes and leading to a decline in yield. This diversion must occur as soon as K in the vacuole begins to decrease so growth is affected at tissue K concentrations at which cytoplasmic K is still unaffected.

There is evidence that cytoplasmic and vacuolar K concentrations behave in the way predicted. Leigh *et al.* (1986) used the technique of X-ray microanalysis to examine the distribution of K in mesophyll cells of barley leaves grown in nutrient solutions containing different K concentrations. They found that in plants grown at a high K concentration, K was detectable in both vacuole and cytoplasm. However, in plants grown at a low K concentration, it was undetectable in the vacuole but was still detectable in the cytoplasm. Thus the differential behaviour of K in the two compartments was confirmed.

Results with ^{31}P nuclear magnetic resonance have demonstrated that P_i concentrations in the vacuole and cytoplasm behave in a similar way to that described for K. Removal of the external P supply from pea roots caused a large decline in total tissue P_i content which was due entirely to the depletion of the vacuolar pool (Lee & Ratcliffe, 1983). Despite this, growth declined by only 2% indicating that growth was insensitive to changes in vacuolar P_i level. In contrast, growth of cultured cells of *Acer pseudoplatanus* stopped once the cytoplasmic P_i concentration fell below an unspecified threshold (Rebeille *et al.*, 1983). Thus, as with K, the response of growth to shortages of P appears to be largely influenced by cytoplasmic P_i concentration rather than by that in the vacuole.

Similar considerations probably also apply to NO_3^-. In general, NO_3^- only accumulates when the N requirements of growth processes are satisfied and, once this point is reached, tissue NO_3^- levels can vary widely with little effect on growth (Hylton *et al.*, 1964; Zhen & Leigh, 1991). It is known that the majority of the NO_3^- is located in the vacuole (Martinoia, Heck & Wiemken, 1981; Granstedt & Huffaker, 1982). Therefore, it can be surmised that variation in vacuolar NO_3^- has little effect on growth rate. The relation between cytoplasmic NO_3^- and growth cannot be elucidated at this time because there are no techniques that readily allow the NO_3^- concentration in this compartment to be monitored. However, NO_3^--sensitive microelectrodes that can be used to impale plant cells have been developed (A.J. Miller, personal communication) and these should allow progress to be made on this problem.

The importance of compartmentation between cells

The majority of the foregoing has been based on the premise that knowledge of the behaviour of nutrient concentrations in individual cells can be used to interpret the events at the whole plant level. In effect, the whole plant has been treated as a collection of identical cells. This is clearly an oversimplification as different cell types, and even adjacent cells of the same type, are known to differ greatly in their composition (e.g. van Steveninck *et al.*, 1982; Hodson & Sangster, 1988; Huang & van Steveninck, 1989*b*). However, the importance of these intercellular differences has never been clearly established but evidence has begun to emerge that there may be a pattern to the way in which different nutrients are distributed between different cell types and that these could play a role in adaption to stress.

The evidence has come from the application of X-ray microanalysis to different cell types in leaves, particularly those from barley. Leigh *et al.* (1986) found that although K-deficiency caused an increase in the Ca level in barley leaves, this element was undetectable in mesophyll cells, instead all detectable Ca was located in the vacuoles of epidermal cells. More detailed analysis (Storey & Leigh, unpublished results) confirmed this and showed that Cl and P were also differentially compartmented between epidermal and mesophyll cells (Table 4). Essentially all of the P that was detectable was found in the mesophyll cells whereas Cl and Ca were found in the

Table 4. *Distribution of P, Cl, and Ca between epidermal and mesophyll cells in barley leaves, determined by X-ray microanalysis.*

Cell type	Number of cells analysed	Per cent of cells with detectable level of element[a]		
		P	Cl	Ca
Mesophyll	97	71	0	0
Epidermis				
Adaxial	127	0	62	72
Abaxial	109	0	67	61

[a] Detectable level of element indicated by a peak : background ratio for that element of > 0.03.

epidermis. This indicates that as nutrients move through the leaf in the transpiration stream, the mesophyll cells selectively absorb particular ions, leaving others to be absorbed by the epidermis. The selective compartmentation of Cl in the epidermis has also been reported for barley (Huang & van Steveninck, 1989*b*) and for wheat (Hodson & Sangster, 1988).

The physiological consequences of this differential distribution between cells remain to be established. However, in barley it is maintained in salt stress (Huang & van Steveninck, 1989*b*) suggesting that exclusion of Cl from the mesophyll could be a mechanism for reducing the Cl load in these cells. Thus the ability of the epidermis to act as a sink for Cl could be an important mechanism for increasing the salt tolerance of plants. In barley, the less salt-sensitive variety, Clipper, had higher Cl concentrations in its mesophyll cells than the more tolerant variety, California Mariout (Huang & van Steveninck, 1989*b*). It would be interesting to know whether differences in the ability of different plant species to selectively accumulate Cl in the epidermis can explain differences in their responses to salt. Sodium does not seem to be excluded from the mesophyll in the same way as Cl. In K-deficient barley, Na is accumulated to relatively high concentrations in the mesophyll before it appears in the epidermis (Leigh *et al.*, 1986; Storey & Leigh, unpublished results). This may indicate that this ion can be more effectively compartmented in the vacuole of mesophyll cells than Cl and so it does not need to be so stringently excluded from these cells.

The finding that some ions are excluded from the mesophyll suggests that some control is exerted over the composition of these cells. This is also suggested by X-ray microanalysis measurements of K levels in these cells compared with the epidermis. Fig. 4 shows a comparison of the peak : background ratios (a relative measure of K concentration, see Huang

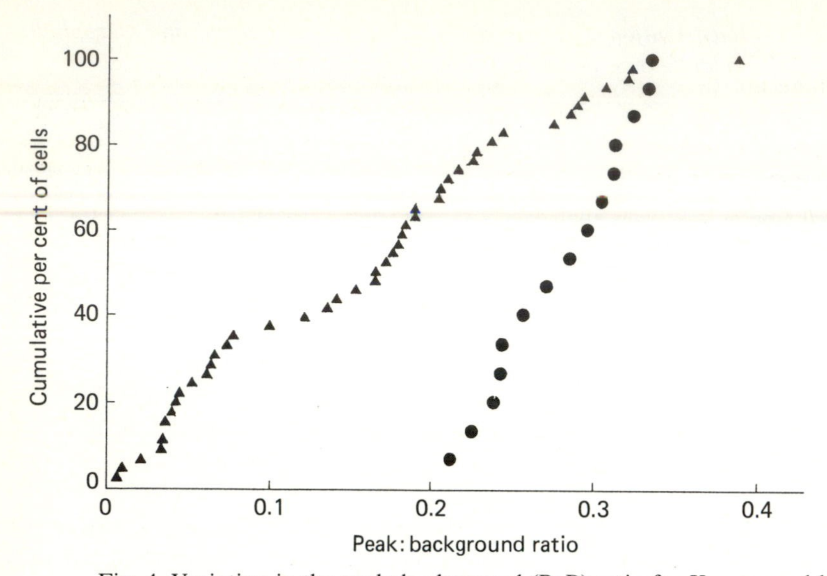

Fig. 4. Variation in the peak: background (P: B) ratio for K measured by X-ray microanalysis of epidermal (▲) or mesophyll (●) cells in a bulk-hydrated frozen sample from a single barley leaf. Data are plotted as the cumulative per cent of cells having P: B equal to or less than the value indicated.

& van Steveninck, 1989*a*, *b*) calculated from X-ray microanalysis spectra of mesophyll and epidermal cells from the same barley leaf. The data are plotted as the cumulative percentage of cells having particular peak: background ratios. It can be seen that the spread of values for the epidermis is very wide whereas the range is much more restricted for the mesophyll. These results are consistent with the idea that the mesophyll K concentration is regulated more highly than that in the epidermis.

The finding that nutrient ions can be differentially distributed between different cell types has implications for the interpretation of responses of crops to nutrient stresses. In general, responses are interpreted on the basis of total tissue concentrations of nutrients (see above). However, if nutrients are differentially distributed, then such interpretations will have little validity because relating changes in physiological processes to average tissue nutrient concentrations is irrelevant if the processes and the nutrients are not located in the same cells. Instead knowledge is needed of the way in which distributions of nutrients between different cell types change in response to stress and whether intercellular compartmentation breaks down under particular conditions.

Conclusion

Plants are not simply sponges that absorb nutrients from their surroundings in an uncontrolled way. Instead they regulate their composition at a number of different levels and for a number of different reasons. In this chapter, evidence has been given for control of ion, particularly K, concentrations in cytoplasm, in mesophyll cells, and in whole shoots. By taking account of this regulation in the interpretation of plant responses to changes in nutrient supply it is possible to gain greater insight into the underlying events and to establish those processes that need to be modified to improve the responses of crops to nutrient stresses.

References

Ahmad, N. & Wyn Jones, R.G. (1982). Tissue distribution of glycinebetaine, proline and inorganic ions in barley at different times during the plant growth cycle. *Journal of Plant Nutrition*, **5**, 195–205.

Asher, C.J. & Ozanne, P.G. (1967). Growth and potassium contents of plants in solution cultures maintained at constant potassium concentrations. *Soil Science*, **103**, 155–61.

Barraclough, P.B. (1989). Root growth, macro-nutrient uptake dynamics and soil fertility requirements of a high-yielding winter oilseed rape crop. *Plant and Soil*, **119**, 59–70.

Bates, T.E. (1971). Factors affecting critical nutrient concentrations in plants and their evaluation: a review. *Soil Science*, **112**, 116–30.

Chapman, H.D. (1966). *Diagnostic Criteria for Plants and Soils*. Davis: University of California, Division of Agricultural Sciences.

Dainty, J. (1963). The structure and possible function of the vacuole. In *Plant Cell Organelles*, ed. J.B. Pridham, pp. 40–6. London: Academic Press.

Evans, H.J. & Sorger, G.J. (1966). Role of mineral elements with the emphasis on univalent cations. *Annual Review of Plant Physiology*, **17**, 47–76.

Flowers, T.J., Troke, P.F. & Yeo, A.R. (1977). The mechanisms of salt tolerance in halophytes. *Annual Review of Plant Physiology*, **28**, 89–121.

Gibson, T.S., Spiers, J. & Brady, C.J. (1984). Salt-tolerance in plants. II. *In vitro* translation of m-RNAs from salt-tolerant and salt-sensitive plants on wheat germ ribosomes. Responses to ions and compatible solutes. *Plant, Cell and Environment*, **7**, 579–87.

Gorham, J. & Wyn Jones, R.G. (1983). Solute distribution in *Suaeda maritima*. *Planta*, **157**, 344–9.

Granstedt, R.C. & Huffaker, R.C. (1982). Identification of the leaf vacuole as a major nitrate storage pool. *Plant Physiology*, **70**, 410–13.

Hajibagheri, M.A., Flowers, T.J., Collins, J.C. & Yeo, A.R. (1988). A comparison of the methods of X-ray microanalysis, compartmental analysis and longitudinal ion profiles to estimate cytoplasmic ion concentrations in two maize varieties. *Journal of Experimental Botany*, **39**, 279–90.

Harvey, D.M.R., Hall, J.L., Flowers, T.J. & Kent, B. (1981). Quantitative ion localisation within *Suaeda maritima* leaf mesophyll cells. *Planta*, **151**, 555–60.

Hodson, M.J. & Sangster, A.G. (1988). Observations on the distribution of mineral elements in the leaf of wheat (*Triticum aestivum* L.), with particular reference to silicon. *Annals of Botany*, **62**, 463–71.

Holleman, J.M. & Key, J.L. (1967). Inactive and protein precursor pools of amino acids in soybean hypocotyl. *Plant Physiology*, **58**, 656–62.

Huang, C.X. & van Steveninck, R.F.M. (1988). Effect of moderate salinity on patterns of potassium, sodium and chloride accumulation in cells near the root tip of barley: role of differentiating metaxylem vessels. *Physiologia Plantarum*, **73**, 525–33.

Huang, C.X. & van Steveninck, R.F.M. (1989*a*). Longitudinal and transverse profiles of K^+ and Cl^- concentration in 'low-' and 'high-salt' barley roots. *New Phytologist*, **112**, 475–80.

Huang, C.X. & van Steveninck, R.F.M. (1989*b*). Maintenance of low Cl^- concentrations in mesophyll cells of leaf blades of barley seedlings exposed to salt stress. *Plant Physiology*, **90**, 1440–3.

Hylton, L.O., Williams, D.E., Ulrich, A. & Cornelius, D.R. (1964). Critical nitrate concentrations for growth of Italian ryegrass. *Crop Science*, **4**, 16–19.

Jarrell, W.M. & Beverly, R.B. (1981). The dilution effect in plant nutrition studies. *Advances in Agronomy*, **34**, 197–224.

Kenyon, W.H., Kringstad, R. & Black, C.C. (1978). Diurnal changes in the malic acid content of vacuoles isolated from leaves of the Crassulacean acid metabolism plant, *Sedum telephium*. *FEBS Letters*, **94**, 281–3.

Koyro, H.-W. & Stelzer, R. (1988). Ion concentrations in the cytoplasm and vacuoles of rhizodermis cells from NaCl treated *Sorghum*, *Spartina* and *Puccinellia* plants. *Journal of Plant Physiology*, **133**, 441–6.

Lawlor, D.W. (1991). Concepts of nutrition in relation to cellular processes and environment. In *Plant Growth: Interactions with Nutrition and Environment*, eds. J.R. Porter & D.W. Lawlor, Society for Experimental Biology, Seminar Series 43, pp. 1–32. Cambridge University Press.

Lee, R.B. & Ratcliffe, R.G. (1983). Phosphorus nutrition and the intracellular distribution of inorganic phosphate in pea root tips: A quantitative study using ^{31}P-NMR. *Journal of Experimental Botany*, **34**, 1222–44.

Leigh, R.A., ap Rees, T., Fuller, W.A. & Banfield, J. (1979). The location of acid invertase activity and sucrose in the vacuoles of storage roots of beetroot (*Beta vulgaris* L.). *Biochemical Journal*, **178**, 539–47.

Leigh, R.A., Chater, M., Storey, R. & Johnston, A.E. (1986). Accumulation and subcellular distribution of cations in relation to the growth of potassium-deficient barley. *Plant, Cell and Environment*, **9**, 595–604.

Leigh, R.A. & Johnston, A.E. (1983*a*). Concentrations of potassium in the dry matter and tissue water of field-grown spring barley and their relationships to grain yield. *Journal of Agricultural Science, Cambridge*, **101**, 675–85.

Leigh, R.A. & Johnston, A.E. (1983*b*). The effects of fertilizers and drought on the concentrations of potassium in the dry matter and tissue water of field-grown spring barley. *Journal of Agricultural Science, Cambridge*, **101**, 741–8.

Leigh, R.A. & Johnston, A.E. (1985). Nitrogen concentrations in field-grown spring barley: an examination of the usefulness of expressing concentrations on the basis of tissue water. *Journal of Agricultural Science, Cambridge*, **105**, 397–406.

Leigh, R.A. & Johnston, A.E. (1986). An investigation of the usefulness of phosphorus concentrations in tissue water as indicators of the phosphorus status of field-grown spring barley. *Journal of Agricultural Science, Cambridge*, **107**, 329–33.

Leigh, R.A. & Wyn Jones, R.G. (1984). A hypothesis relating the critical potassium concentration for growth to the distribution and functions of this ion in the plant cell. *New Phytologist*, **97**, 1–13.

Leigh, R.A. & Wyn Jones, R.G. (1986). Cellular compartmentation in plant nutrition: The selective cytoplasm and the promiscuous vacuole. In *Advances in Plant Nutrition*, Vol. 2, eds. P.B. Tinker & A. Läuchli, pp. 249–79. New York: Praeger.

Lips, S.H. & Beevers, H. (1966). Compartmentation of organic acids in corn roots. I. Differential labeling of 2 malate pools. *Plant Physiology*, **41**, 709–12.

Lubin, M. & Ennis, H.L. (1964). The role of intracellular potassium in protein synthesis. *Biochimica et Biophysica Acta*, **80**, 614–31.

MacRobbie, E.A.C. (1970). The active transport of ions in plant cells. *Quarterly Review of Biophysics*, **3**, 251–94.

Martinoia, E., Heck, U. & Wiemken, A. (1981). Vacuoles as storage compartments for nitrate in barley leaves. *Nature*, **289**, 292–4.

Memon, A.R., Siddiqi, M.Y. & Glass, A.D.M. (1985). Efficiency of K^+ utilization by barley varieties: Activation of pyruvate kinase. *Journal of Experimental Botany*, **36**, 79–90.

Miller, A.J. & Sanders, D. (1987). Depletion of cytosolic free calcium induced by photosynthesis. *Nature*, **326**, 397–400.

Mott, R.L. & Steward, F.C. (1972). Solute accumulation in plant cells. I. Reciprocal relations between electrolytes and nonelectrolytes. *Annals of Botany*, **36**, 915–37.

Oaks, A. & Bidwell, G.S. (1970). Compartmentation of intermediary metabolites. *Annual Review of Plant Physiology*, **21**, 43–66.

Perry, C.A., Tomos, A.D., Leigh, R.A., Wyse, R.E. & Hall, J.L. (1987). The regulation of turgor pressure during sucrose mobilisation and salt accumulation by excised storage-root tissue of red beet. *Planta*, **170**, 353–61.

Pierce, W.S. & Higinbotham, N. (1970). Compartmentation and fluxes of K^+, Na^+, and Cl^- in *Avena* coleoptile cells. *Plant Physiology*, **46**, 666–73.

Pitman, M.G., Mowat, J. & Nair, H. (1971). Interpretation of the processes for accumulation of salt and sugar in barley roots. *Australian Journal of Biological Sciences*, **24**, 619–31.

Rebeille, F., Bligny, R., Martin, J.-P. & Douce, R. (1983). Relationship between the cytoplasm and vacuole phosphate pool in *Acer pseudoplatanus* cells. *Archives of Biochemistry and Biophysics*, **225**, 143–8.

Robinson, S.P. & Downton, W.J. (1984). Potassium, sodium, and chloride content of isolated intact chloroplasts in relation to ionic compartmentation in leaves. *Archives of Biochemistry and Biophysics*, **228**, 197–206.

Sanders, D. & Slayman, C.L. (1989). Transport at the plasma membrane of plant cells: a review. In *Plant Membrane Transport: The Current Position*, eds. J. Dainty, M.I. de Michelis, E. Marré & F. Rasi-Caldogno, pp. 3–11. Amsterdam: Elsevier.

Schnarrenberger, C. & Fock, H. (1976). Interactions among organelles involved in photorespiration. In *Encyclopedia of Plant Physiology, New Series*, vol. 3, eds. C.R. Stocking & U. Heber, pp. 185–234. Berlin: Springer-Verlag.

Schröppel-Meier, G. & Kaiser, W.M. (1988*a*). Ion homeostasis in chloroplasts under salinity and mineral deficiency. I. Solute concentrations in leaves and chloroplasts from spinach plants under NaCl or $NaNO_3$ salinity. *Plant Physiology*, **87**, 822–7.

Schröppel-Meier, G. & Kaiser, W.M. (1988*b*). Ion homeostasis in chloroplasts under salinity and mineral deficiency. II. Solute distribution between chloroplasts and extrachloroplastic space under excess or deficiency of sulfate, phosphate, or magnesium. *Plant Physiology*, **87**, 828–32.

Skulachev, V.P. (1978). Membrane-linked energy buffering as the biological function of the Na^+/K^+ gradient. *FEBS Letters*, **87**, 171–9.

Storey, R., Pitman, M.G., Stelzer, R. & Carter, C. (1983). X-ray microanalysis of cells and cell compartments in *Atriplex spongiosa*. I. Leaves. *Journal of Experimental Botany*, **34**, 778–94.

Thom, M., Maretzki, A. & Komor, E. (1982). Vacuoles from sugarcane suspension cultures. I. Isolation and partial characterization. *Plant Physiology*, **69**, 1315–19.

Ulrich, A. & Hills, F.J. (1967). Principles and practices of plant analysis. In *Soil Testing and Plant Analysis*, Part II, ed. G.W. Hardy, Special Publication No. 2, pp. 11–24. Madison, Wisconsin: Soil Science Society of America.

van Steveninck, R.F.M., van Steveninck, M.E., Stelzer, R. & Läuchli, A. (1982). Variations in vacuolar solutes of *Lupinus luteus* L. leaf tissue shown by electron probe X-ray analysis. *Zeitschrift für Pflanzenphysiologie*, **107**, 91–5.

Warren, R.G. & Johnston, A.E. (1962). Barnfield. *Rothamsted Experimental Station Report for 1961*, pp. 227–47.

Wiebe, H.H. (1978). The significance of plant vacuoles. *Bioscience*, **28**, 327–31.

Wyn Jones, R.G., Brady, C.J. & Spiers, J. (1979). Ionic and osmotic relations in plant cells. In *Recent Advances in the Biochemistry of Cereals*, eds. D.L. Laidman & R.G. Wyn Jones, pp. 63–103. London: Academic Press.

Wyn Jones, R.G. & Pollard, A. (1983). Proteins, enzymes and inorganic ions. In *Encyclopedia of Plant Physiology, New Series, Vol. 15B, Inorganic Plant Nutrition*, eds. A. Läuchli & R.L. Bielski, pp. 528–62. Berlin: Springer-Verlag.

Wyn Jones, R.G., Storey, R., Leigh, R.A., Ahmad, N. & Pollard, A. (1977). A hypothesis on cytoplasmic osmoregulation. In *Regulation of Cell Membrane Activities in Plants*, eds. E. Marré & O. Ciferri, pp. 121–36. Amsterdam: North-Holland.

Zhen, R.G. & Leigh, R.A. (1991). Nitrate accumulation in relation to growth and tissue N concentrations. In *Proceedings of the XIth International Plant Nutrition Colloquium, Amsterdam, 1989*. Dordecht: Kluwer Academic Publishers (in press).

NORMAN TERRY AND I. MADHUSUDANA RAO

Nutrients and photosynthesis: iron and phosphorus as case studies

Introduction

Although many of the 13 mineral elements known to be required for plant growth have been shown to influence photosynthesis in some way, relatively few have been comprehensively studied from the molecular to the whole plant level. Three mineral nutrients which have been studied in depth are iron, phosphorus and nitrogen. The relationship of photosynthesis to nitrogen supply was the subject of an excellent review by Evans (1989); since we have extensively investigated iron and phosphorus, these were selected as case studies.

These two elements are particularly appropriate to serve as case studies for exploring the interactions of photosynthesis with nutrient supply. Iron is a micronutrient which has a major effect on the light reactions of photosynthesis – a fact which led to the use of iron deficiency as an experimental tool to study photosynthetic limitation (Terry, 1980; Taylor & Terry, 1984). Phosphorus, on the other hand, is a macronutrient which is more involved with the reductive pentose phosphate (RPP) pathway (Calvin cycle) of photosynthesis (Brooks, 1986; Rao & Terry, 1989; Rao, Arulanantham & Terry, 1989). Phosphorus is important in other respects; the level of orthophosphate (P_i) in leaves probably regulates photosynthesis

The following abbreviations are used in this chapter: ADPG, adenosine 5′-diphosphoglucose; C_i, intercellular CO_2 partial pressure; CP, chlorophyll-protein; cyt, cytochrome; DHAP, dihydroxyacetone phosphate; DPGA, diphosphoglycerate; ER, Rubisco enzyme–RuBP complex; F6P, fructose-6-phosphate; FBP, fructose 1,6-bisphosphate; F2,6BP, fructose 2,6-bisphosphate; GAP, glyceraldehyde phosphate; G3P, glyceraldehyde-3-phosphate; G1P, glucose-1-phosphate; G6P, glucose-6-phosphate; IMP, inosine 5′-monophosphate; K_m, Michaelis constant; LHC, light harvesting chlorophyll a/b protein; PFD, photon flux density; 6PG, 6-phosphogluconate; PGA, 3-phosphoglycerate; P_i, orthophosphate; R5P, ribose-5-phosphate; RPP, reductive pentose phosphate; Ru5P, ribulose-5-phosphate; RuBP, ribulose 1,5-bisphosphate; Rubisco, ribulose 1,5-bisphosphate carboxylase–oxygenase; SBP, sedoheptulose 1,7-bisphosphate; S7P, sedoheptulose-7-phosphate; triose-P, glyceraldehyde-3-phosphate + dihydroxyacetone phosphate; UDPG, uridine 5-diphosphoglucose; UTP, uridine 5′-triphosphate.

and carbon partitioning through the P_i-translocator located on the inner membrane of the chloroplast envelope. Accordingly, the concentration of P_i in the cytosol controls the partitioning of newly formed assimilates between sucrose and starch synthesis (Heldt *et al.*, 1977; Walker, 1980; Leegood, Walker & Foyer, 1985; Stitt, Huber & Kerr, 1987). The effects of iron and phosphorus supply on photosynthesis are discussed separately in the following two sections.

Iron supply and photosynthesis

Iron and thylakoids

Iron has an important role in the formation of thylakoid membranes which determine the effects of iron deficiency on photosynthesis. Iron-deficient plants, characteristically, develop chloroplasts with few thylakoids (Fig. 1). In sugar beet, iron deficiency markedly reduces the amount of thylakoid membranes per chloroplast without appreciably affecting other regions of the cell. This is evident in the electron micrographs of Platt-Aloia, Thomson & Terry (1983) who found that visible effects of iron deficiency were confined to the chloroplasts and that even iron-containing organelles such as peroxisomes and mitochondria were unaffected structurally by iron deficiency.

The quantitative loss of thylakoid membranes with iron deficiency has been assessed using measurements of membrane constituents such as thylakoid lipids and proteins. Thylakoid membranes are rich in galactolipids which constitute 70% of the non-pigment lipids (Hiller & Goodchild, 1981). Iron deficiency reduced thylakoid galactolipids by 75% and thylakoid total protein by 60%, while reducing total chlorophyll content by 90% (Nishio, Taylor & Terry, 1985*a*). Chlorophyll a, chlorophyll b and carotene decrease in proportion with iron deficiency. Xanthophyll decreases to a lesser extent so that iron-deficient leaves are characteristically richer in xanthophyll (Bolle-Jones & Notton, 1953; Terry, 1980).

Iron is a constituent of many electron carriers in the electron transport chain (Hewitt, 1983; Hipkins, 1983; Sandmann & Malkin, 1983). The water-splitting unit associated with photosystem II, which is located mainly in the appressed thylakoid membranes (and in the granal-stromal partition regions), has several iron-containing components, as does photosystem I. The loss of thylakoids with iron deficiency is associated with decreased amounts of electron carriers. In sugar beet, iron deficiency diminishes cytochrome f per area (Spiller & Terry, 1980), as well as the Fe–S centres associated with photosystem I (Spiller, 1979). Ferredoxin amounts decrease in iron deficient *Vigna sinensis* L. (Marsh, Evans & Matrone, 1963) and in lemon (*Citrus limonum*) (Alcaraz *et al.*, 1985). The latter research shows that iron deficiency reduced chlorophyll to a much greater extent than it

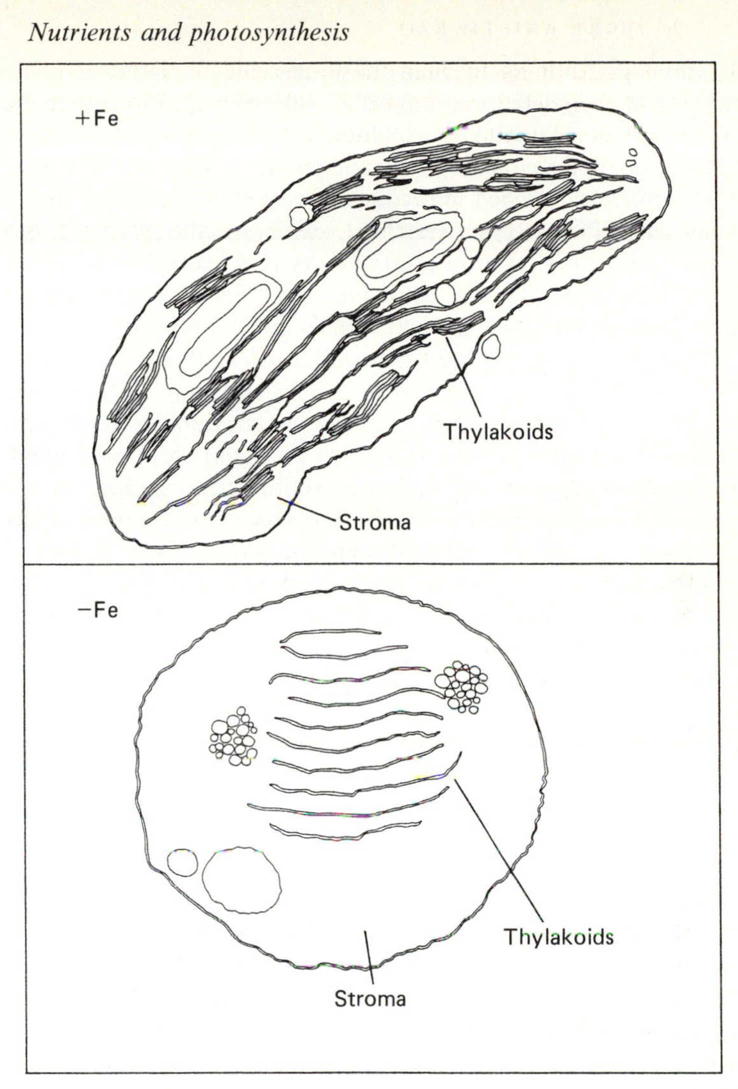

Fig. 1. Effect of iron stress on chloroplast structure. The upper drawing depicts a chloroplast from an iron-sufficient control sugar beet leaf. The lower drawing, which depicts a chloroplast from a severely iron-stressed plant, shows a small number of thylakoids and a relatively large stroma (after Spiller & Terry, 1980).

decreased ferredoxin, suggesting that iron deficiency exerts a greater effect on thylakoid constituents (chlorophyll) than on stromal components (ferredoxin) (Terry & Abadia, 1986).

Iron deficiency reduces the number of photochemical units per leaf area. Based on the observation that iron deficiency diminished the amounts of chlorophyll, P_{700} and cytochrome f to similar extents, Spiller & Terry (1980) proposed that iron deficiency reduced the number of 'photosynthetic units' (consisting of approximately 400 chlorophyll molecules per cytochrome f or per P_{700}). Today, the concept of a single photosynthetic unit embracing two photosystems has changed in favour of two separate photochemical units, photosystem II and photosystem I, with specific pigment molecules organised with each reaction centre (Terry & Abadia 1986).

Using gel electrophoresis techniques, the effects of iron deficiency on specific chlorophyll-proteins and other polypeptides of the thylakoid membrane have been assessed (Machold, 1971; Price, Ortiz & Gaynor, 1978; Guikema & Sherman, 1984). Using this approach, Nishio *et al.* (1985*a*, *b*) showed that iron deficiency reduces the number of photochemical units in the membrane compared to other proteins (e.g. structural proteins). Iron deficiency appears to diminish photosystem I electron transport activity more than photosystem II. Nishio *et al.* (1985*b*) found photosystem II activity was slightly increased and photosystem I activity slightly decreased (per chlorophyll) by iron deficiency and that CP1 was more diminished by iron deficiency than CPa.

Light absorption and photochemistry

Research with sugar beet has shown that iron deficiency affects light absorption much less than electron transport, despite the large loss of pigment (especially chlorophyll) associated with the decrease in thylakoids (Fig. 2) (Terry, 1980). Absorptance (the proportion of incoming light energy absorbed by the leaf) remains high (about 50%) even in leaves which have lost over 90% of their chlorophyll (Fig. 2*a*) (Terry, 1980). There are several possible reasons for this. Firstly, there is more light absorption by carotenoids in iron-deficient leaves (xanthophylls are increased in proportion to the decreased chlorophyll concentration). Secondly, due to light scattering, leaves absorb light more efficiently than pigment solutions of comparable concentration.

Photochemical conversion of absorbed light proceeds efficiently in iron-deficient leaves at low photon flux densities (PFD). Photosynthetic quantum yields measured for attached leaves illuminated with red light in the absence of oxygen were about 0.1 mol CO_2 fixed per mol photons absorbed and did not vary with iron deficiency (Terry, 1980). As the incident photon flux increased, however, the iron-deficient leaves were less able to utilise the incoming photons due to the decreased capacity for photosynthetic electron transport (Fig. 2*b*) (Terry, 1980; Nishio *et al.*, 1985*b*). This is clearly a consequence of the reduction in reaction centres and electron carriers associated with the loss of thylakoids.

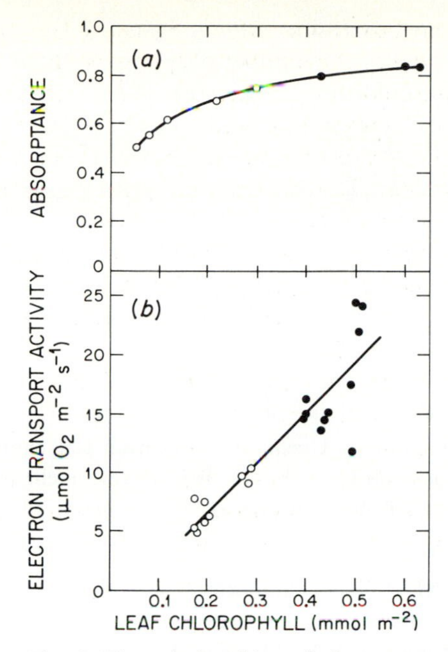

Fig. 2. The relationships of absorptance (*a*) and photosynthetic electron transport activity (*b*) with leaf chlorophyll content in Fe-deficient (○) and Fe-sufficient (●) plants. Absorptance is the proportion of incident photon flux density absorbed by leaves. Electron transport activity was measured from $H_2O \rightarrow FeCN$, monitoring O_2 evolution with an O_2 electrode.

Specificity and reversibility

Electron microscopy revealed that iron deficiency diminishes the thylakoid system with no apparent effect on other parts of the cell. That the effects of iron stress are specific to the thylakoids is supported by other evidence (Terry, 1980). Iron deficiency had no effect on the number of chloroplasts per cell or per unit leaf area and chloroplasts apparently replicated at normal rates under iron stress. Leaf structural attributes such as the number of cells per unit leaf area and average leaf cell volume were unaffected by iron stress. In fact, leaves grew almost as fast under iron deficiency as leaves of control plants, and, at the same stage of leaf development, had the same thickness, fresh weight per area and percentage intercellular space as normal leaves (Terry, 1979).

Whereas iron stress decreased the thylakoid components of chloroplasts by 90 %, ribulose bisphosphate (RuBP) carboxylase–oxygenase (Rubisco) activity on the other hand was diminished by only 30 % (Terry, 1980). The total extractable activities of other RPP pathway enzymes, fructose 1,6-

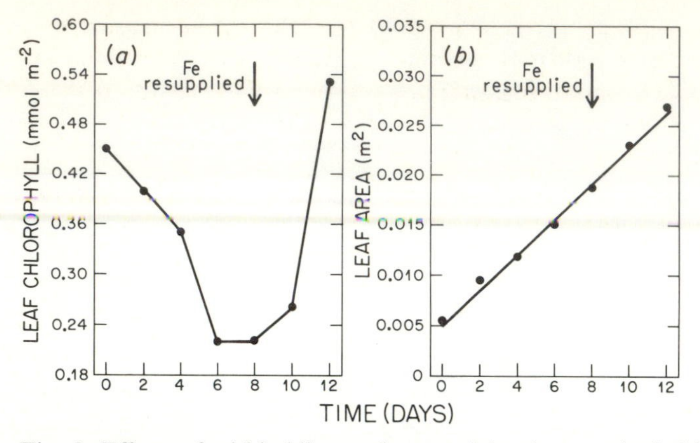

Fig. 3. Effects of withholding and resupplying iron on leaf chlorophyll content (*a*) and leaf expansion (*b*) of leaves. Iron was withheld from the treated plant on day 0 and resupplied on day 8 (after Terry, 1979).

bisphosphatase (FBPase), NADP-glyceraldehyde-3-phosphate (G3P) dehydrogenase, and ribulose-5P (Ru5P) kinase, were not decreased by iron deficiency (Taylor, Terry & Huston, 1982). These and other data show that iron deficiency had much less effect on the RPP pathway enzymes than on thylakoids.

The effect of iron stress on the thylakoid system was not only specific but reversible (Fig. 3) (Terry, 1979). When iron was supplied after eight days of iron deficiency, the chlorophyll content of the leaf returned to normal within 96 hours (Fig. 3*a*). Leaf growth rate remained constant irrespective of iron resupply (Fig. 3*b*). Since the amount of chlorophyll per leaf did not change during iron deficiency, it appears that the thylakoid material was conserved and diluted during leaf growth with chloroplast replication resulting in less thylakoid material per chloroplast. Electron transport components such as Q, P_{700}, cytochrome f, lamellar iron and lamellar manganese also increased when iron was supplied again (Nishio & Terry, 1983; Young & Terry, 1983).

Iron stress as an experimental technique to study photosynthetic limitation *in vivo*

By comparing the gas exchange of leaves from iron-sufficient and iron-deficient plants, it is possible to determine how changes in photosynthetic electron transport capacity influence photosynthetic rate in specific light and CO_2 environments (Terry, 1983). The rate of photosynthesis, measured in an optimum environment of about 100 Pa partial pressure of external

CO_2, was linearly related to chlorophyll content (a measure of photosynthetic electron transport capacity; Terry, 1980). The data at 30 Pa of external CO_2 partial pressure are particularly interesting because they are closest to photosynthesis under field conditions. The light-saturated rate of photosynthesis per area increased by 36% with an increase in chlorophyll (photochemical capacity) over the range of 0.45 to 0.73 mmol m^{-2} (Fig. 4). Thus, at 30 Pa CO_2, the rate of photosynthesis under field conditions may be co-limited by photosynthetic electron transport capacity and ambient CO_2.

The increase in photosynthesis per unit leaf area with increase in chlorophyll content at light saturation is attributed to increased efficiency of photochemical conversion of absorbed light, not to an increase in light absorption (Terry, 1980). At irradiances that are limiting one might expect that an increase in chlorophyll content would enhance photosynthesis per unit area by increasing the quantity of light (PFD) absorbed. However, even at limiting PFD the improvement in light absorption of photosynthetically active radiation (PAR) is far less important than the increase in efficiency of photochemical conversion. For example, photosynthesis per unit leaf area at the limiting PFD of 500 μmol m^{-2} s^{-1} increased six-fold with increase in chlorophyll content from 0.056 to 0.67 mmol m^{-2}. Absorption of white light, however, increased much less, that is, from 51% to 81% over the same range of chlorophyll content.

Iron stress effects on thylakoids: how are they mediated on CO_2 fixation?

Taylor & Terry (1984) showed that the initial slope of the photosynthesis (A)/intercellular CO_2 partial pressure (C_i) curve decreased in proportion to the decrease in photosynthetic electron transport capacity induced by iron deficiency. This shows that photosynthetic rate was co-limited by photochemical energy supply and CO_2 concentration, even at very low and limiting C_i values. How might co-limitation between photochemical energy supply and CO_2 concentration be mediated in biochemical terms? Farquhar and his associates have proposed that, at low C_i, the total (free and bound) RuBP concentration is greater than the concentration of active sites of Rubisco (Farquhar, 1979; Farquhar, von Caemmerer & Berry, 1980; Farquhar & von Caemmerer, 1982). Thus, the formation of the Rubisco enzyme–RuBP complex (ER) is occurring at its maximal velocity and photosynthetic rate is determined solely by the rates of carboxylation and oxygenation of the ER complex. For co-limitation to occur, photochemical energy supply would have to influence either the level of activation of Rubisco or the concentration of RuBP. Either way, a reduction in

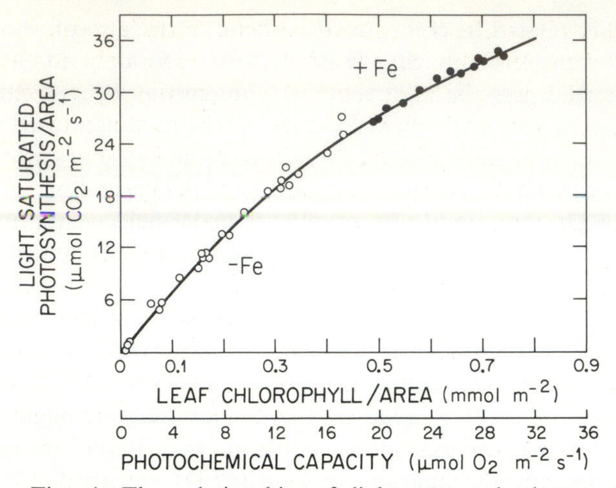

Fig. 4. The relationship of light-saturated photosynthesis with leaf chlorophyll content in Fe-deficient (○) and Fe-sufficient (●) sugar beet plants. Chlorophyll is considered to be an index of photochemical capacity (Terry, 1980). To illustrate further this point, an additional abscissa (photochemical capacity) is shown. This represents the rates of photosynthetic electron transport at comparable chlorophyll contents. The rate of photosynthesis was measured at an ambient CO_2 partial pressure of 30 Pa (after Terry, 1983).

photochemical energy supply would lead to a decrease in the rate of ER formation and, with it, photosynthetic rate (Terry & Farquhar, 1984; Taylor & Terry, 1984).

The decrease in photosynthesis with diminished photosynthetic electron transport capacity in iron-deficient leaves was probably mediated via a reduction in RuBP regeneration. Arulanantham, Rao & Terry (1990) showed from estimates of the Rubisco binding site concentration and steady-state pool sizes of RuBP that the RuBP content was below the binding site concentrations in iron-stressed leaves. In control leaves RuBP levels were almost double the binding site concentration, consistent with previous reports; see review by Woodrow & Berry (1988). Furthermore, the ratio of RuBP/triose-P decreased with increasing severity of iron deficiency, suggesting that the regeneration of RuBP from triose-P (Machler *et al.*, 1988; Rao *et al.*, 1989) was increasingly impaired as the photochemical capacity of the leaf decreased with Fe deficiency (Arulanantham, 1989).

The decrease in initial activity of Rubisco with increasing iron deficiency was not sufficient to decrease photosynthetic rate (Taylor & Terry, 1986; Arulanantham *et al.*, 1990). Leegood *et al.* (1985) suggested that the *in vitro* initial activity of Rubisco measured with saturating CO_2 is usually four to five times higher than the net photosynthetic rate (at ambient CO_2 concentration) of the leaf. We calculated that in moderately and severely deficient leaves, the *in vitro* initial activity of Rubisco was eight to ten times higher than the photosynthetic rate of the leaf (Arulanantham *et al.*, 1990).

There was no evidence from our work that RuBP regeneration or photosynthetic rate was limited by the supply of ATP and/or NADPH. Considering the large reductions in amount of thylakoid and therefore in the capacity of photosynthetic electron transport with moderate and severe Fe deficiency, the decreases in ATP and NADPH were remarkably small (Arulanantham *et al.*, 1990). Giersch & Robinson (1987) used the formation of triose-P from PGA as a measure of ATP and/or NADPH limitation of the RPP pathway. Arulanantham *et al.* (1990) estimated the triose-P/PGA ratio and found that this ratio increased with Fe deficiency, suggesting that there was no ATP/NADPH limitation in the capacity for PGA reduction to triose-P. Furthermore, since ATP did not limit the formation of triose-P from PGA it is unlikely that it would limit the phosphorylation of Ru5P by Ru5P kinase because the K_m for ATP is much lower for Ru5P kinase than for PGA kinase (Leegood *et al.*, 1985).

Since decreased photochemical capacity did not appear to affect RuBP regeneration via the supply of ATP and/or NADPH, it might have affected RuBP regeneration via the activities of enzymes which catalyse the formation of RuBP in the RPP pathway. The enzyme most affected by decreased photochemical capacity was Ru5P kinase, the initial activity of which was diminished by 60% under severe Fe deficiency (Arulanantham *et al.*, 1990), suggesting that the reduction in RuBP regeneration is due to decreased activity of Ru5P kinase (Arulanantham, 1989).

Phosphorus supply and photosynthesis

Growth, carbon fixation and carbon export

When phosphorus is supplied in limiting amounts to sugar beet, it has a much greater impact on growth than on photosynthesis (Rao, Abadia & Terry, 1986; 1987*a*, *b*; Rao & Terry, 1989). In our research we typically supply P_i at 1 mM concentration to control plants grown hydroponically in half-strength Hoagland's solution. Plants subjected to limiting P_i ('low-P') are grown at 0.05 mM P_i in half-strength Hoagland's solution. In sugar beet, the total and acid soluble phosphorus concentrations in leaves decreased by about 90% in a period of two weeks of low-P treatment (Rao & Terry, 1989). Over the same period, plant dry weight decreased 60% and photosynthesis per unit area of leaf 35% (or less, depending on the PFD). Similar results were obtained with soybean (Fredeen, Rao & Terry, 1989). This effect of P on growth was primarily mediated through the leaf surface,

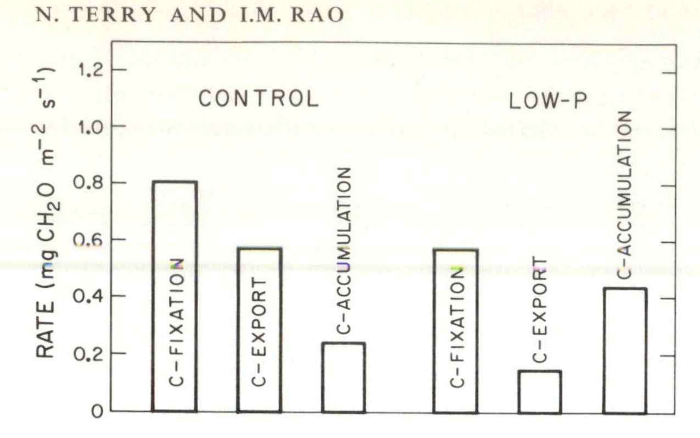

Fig. 5. Rates of photosynthetic carbon fixation, export and accumulation as influenced by P supply to sugar beet leaves. Measurements were made at light saturation and at an ambient CO_2 partial pressure of 30 Pa. Values represent averages over four time points measured from 0 to 8 hours (after Rao *et al.*, 1990).

particularly the expansion of new leaves so that total leaf area decreased by 76% (Rao & Terry, 1989). Fibrous root growth by comparison was much less affected (Rao & Terry, 1989).

Low-P treatment affected photosynthesis much less at low than at high PFD. Photosynthesis at very low PFD, such as that used to measure quantum yield, was virtually unaffected by low-P (Abadia, Rao & Terry, 1987). Low-P treatment decreased leaf conductance which resulted in slightly lower C_i in low-P leaves at each PFD level (Rao & Terry, 1989). More importantly, the rate of photosynthesis decreased significantly with low-P at all values of C_i. These results indicate that the major effect of low-P on photosynthesis was on the enzymatic reactions occurring within the chloroplast (Rao & Terry, 1989) and is consistent with the findings of Brooks (1986) with spinach.

The low-P treatment impaired the export of newly-fixed carbon out of the leaf blade (Fig. 5) more than it decreased photosynthesis per unit leaf area so that photosynthate accumulated. Dry weight per unit area of leaf laminae increased by 30% (Rao & Terry, 1989). Over an eight-hour period at light saturation, photosynthesis was reduced by about 30% by low-P while export was reduced by as much as 87% (Fig. 5) (Rao, Fredeen & Terry, 1990). Radin & Eidenbock (1986) obtained similar results with cotton.

It is unclear why low-P had such a pronounced effect on carbon export at light saturation. It is conceivable that low-P acts in the leaf, e.g. at the phloem loading step (see Grange, 1987). The loading of sugars into the phloem has a large requirement for ATP (Sovonick, Geiger & Fellows,

1974; Giaquinta, 1983), and there may be insufficient ATP in the low-P leaves to maintain export at a rapid rate (ATP levels were reduced as much as 60% in low-P leaves (Fig. 6)) (Rao *et al.*, 1990). However, it seems more likely that carbon export from the leaf blade was reduced because of insufficient sinks for carbohydrate (Rao *et al.*, 1990).

Phosphorus and the light reactions

Abadia *et al.* (1987) found that the low-P treatment had relatively small effects on photosynthetic electron transport and even smaller effects on photosynthetic quantum yield (Table 1). Low-P did not affect thylakoid polypeptide composition, PSI electron transport/area of leaf, number of PSII reaction centres/area (measured as atrazine-binding sites), Cyt b_{559}/area and CPa Chl/area, and had only small effects on Cyt f/area, Chl a/b ratio, PSII electron transport/area and the ratio of PSII/PSI fluorescence at 77K. Low-P appeared to modify thylakoid membrane structure by increasing the amounts of PSI and light harvesting Chl-protein complexes per area, and the amounts of the cytochromes f and b_{563}, without affecting the PSII components (including CPa, atrazine binding sites and Cyt b_{559}). However, the effects of low-P treatment on thylakoid function were relatively mild and were easily reversible within a few hours of increasing the supply of P_i to low-P plants. We conclude, as did Brooks (1986), that the photosynthetic rate of low-P leaves was not limited by photochemical capacity.

Phosphorus, the RPP pathway and photosynthetic rate

One of the most pronounced effects of low-P was to reduce markedly the sugar phosphates (Rao *et al.* 1987*a*; 1989) (Fig. 6). RuBP decreased by as much as 69%, but PGA decreased more. Other sugar phosphates, FBP, F6P, and G6P were also appreciably decreased. Low-P treatment also affected the total and initial activities of several RPP pathway enzymes (Rao & Terry, 1989). It significantly increased the total activities of FBPase, FBP aldolase, and transketolase, had no significant effect on Ru5P kinase, Rubisco, SBPase and triose phosphate isomerase and decreased PGA kinase and NADP-G3P dehydrogenase. The initial activities of NADP-G3P dehydrogenase, PGA kinase and Ru5P kinase were significantly decreased by 77, 32 , and 31 %, respectively, by low-P treatment. Low-P did not affect the initial activity of Rubisco.

Phosphorus supply appears to affect photosynthesis through RuBP regeneration in a similar manner to iron. Our calculations show that RuBP concentrations in low-P leaves were about half of the concentration of

Table 1. *Effect of low-P treatment on certain photochemical characteristics of five-week-old sugar beet plants (after Abadia et al., 1987).*

Characteristics	Treatment		
	Control	Low-P	% control
Chl (μmol m^{-2}) in:			
chlorophyll-protein-1	86	111	129
LHCII	213	268	126
CPa	69	70	101
Cytochromes (μmol m^{-2});			
Cyt f	0.71	0.8	113
Cyt b_{559}	1.2	1.2	100
Cyt b_{563}	2.82	3.4	121
Atrazine binding sites (μmol m^{-2})	2.73	2.94	108
Chl a/b (mol mol^{-1})	3.7	3.5	95
Chl (mmol m^{-2})	0.48	0.59	123
Electron transport (μmol O_2 m^{-2} s^{-1})			
PSI	112	115	103
PSII	43.6	38.2	88
Photosynthetic quantum yield (mol CO_2 mol $photons^{-1}$)	0.12	0.11	92

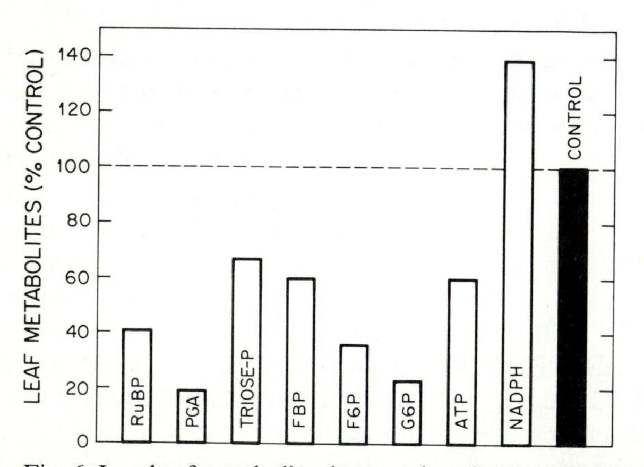

Fig. 6. Levels of metabolites in sugar beet leaves as influenced by P supply. Values represent averages over five time points during a 16 hour light (500 μmol m^{-2} s^{-1}) period (after Rao *et al.*, 1989).

active sites of Rubisco (Rao *et al.*, 1989) but in the control there was approximately 1.4 times more RuBP than the concentration of active sites. Low-P treatment substantially decreased RuBP levels in spinach (Brooks, 1986; Brooks, Woo & Wong 1988), barley (Sicher & Kremer, 1988) and soybean (Fredeen *et al.*, 1990).

If photosynthetic rate is decreased by slow RuBP regeneration, how is RuBP regeneration controlled by low-P? One possibility is that RuBP regeneration is limited via effects of P on the supply of NADPH and ATP to the stromal enzymes of the RPP pathway. This possibility seems unlikely. Leaf NADPH content (Fig. 6), and the NADPH/$NADP^+$ ratio, increased with low-P treatment, suggesting that there was an excess production of reducing power relative to its consumption in the reactions of the RPP pathway (Rao *et al.*, 1989). Although the absolute amount of ATP was reduced by low-P (Fig. 6), ATP/ADP ratios were not. Furthermore, low-P did not decrease triose-P levels and substantially increased the triose-P/PGA ratio. This suggests that the decreased amounts of ATP in low-P leaves did not impair the capacity of leaves to phosphorylate PGA to triose-P (see Giersch & Robinson, 1987). If the supply of ATP was not limiting the reduction of PGA, it is also unlikely that it would limit the phosphorylation of Ru5P (Leegood *et al.*, 1985).

The main reason for the small amounts of ATP in low-P leaves was that net adenylate accumulation was reduced (Rao *et al.*, 1990). Similar affects of low-P were observed with *Lemna gibba* (Thorsteinsson & Tillberg, 1987) and soybean (Fredeen *et al.*, 1990). The decreased adenylate content observed in low-P leaves may be due to the combined effect of lower rates of adenylate synthesis and higher rates of adenylate degradation (Miginiac-Maslow, Nguyen & Hoarau, 1986). Adenylates were synthesised via the IMP pathway which utilises ribose-5-P for its precursor. Since low-P seems to decrease sugar phosphates to a marked degree, especially those involved in RuBP regeneration from triose-P, it seems likely that the reduction in adenylates may have been attributable to a reduction in ribose-5-P (Rao *et al.*, 1989).

Since low-P treatment did not appear to affect RuBP regeneration via the supply of ATP and/or NADPH, how might low-P impair RuBP regeneration? The decrease in RuBP regeneration could have resulted from smaller initial activities of PGA kinase and NADP-G3P dehydrogenase. This seems unlikely in view of the high levels of triose-P and the large ratio of triose-P/PGA. RuBP regeneration may have been limited by the extent of activation of Ru5P kinase *in vivo* since low-P treatment decreased the initial activity of this enzyme appreciably (Rao & Terry, 1989). We believe, however, that the levels of RuBP and other sugar phosphates, FBP, F6P and G6P, were diminished as a result of the increased synthesis of starch.

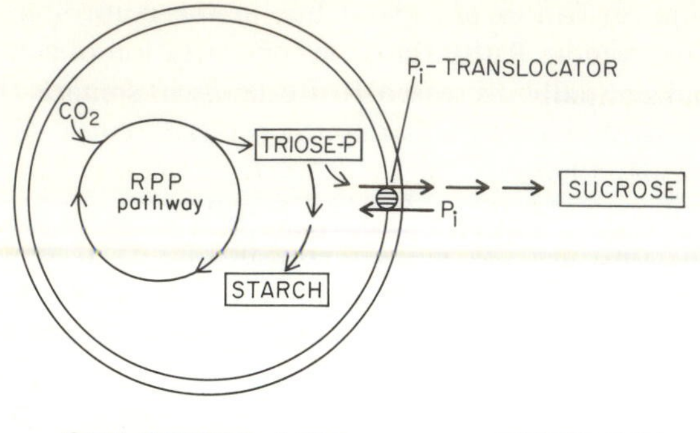

Fig. 7. Simplified scheme showing the partitioning of triose-P between starch and sucrose syntheses. The export of triose-P from the chloroplast to the cytosol and the import of P_i from cytosol to chloroplast are mediated by the P_i translocator.

Low-P plants exhibited increased activities of enzymes in the starch synthesis pathway, e.g. FBPase, FBP aldolase (Rao & Terry, 1989) and ADPG pyrophosphorylase (Rao *et al*., 1987*b*; 1990).

The role of P_i in carbon partitioning

Several studies in recent years have indicated that the concentration of P_i in leaves may regulate photosynthesis and carbon partitioning *in vivo*. Leaf P_i concentration probably influences photosynthesis through the P_i-translocator, an antiporter located in the inner membrane of the chloroplast envelope (Fig. 7). The P_i-translocator facilitates the counter-exchange of P_i, triose-P and PGA (Heldt *et al*., 1977; Walker, 1980). It permits the transport of triose-P from the stroma to the cytosol in a one-to-one stoichiometric exchange for P_i (Flügge, 1987). Since one molecule of P_i must be available for incorporation into triose-P for every three molecules of CO_2 fixed, a significant flux of P_i must be provided to maintain carbon fixation. Much of this P_i is believed to be released by the synthesis of sucrose from triose-P, i.e. during the action of cytosolic FBPase and sucrose-P phosphatase, and pyrophosphatase after UDPG formation. Also, some P_i will be released within the stroma as triose-P is utilised in starch synthesis, which is, however, slower than the maximal rate of CO_2 fixation by a factor of three to four (Heldt *et al*., 1977).

Since P_i, triose-P and PGA are exchanged through the P_i-translocator, changes in the P_i concentration outside the chloroplast may affect the RPP pathway by changing the levels of phosphorylated intermediates within the chloroplast (Walker, 1980). In addition, P_i may also affect the activation of RPP pathway enzymes (see Leegood *et al.*, 1985, for review). According to the *in vitro* studies, low levels of cytosolic P_i decrease the export of triose-P from the chloroplast to the cytosol; this in turn causes: (1) a build-up of RPP pathway intermediates, (2) increased starch synthesis, and (3) decreased sucrose synthesis (Heldt *et al.*, 1977; Walker, 1980).

Plants grown with limited P supply may lose 90% or more of their total leaf P and soluble P_i compared to control plants. It is reasonable to assume that deficiency will also have decreased cytosolic P_i concentration although direct evidence on this point is lacking (Rao *et al.*, 1989, calculated that nonvacuolar P_i in low-P plants decreased by 23%). Assuming that a decrease in cytosolic P_i does occur in low-P plants, then we might expect to see increased concentrations of chloroplast sugar phosphates, increased starch and less sucrose.

Instead of increased chloroplast sugar phosphates with low-P, there was a large reduction in RuBP, which is located in the chloroplast, as well as in other sugar phosphates, part of which are also found in chloroplasts, i.e. PGA, FBP, F6P, and G6P (Brooks, 1986; Dietz & Foyer, 1986; Rao *et al.*, 1987*a*; 1989; Sicher & Kremer, 1988; Fredeen *et al.*, 1990). The decreased pool sizes of sugar phosphates with low-P was in part due to increased phosphatase activities (e.g. acid and alkaline phosphatases) (Rao *et al.*, 1990). However, it also seems likely that the leaf sugar phosphate pools were depleted in response to the increased flow of carbon towards starch and sucrose synthesis (Rao *et al.*, 1990). Other studies (in which leaf P_i was decreased) have shown increased starch/sucrose ratios, e.g. in soybean (Foyer & Spencer, 1986; Fredeen, 1988; Fredeen *et al.*, 1989) and pea (Foyer & Spencer, 1986). With sugar beet, however, starch/sucrose ratios were substantially decreased, not increased by low-P treatment (Rao *et al.*, 1990). Fleck *et al.* (1987) also observed decreased starch/sucrose ratios in P-deficient wheat. The decrease in starch/sucrose ratio with low-P treatment in our work was due mainly to the accumulation of sucrose. Sucrose levels were four- to six-fold higher in low-P leaves compared to the control leaves. Starch levels increased to only a small extent (Rao *et al.*, 1990).

Our results suggest that low-P increased the enzymatic capacities for both starch and sucrose syntheses (Rao *et al.*, 1990) (Fig. 8). For example, ADPG pyrophosphorylase activity, a key regulatory enzyme in the starch synthesis pathway, was increased substantially (57%) by low-P. Low-P also increased the activities of FBP aldolase (53%) and chloroplastic FBPase (62%), two other enzymes in the starch synthesising pathway. With regard to the sucrose synthesising pathway, three key enzymes were substantially increased by low-P: sucrose phosphate synthase (97%), cytosolic FBPase

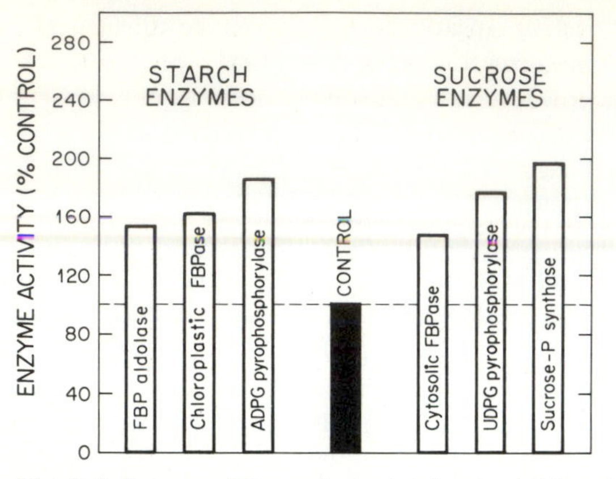

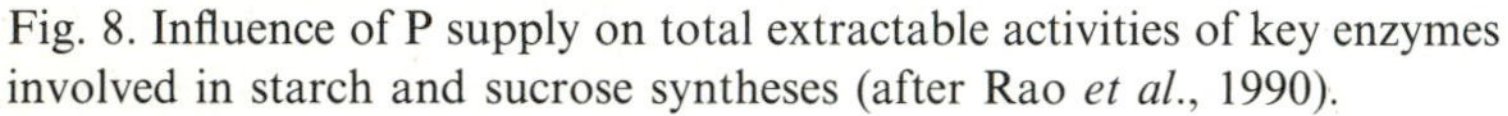
Fig. 8. Influence of P supply on total extractable activities of key enzymes involved in starch and sucrose syntheses (after Rao *et al.*, 1990).

(47 %) and UDPG pyrophosphorylase (76 %). Similar effects of low-P on sucrose phosphate synthase were reported by Sicher & Kremer (1988) for barley seedlings.

Since low-P increased the accumulation of both starch and sucrose, and decreased chloroplast sugar phosphates, it seems unlikely that carbohydrate partitioning was mediated through the P_i-translocator. Another factor which is thought to influence starch/sucrose partitioning is the level of the signal metabolite, fructose 2,6-bisphosphate (F2,6BP). F2,6BP is a potent inhibitor of cytosolic FBPase, which catalyses an important initial step in the sucrose synthesis pathway (Stitt *et al.*, 1987). Low-P treatment decreases F2,6BP in both light and dark and this could have increased carbon flow to sucrose.

Regardless of the partitioning of fixed carbon between the different storage carbohydrates, low-P treatment clearly increased the pool sizes of unphosphorylated carbon compounds (starch, sucrose, glucose) (Rao *et al.*, 1990) and may also have increased structural, as well as non-structural, carbohydrates. Low-P increased leaf dry weight per unit area by 30 %. Only 9 % (or less) of the increase in dry weight in low-P leaves was due to starch; the remainder of the increase in dry weight may have been due to other carbon compounds such as cell wall polymers (cellulose and hemicelluloses). The glucosyl donor for cellulose synthesis is probably UDPG, while pectic substances and hemicelluloses are formed from nucleoside diphospho-sugars that are made by direct interconversion from UDPG (for review see ap Rees, 1987). The formation of UDPG for structural polysaccharide synthesis in leaves occurs mainly via UDPG pyrophosphorylase (ap Rees, 1987). Since our research shows that the activities of cytosolic FBPase and

UDPG pyrophosphorylase increased, it seems likely that the synthesis of UDPG would be greater under low-P conditions, resulting in an accumulation of structural carbohydrates.

Adaptation to low-P environments

There is evidence that plants adapt to low-P environments, a situation frequently experienced under natural conditions. For example, plants have increased activity of phosphatases in response to P deficiency (Bieleski, 1973; Besford, 1979; Bouma, 1983; Elliott & Lauchli, 1986). The increases in phosphatases in roots help to release phosphate bound up in root cell walls and perhaps even in the soil solution (Szabo-Nagy, Olah, & Erdei, 1987). In leaves it is likely that their major role is to increase the availability of phosphate for photosynthesis and other important functions. This is especially evident with sugar beet since several phosphatases were substantially increased by low-P treatment (Rao *et al.*, 1990).

We speculate that the physiological and metabolic changes which occur in response to P deficiency (Fig. 9) may be part of an adaptive strategy. Thus, as P_i limits growth, changes in enzymatic pathways occur such that phosphate tied up in phosphorylated intermediates, and the amounts of phosphate-free carbon compounds are increased. This increases the amount of P_i available for photosynthesis and other essential functions. The storage of carbon in non-phosphorylated compounds such as starch, sucrose and glucose provides carbon reserves for growth if and when P subsequently becomes available.

The P_i translocator might also form part of an adaptive stratagem to P deficiency since it functions as a mechanism for providing the chloroplast with a continuing supply of orthophosphate under conditions of low-P. Since the P_i-translocator ensures the uptake of one P_i for every triose-P exported from the chloroplast, the P_i content of the chloroplast is maintained at levels which permit substantial rates of photosynthesis, thereby permitting carbohydrate accumulation and also preventing potential damage to the photosynthetic apparatus from photoinhibition.

Concluding remarks

In this chapter, we have explored the roles of iron and phosphorus from the molecular to the whole plant level. The responses of sugar beet plants to deficiency of these two elements are compared directly in Table 2, which shows how various physiological and biochemical parameters change in response to moderate deficiency and highlights the fact that Fe deficiency primarily affects the light-harvesting and electron transport system while P deficiency effects are centred on carbon assimilation, partitioning and

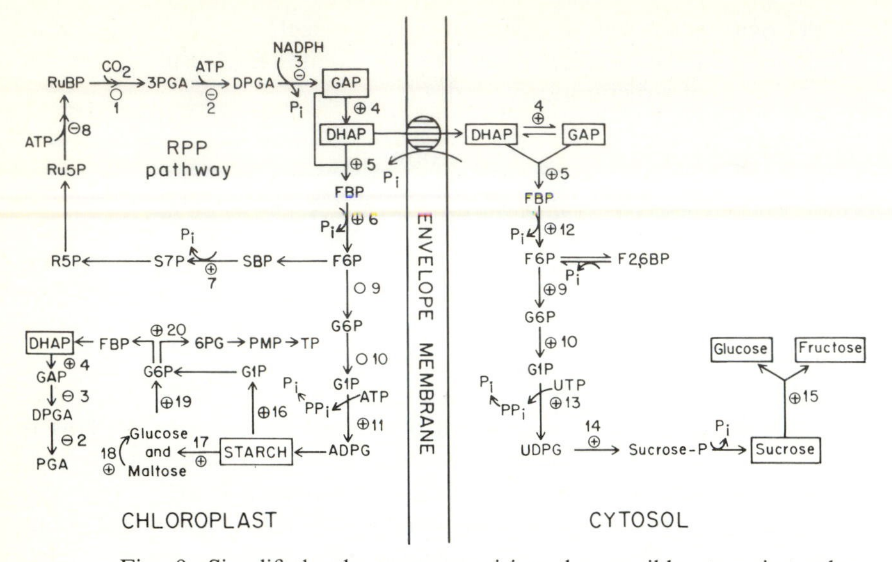

Fig. 9. Simplified scheme summarising the possible steps in carbon metabolism that may be affected by P supply in sugar beet leaves. Specific enzymes are (1) Rubisco; (2) PGA kinase; (3) NADP-G3P dehydrogenase; (4) triose-P isomerase; (5) FBP aldolase; (6) chloroplastic FBPase; (7) SBPase; (8) Ru5P kinase; (9) phosphohexoisomerase; (10) phosphoglucomutase; (11) ADPG-pyrophosphorylase; (12) cytosolic FBPase; (13) UDPG-pyrophosphorylase; (14) sucrose-P synthase; (15) invertase; (16) phosphorylase; (17) β-amylase; (18) maltase or maltose phosphorylase; (19) hexokinase; and (20) NADP-G6P dehydrogenase. The enzymes which have been shown to be increased in activity by low-P treatment are designated as ⊕, decreased as ⊖, and not significantly affected as ○.

export. Such multidisciplinary studies are useful in that they provide information about how the photosynthetic system is influenced by the supply of a specific nutrient. More importantly, however, such studies can provide information about the photosynthetic process itself. From iron stress studies we have learned that photosynthetic rate may be co-limited by photosynthetic electron transport capacity even under CO_2-limited conditions. From studies of the effects of phosphorous, we have developed an alternative view of the role of the P_i-translocator in photosynthesis, i.e., that it may serve as part of a phosphorus conservation mechanism which is triggered in plants trying to cope with periods of phosphorus deprivation. From iron (Arulanantham *et al.*, 1989), phosphorus (Rao *et al.*, 1990) and nitrogen (Machler *et al.*, 1988) studies we have learned that the rate of photosynthesis is commonly limited by the rate of RuBP regeneration,

Table 2. *Summary of the effects of Fe and P deficiencies on several components of the photosynthetic system.*

Components	Low-Fe	Low-P
Light-harvesting and electron transport		
Leaf chlorophyll	−	+
Chl in:		
CP1	−	+
LHCII	−	+
CPa	−	0
Cytochrome f	−	+
PSII reaction centres	−	+
Electron transport:		
PSI	−	+
PSII	−	−
Carbon metabolism		
Enzyme (total activity):		
Rubisco	0	+
PGA kinase	−	−
NADP-G3PD	0	−
FBPase	0	+
SBPase	0	+
Ru5P kinase	0	0
Leaf metabolites:		
RuBP	−	−
PGA	−	−
Triose-P	−	−
FBP	−	−
F6P	−	−
ATP	−	−
NADPH	−	+
Starch	−	+
Sucrose	−	+
Glucose	−	+
Photosynthesis, carbon export and leaf expansion		
Photosynthesis/area	−	−
Carbon export	n.a.	−
Leaf expansion	0	−

Components that increase with deficiency are shown as +, those that decrease as −, and those that show no change as 0 (n.a., data not available).

although the mechanisms by which this limitation is caused may be quite different for each element. We believe that further studies of the relationship of photosynthesis with mineral nutrient supply will continue to yield valuable information at the fundamental as well as agronomic levels. It is essential that such studies combine molecular, biochemical and whole-plant approaches to properly elucidate photosynthesis/nutrient interactions in plant ecology.

References

Abadia, J., Rao, I.M. & Terry, N. (1987). Changes in leaf phosphate status have only small effects on the photochemical apparatus of sugar beet leaves. *Plant Science*, **50**, 49–55.

Alcaraz, C.F., Hellin, E., Sevilla, F.& Martinez-Sanchez. (1985). Influence of the leaf iron contents on the ferredoxin levels in citrus plants. *Journal of Plant Nutrition*, **8**, 603–11.

ap Rees, T. (1987). Hexose phosphate metabolism by nonphotosynthetic tissues of higher plants. In *The Biochemistry of Plants. A Comprehensive Treatise*, Vol. 14, ed. J. Preiss, pp. 1–33. New York: Academic Press.

Arulanantham, A.R. (1989). *Studies on the physiology and biochemistry of iron stress in Beta vulgaris L.* Ph.D. Thesis, University of California, Berkeley.

Arulanantham, A.R., Rao, I.M. & Terry, N. (1990). Limiting factors in photosynthesis. VI. Regeneration of ribulose 1,5-bisphosphate limits photosynthesis at low photochemical capacity. *Plant Physiology*, **93**, 1466–75.

Besford, R.T. (1979). Phosphorus nutrition and acid phosphatase activity in leaves of seven plant species. *Journal of the Science of Food and Agriculture*, **30**, 281–5.

Bieleski, R.L. (1973). Phosphate pools, phosphate transport, and phosphate availability. *Annual Review of Plant Physiology*, **24**, 225–52.

Bolle-Jones, E.W. & Notton, B.A. (1953). The relative proportions of the chloroplast pigments as influenced by different levels of iron and potassium supply. *Plant and Soil*, **5**, 87–100.

Bouma, D. (1983). Diagnosis of mineral deficiencies using plant tests. In *Encyclopedia of Plant Physiology, New Series, Vol. 15B, Inorganic Plant Nutrition*, eds. A. Lauchli & R.L. Bieleski, pp. 120 – 46. Berlin: Springer-Verlag.

Brooks, A. (1986). Effects of phosphorus nutrition on ribulose-1,5-bisphosphate carboxylase activation, photosynthetic quantum yield and amounts of some Calvin-cycle metabolites in spinach leaves. *Australian Journal of Plant Physiology*, **13**, 221–37.

Brooks, A., Woo, K.C. & Wong, S.C. (1988). Effects of phosphorus nutrition on the response of photosynthesis to CO_2 and O_2, activation of ribulose bisphosphate carboxylase and amounts of ribulose bisphosphate and 3-phosphoglycerate in spinach leaves. *Photosynthesis Research*, **15**, 133–41.

Dietz, K.-J. & Foyer, C. (1986). The relationship between phosphate status and photosynthesis in leaves; reversibility of the effects of phosphate deficiency on photosynthesis. *Planta*, **167**, 376–81.

Elliott, G.C. & Lauchli, A. (1986). Evaluation of an acid phosphatase assay for detection of phosphorus deficiency in leaves of maize (*Zea mays* L.). *Journal of Plant Nutrition*, **9**, 1469–77.

Evans, J.R. (1989). Photosynthesis and nitrogen relationships in leaves of C_3 plants. *Oecologia*, **78**, 9–19.

Farquhar, G.D. (1979). Models describing the kinetics of ribulose bisphosphate carboxylase oxygenase. *Archives of Biochemistry and Biophysics*, **193**, 456–68.

Farquhar, G.D. & von Caemmerer, S. (1982). Modelling of photosynthetic response to environmental conditions. In *Encyclopedia of Plant Physiology, new Series, Vol. 12B*, Physiological Plant Ecology II, eds. O.L. Lange, P.S. Nobel & C.B. Osmond, pp. 550–87. Berlin: Springer-Verlag.

Farquhar, G.D., von Caemmerer, S. & Berry, J.A. (1980). A biochemical model of photosynthetic CO_2 assimilation in leaves of C_3 species. *Planta*, **148**, 78–90.

Fleck, I, Fransi, A., Florensa, I. & Rafales, M. (1987). Influence of P-nutrition on aging of the flag leaf of wheat after anthesis with respect to ribulose bisphosphate carboxylase activity and sugar content. *Plant Physiology and Biochemistry* **25**, 609–16.

Flügge, U.-I. (1987). Physiological function and physical characteristics of the chloroplast phosphate translocator. In *Progress in Photosynthesis Research*, Vol. 3, ed. J. Biggins, pp. 739–46. The Hague: Martinus Nijhoff/Dr. W. Junk, Publishers.

Foyer, C. & Spencer, C. (1986). The relationship between phosphate status and photosynthesis in leaves. Effects on intracellular P_i distribution, photosynthesis and assimilate partitioning. *Planta*, **167**, 369–75.

Fredeen, A.L. (1988). *Influence of vesicular-arbuscular mycorrhizal-infection and phosphorus nutrition on growth, photosynthesis, and carbon-metabolism in Glycina max*. Ph.D. Thesis, University of California, Berkeley.

Fredeen, A.L., Rao, I.M. & Terry, N. (1989). Influence of phosphorus nutrition on growth and carbon partitioning in *Glycine max. Plant Physiology* **89**, 225–30.

Fredeen, A.L., Raab, T.K., Rao, I.M. & Terry, N. (1990). Effects of phosphorus nutrition on photosynthesis in *Glycine max*. (L.) Merr. *Planta*, **181**, 399–405.

Giaquinta, R.T. (1983). Phloem loading of sucrose. *Annual Review of Plant Physiology*, **34**, 347–87.

Giersch, C. & Robinson, S.P. (1987). Regulation of photosynthetic carbon metabolism during phosphate limitation of photosynthesis in isolated spinach chloroplasts. *Photosynthesis Research*, **14**, 211–27.

Grange, R.I. (1987). Carbon partitioning in mature leaves of pepper: effects of transfer to high or low PFD. *Journal of Experimental Botany*, **38**, 77–83.

Guikema, J.A. & Sherman, L.A. (1984). Influence of iron deprivation on the membrane composition of *Anacystis nidulans*. *Plant Physiology*, **74**, 90–5.

Heldt, H.W., Chon, C.J., Maronde, D., Herold, A., Stankovic, A.Z., Walker, D.A., Kraminer, A., Kirk, M.R. & Heber, U. (1977). Role of orthophosphate and other factors in the regulation of starch formation in leaves and isolated chloroplasts. *Plant Physiology* **59**, 1146–55.

Hewitt, E.J. (1983). Essential and functional metals in plants. In *Metals and Micronutrients: Uptake and Utilization by Plants*, eds. D.A. Robb & W.S. Pierpoint, pp. 277–323. New York: Academic Press.

Hiller, R.G. & Goodchild, D.J. (1981). Thylakoid membrane and pigment organization. In *The Biochemistry of Plants, Vol. 8, Photosynthesis*, eds. M.D. Hatch & N.K. Boardman, pp. 2–46. New York: Academic Press.

Hipkins, M.F. (1983). Metals and photosynthesis. In *Metals and Micronutrients: Uptake and Utilization by Plants*, eds. D.A. Robb & W.S. Pierpoint, pp. 147–68. New York: Academic Press.

Leegood, R.C., Walker, D.A. & Foyer, C.H. (1985). Regulation of the Benson–Calvin cycle. In *Photosynthetic Mechanisms and the Environment*, eds. J. Barber & N.R. Baker, pp. 189–258. Amsterdam, New York, Oxford: Elsevier Science Publishers BV (Biomedical Division).

Machler, F., Oberson, A., Crub, A., & Nosberger, S. (1988). Regulation of photosynthesis in nitrogen deficient wheat seedlings. *Plant Physiology*, **87**, 46–9.

Machold, O. (1971). Lamellar proteins of green and chlorotic chloroplasts as affected by iron deficiency and antibiotics. *Biochimica Biophysica Acta* **238**, 324–31.

Marsh, H.V., Evans, H.J. & Matrone, G. (1963). Investigations of the role of iron in chlorophyll metabolism I. Effects of iron deficiency on chlorophyll and heme content and on the activities of certain enzymes in leaves. *Plant Physiology* **38**, 632–8.

Miginiac-Maslow, M., Nguyen, J. & Hoarau, A. (1986). Adenylate metabolism in phosphate-depleted isolated soybean leaf cells and wheat leaf fragments. *Journal of Plant Physiology*, **123**, 69–77.

Nishio, J.N. & Terry, N. (1983). Iron nutrition-mediated chloroplast development. *Plant Physiology*, **71**, 688–91.

Nishio, J.N., Taylor, S.E. & Terry, N. (1985*a*). Changes in thylakoid galactolipids and proteins during iron nutrition-mediated chloroplast development. *Plant Physiology* **77**, 705–11.

Nishio, J.N., Abadia, J. & Terry, N. (1985*b*). Chlorophyll-proteins and electron transport during iron nutrition-mediated chloroplast development. *Plant Physiology*, **78**, 296–9.

Platt-Aloia, K.A., Thomson, W.W. & Terry, N. (1983). Changes in plastid ultrastructure during iron nutrition-mediated chloroplast development. *Protoplasma*, **114**, 85–92.

Price, C.A., Ortiz, W. & Gaynor, J.J. (1978). Regulation of protein synthesis in isolated chloroplasts of *Euglena gracilis*. In *Chloroplast Development*, eds. G. Akoyunoglou & J.H. Argyroudi-Akoyunoglou, pp. 257–66. The Hague: Elsevier/North-Holland Biomedical Press.

Radin, J.W. & Eidenbock, M.P. (1986). Carbon accumulation during photosynthesis in leaves of nitrogen- and phosphorus-stressed cotton. *Plant Physiology*, **82**, 869–71.

Rao, I.M., Abadia, J. & Terry, N. (1986). Leaf phosphate status and photosynthesis *in vivo*: Changes in light scattering and chlorophyll fluorescence during photosynthetic induction in sugar beet leaves. *Plant Science*, **44**, 133–8.

Rao, I.M., Abadia, J. & Terry, N. (1987*a*). The role of orthophosphate in the regulation of photosynthesis *in vivo*. In *Progress in Photosynthesis Research*, Vol. 3, ed. J. Biggins, pp. 325–8. The Hague: Martinus Nijhoff/Dr. W. Junk, Publishers.

Rao, I.M., Abadia, J. & Terry, N. (1987*b*). Leaf phosphate status and its effects on photosynthetic carbon partitioning and export in sugar beet. In *Progress in Photosynthesis Research*, Vol. 3, ed. J. Biggins, pp. 571–4. The Hague: Martinus Nijhoff/Dr. W. Junk, Publishers.

Rao, I.M., Arulanantham, A.R. & Terry, N. (1989). Leaf phosphate status, photosynthesis and carbon partitioning in sugar beet. II. Diurnal changes in sugar phosphates, adenylates and nicotinamide nucleotides. *Plant Physiology*, **90**, 820–6.

Rao, I.M., Fredeen, A.L. & Terry, N. (1990). Leaf phosphate status, photosynthesis and carbon partitioning in sugar beet. III. Diurnal changes in carbon partitioning and carbon export. *Plant Physiology*, **92**, 29–36.

Rao, I.M. & Terry, N. (1989). Leaf phosphate status and photosynthesis *in vivo* in sugar beet. I. Changes in growth, photosynthesis and Calvin cycle enzymes. *Plant Physiology*, **90**, 814–19.

Sandmann, G. & Malkin, R. (1983). Iron–sulfur centers and activities of photosynthetic electron transport chain in iron-deficient cultures of the blue-green algae *Aphanocapsa*. *Plant Physiology*, **73**, 724–8.

Sicher, R.C. & Kremer, D.F. (1988). Effects of phosphate deficiency on assimilate partitioning in barley seedlings. *Plant Science*, **57**, 9–17.

Sovonick, S.A., Geiger, D.R. & Fellows, R.J. (1974). Evidence for active phloem loading in the minor veins of sugar beet. *Plant Physiology*, **43**, 886–91.

Spiller, S.C. (1979). *The influence of iron stress on the photosynthetic apparatus of Beta vulgaris*. Ph.D. Thesis, University of California, Berkeley, USA.

Spiller, S.C. & Terry, N. (1980). Limiting factors in photosynthesis. II. Iron stress diminishes photochemical capacity by reducing the number of photosynthetic units. *Plant Physiology*, **65**, 121–5.

Stitt, M., Huber, S. & Kerr, P. (1987). Control of photosynthetic sucrose formation. In *The Biochemistry of Plants. A Comprehensive Treatise*, Vol. 10, eds. M.D. Hatch & N.K. Boardman, pp. 327–409. New York: Academic Press.

Szabo-Nagy, A., Olah, Z. & Erdei, L. (1987). Phosphatase induction in roots of winter wheat during adaptation to phosphorus deficiency. *Physiologia Plantarum*, **70**, 544–52.

Taylor, S.E. & Terry, N. (1984). Limiting factors in photosynthesis. V. Photochemical energy supply colimits photosynthesis at low values of intercellular CO_2 concentration. *Plant Physiology*, **75**, 82–6.

Taylor, S.E. & Terry, N. (1986). Variation in photosynthetic electron transport capacity *in vivo* and its effects on the light modulation of ribulose bisphosphate carboxylase. *Photosynthesis Research*, **8**, 249–56.

Taylor, S.E., Terry, N. & Huston, R.P. (1982). Limiting factors in photosynthesis. III. Effects of iron nutrition on the activities of three regulatory enzymes of photosynthetic carbon metabolism. *Plant Physiology*, **70**, 1541–3.

Terry, N. (1979). The use of mineral nutrient stress in the study of limiting factors in photosynthesis. In *Photosynthesis and Plant Development*, eds. R. Marcelle, H. Clijsters & M. van Poucke. The Hague, Boston, London: Dr. W. Junk, Publishers.

Terry, N. (1980). Limiting factors in photosynthesis I. Use of iron stress to control photochemical capacity *in vivo*. *Plant Physiology*, **65**, 114–20.

Terry, N. (1983). Limiting factors in photosynthesis. IV. Iron stress-mediated changes on light-harvesting and electron transport capacity and its effects on photosynthesis *in vivo*. *Plant Physiology*, **71**, 855–60.

Terry, N. & Abadia, J. (1986). Function of iron in chloroplasts. *Journal of Plant Nutrition*, **9**, 609–46.

Terry, N. & Farquhar, G.D. (1984). Photochemical capacity and photosynthesis. In *Control of Crop Productivity*, ed. C.J. Pearson. Cambridge University Press.

Thorsteinsson, B. & Tillberg, J.-E. (1987). Carbohydrate partitioning, photosynthesis and growth in *Lemna gibba* G3. II. The effects of phosphorus limitation. *Physiologia Plantarum*, **71**, 271–6.

Walker, D.A. (1980). Regulation of starch synthesis in leaves – The role of orthosphosphate. In *Physiological Aspects of Crop Productivity*, pp. 195–207. Bern: Proc. 15th Colloq. Intl. Pot. Inst.

Woodrow, I.E. & Berry, J.A. (1988). Enzymatic regulation of photosynthetic CO_2 fixation in C_3 plants. *Annual Review of Plant Physiology and Plant Molecular Biology*, **39**, 533–94.

Young, T.F. & Terry, N. (1983). Kinetics of iron transport into the leaf symplast during recovery from iron stress. *Canadian Journal of Botany*, **61**, 2496–9.

G.R. STEWART

The comparative ecophysiology of plant nitrogen metabolism

Introduction

The interactions between the processes of nitrogen acquisition, assimilation and allocation in plant tissues and the transformations of nitrogen in the environment are complex. It is generally held that in soils of low pH and low temperatures, the ammonium ion is produced and may accumulate whereas nitrification, the production of the nitrate (NO_3^-) ion, is favoured by neutral or alkaline pH and by warmer temperatures (Robertson, 1982). Soil aeration can also exert a strong influence on the microbial conversion of ammonium ions to nitrate ions and in some relatively warm soils of alkaline pH but where aeration is poor, ammonium can be the source of nitrogen for plant growth (Rabinovitch-Vin, 1983). In many climax forests it has been difficult to relate rates of nitrification to abiotic environmental factors and biotic inhibition of nitrification (allelopathic interactions) probably plays a decisive role in determining the source of nitrogen available to plants (Sprent, 1980).

The assimilation of ammonium ions, dinitrogen (N_2) and nitrate has different potential costs with respect to energy (ATP and reductant) and water, with ammonium being the least costly nitrogen source (Raven, 1985). However, in many plant communities nitrate ions appear to be the nitrogen source generally utilised (Lee & Stewart, 1978; Stewart & Orebamjo, 1983; Al Gharbi & Hipkin, 1984). Nitrogen acquisition, because of its accumulation in cell proteins must, of necessity, be closely linked to growth and other physiological processes, particularly photosynthesis. Mooney and his co-workers (Field & Mooney, 1986) have shown that photosynthetic rates in many species are directly proportional to leaf nitrogen concentrations.

Here I discuss nitrogen acquisition in terms of the availability and forms of nitrogen in different habitats, the assimilation of nitrogen in relation to sites and routes of metabolism within the plant and the allocation of assimilated nitrogen to the functional utilisation of nitrogen compounds.

Nitrogen acquisition

In both the ecological and agricultural literature much emphasis is placed on the limiting role played by nitrogen availability in various plant communities. Indeed, the proposition that nitrogen, more than any other soil nutrient, limits plant growth has almost acquired the status of a dogma although Jeffrey (1987) has cautioned against making a 'nitrogen ecology'. Ribulose bisphosphate carboxylase–oxygenase is the primary carboxylating enzyme of the photosynthetic CO_2 fixation cycle and is the single most important sink for nitrogen in plants. Together with other cycle enzymes and the pigment–protein complexes of the thylakoid membranes the photosynthetic system is the primary nitrogen demand in plants (Lawlor, 1991; Terry & Rao, 1991; both in this volume). Its clear that photosynthetic activity is very much dependent upon plant nitrogen status and that the conversion of inorganic nitrogen to the organic form depends on photosynthesis. Given this, there is surprisingly little experimental evidence in support of the basic tenet of the dogma and such as there is comes in large part from studies of the nitrogen requirements of fast growing crop species (Greenwood, 1986). Other studies have investigated the effects of the addition of large amounts of nitrogen to natural vegetation and where an increase in biomass is obtained this is generally accompanied by a change in species composition (Willis, 1961). A frequently observed response is the appearance of fast growing species which out-compete the 'native' species. Caution needs to be exercised in interpreting such experiments since what they show is that plants of so-called nitrogen limited environments get along very well with the levels of available nitrogen normally present. This could reflect the fact that such plants have either an intrinsically low demand for nitrogen or have special features which enable them to acquire sufficient supplies even though the absolute availability of nitrogen is low.

Various plant characteristics have been suggested to be important in the acquisition of nitrogen from soils of low N availability. Many of these are concerned with root morphology and Fitter (1987) and Robinson & Rorison (1983) have discussed the relation between root architecture and the efficiency of nutrient capture. The absorptive capacity of the root system is increased by increasing fineness of the ultimate roots and by long root hairs. Fitter suggests that although root systems with a herringbone structure are the most efficient in exploring soil, they are least transport efficient and most expensive in terms of carbon allocation. Such root architecture might be most effective in soils where nitrogen sources are distributed heterogeneously. Lamont (1982) has considered the role played by specialised types of root morphology, particularly root clusters, in the acquisition of nutrient from soils in which the nutrient concentration limits plant growth. Four categories of this type of specialised root system, composed of bunches of hairy rootlets, are recognised: proteoid, dauciform, capillaroid and leaf-base clusters. Most studied are the proteoid roots found in all species belonging to the Proteaceae; Lamont (1982) reported that they can be induced by nitrogen limitation. The dauciform root system, which is found in several genera of sedges, consists of some 30 carrot-shaped rootlets, covered in long root hairs. Capillaroid roots, characteristic of genera in the Restionaceae, have sponge-like properties with respect to their water holding capacity and again they consist of clusters of rootlets with long root hairs. The leaf bases of the grass tree, *Kingia australis* form an important reservoir of nutrients such as nitrogen, phosphorus and potassium. This species produces aerial roots from the stem apex which descend between the true stem and the leaf bases. These roots send clusters of hairy laterals into the leaf bases from which they absorb stored nutrients (Lamont, 1982).

Another type of specialised root which may play a role in nitrogen acquisition is the mycorrhizae, of which the principal types are vescicular–arboreal (VA), ericoid, orchid and ectomycorrhiza. Fungi forming VA mycorrhiza are characterised by a lack of host specificity while the fungi of the other three types are much more host-specific. Although the role of mycorrhizal fungi in the phosphorus nutrition of their hosts is well established (Tinker, 1980) their significance with respect to nitrogen nutrition is less clear. Studies of the uptake of nitrate and ammonium ions by ectomycorrhizal roots suggest some enhancement of uptake (Alexander, 1985). Mycorrhizal fungal hyphae increase the effective volume of soil exploited by plant roots. Also ericoid mycorrhizal roots may increase nitrogen uptake by the plant through the exploitation of otherwise unavailable organic nitrogen sources in the soil or litter (Abuzinadah & Read, 1986*a*, *b*; Abuzinadah, Finlay & Read, 1986). The ability of some mycorrhizal fungi to hydrolyse protein in the soil may be of particular importance for the nutrition of woody plants growing in soil of boreal or arctic regions where cool temperatures may limit nitrogen mineralisation (Abuzinadah & Read, 1986*b*).

A group of plants often considered to have access to novel sources of nitrogen are carnivorous species. Chandler & Anderson (1976) showed that nitrogen concentration increased in insect-fed plants of *Drosera* and that growth was greater when nitrogen was supplied as insects compared with nitrate ions, although the contribution of other ions cannot be ignored. It has been estimated that between 20% (Pate & Dixon, 1978) and 100% (Watson, Matthiesean & Springett, 1982) of the nitrogen in carnivorous species can be derived from their insect catch.

A further group of plants which may be able to exploit alternative sources of nitrogen are parasites which can tap into the transport system of their host. The acquisition of nitrogen may be a key factor regulating transport processes between, for example, plants and their mistletoe parasites. The high transpiration rates observed in most mistletoes are essential to satisfy their nitrogen requirements (Schulze & Ehleringer, 1984; Schulze, Turner & Glatzel, 1984). Moreover, Ehleringer *et al.* (1985) suggest that nitrogen concentrations in the host xylem sap regulate stomatal conductance in the parasite. The form of nitrogen utilised by parasitic plants will be determined by both the source of nitrogen available to the host and the biochemical characteristics of its nitrogen metabolism. McNally & Stewart (1987) found that the nitrogen contents of parasitic species of both stem and root, were similar to those of their hosts. Glatzel & Balasubramaniam (1987) have shown that the nitrogen content of the mistletoe, *Dendrophtoe falcata* is very closely related to that of the host species it parasitises.

The plants which most clearly have access to non-soil sources of nitrogen are those forming symbiotic associations with nitrogen fixing prokaryotes. Many such species have an important role in nitrogen acquisition in the early phases of ecological succession (Silvester, 1977). In addition, they play an important role in the savanna flora of Africa, the Far East and South America, which are characterised by a high proportion of species with the potential for nitrogen fixation (Cole, 1986).

At the physiological level, studies of nitrate uptake suggest that some species characteristic of nutrient-poor soils can increase their capacity for nitrate uptake when grown under nitrate limiting conditions. In contrast, nitrophilous species increase their uptake capacity in response to increased nitrate availability (Van der Dijk *et al.*, 1982). The results shown in Table 1 suggest that it is only nitrogen fixing species which are able to acquire additional nitrogen when growing in nitrogen limited environments. The leaf nitrogen contents of leguminous species are markedly higher than those of non-legumes but there is little indication that species with the proteoid root morphology, or the parasitic mode of nutrition, have an enhanced capacity to acquire nitrogen. However, dynamic studies which measure rates of nitrogen acquisition in relation to the growth of individual species are needed to establish the basis for differences in the type of nitrogen nutrition. Clearly, the annual demand for nitrogen will vary between species depending upon their morphology and growth potential. Also, parasitic plants which are holoparasitic will have a small nitrogen requirement since the investment in photosynthetic machinery will be much less than in autotrophic plants. Even in some hemiparasitic species, low rates of photosynthesis and carbon input from the host may reduce the

Table 1. *Comparison of the nitrogen content of leguminous and non-leguminous species from plant communities of eastern Australia.*

	Nitrogen (% dry weight)			
Site	Legumes	Non-legumes	Proteoid	Parasites
Evergreen vine-forest	2.2	2.4	2.0	2.1
Deciduous vine-forest	2.1	2.0	—	2.1
Savanna woodland	1.9	1.7	—	1.9
Heathy open-forest	1.7	1.1	0.8	0.9

Analyses were carried out on mature leaves (Stewart *et al.* unpublished data).

demand for photosynthate and hence the need for nitrogen allocated to the photosynthetic machinery. For example, the root hemiparasites, *Striga asiatica* and *S. hermonthica*, which obtain at least 35% of their carbon from the host (Press *et al.*, 1987) both have low rates of photosynthesis (Shah, Smirnoff & Stewart, 1987).

Nitrogen assimilation

Nitrate absorbed by plant roots is first reduced to ammonium before incorporation into amino acids and this reduction is catalysed by the enzymes nitrate and nitrite reductase (Fig. 1). The capacity for nitrate reduction is widespread among higher plants (Lee & Stewart, 1978; Smirnoff, Todd & Stewart, 1984; Al Gharbi & Hipkin, 1984: Stewart *et al.*, 1986) and except for a few species, such as those of the genus *Erythrina* (Stewart & Orebamjo, 1979) the properties (K_m and pyridine nucleotide specificity) of nitrate reductase show little variation between species.

Differences between species with respect to the site of nitrogen assimilation have attracted much speculation over many years. Dinitrogen fixation generally occurs in the root system, with the exception of a small number of species having stem or petiole nodules. It is thought that problems of pH maintenance restrict the primary assimilation of ammonia to plant roots (Raven & Smith, 1976). Briefly, the argument is as follows, the stoichiometric production of protons for each ammonium ion assimilated requires that the protons are excreted into the soil solution as they cannot be neutralised in biochemical processes (Eqn 1).

$$NH_4^+ \longrightarrow NH_3 + H^+ \quad (1)$$

In contrast, the reduction of nitrate ions (Eqn 2) produces hydroxyl ions

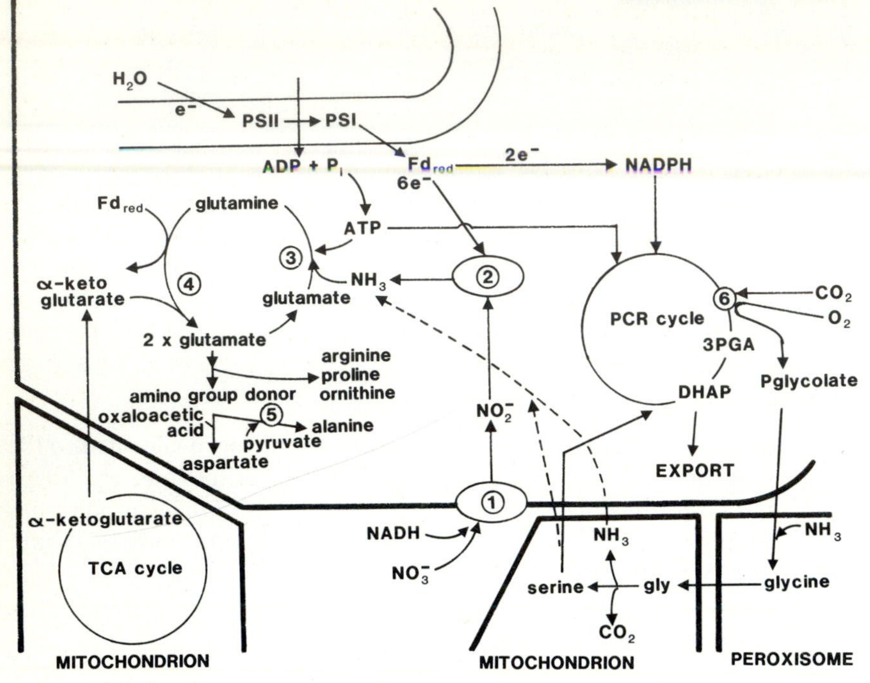

Fig. 1. Nitrate reduction in photosynthetic cells shown schematically, with the central role of nitrate reductase, the glutamine/glutamate cycle of ammonia assimilation and the links to the carbon reduction cycle and associated photorespiratory nitrogen flux indicated. Enzymes identified in the figure are: (1) nitrate reductase; (2) nitrite reductase; (3) glutamine synthetase; (4) GOGAT; (5) pyruvate aminotransferase; (6) ribulose bisphosphate carboxylase–oxygenase (after Lawlor, 1987).

but, through the operation of the biochemical pH stat, intracellular pH can be regulated by organic acid synthesis.

$$NO_3^- + 6H^+ + 6e^- \longrightarrow NH_3 + 3OH^- \qquad (2)$$

In general, small concentrations of ammonium ions are present in the xylem fluid, suggesting that primary ammonium assimilation is, indeed, largely restricted to the root system.

Plants are normally divided into three groups on the basis of the contribution that their root systems make to the reduction of nitrate ions. At one extreme are species in which nitrate reduction is confined to the roots and at the other extreme are those species in which nitrate reduction occurs only in the shoots. However, there are many species in which nitrate reduction occurs in both roots and shoots, although the relative distribution of nitrate reduction between the two varies considerably depending on species and may often change in response to variations in nitrate availability (Smirnoff & Stewart, 1985*a*). Until recently it was generally held that many woody species assimilated nitrate in their roots (Bray, 1983). However, several studies have shown that nitrate reduction is widely distributed in the leaves of woody plants (Stewart & Orebamjo, 1983; Smirnoff *et al.*, 1984; Al Gharbi & Hipkin, 1984; Stewart, Hegarty & Specht, 1988).

Andrews (1986) suggested a number of generalisations regarding the relation between the site of nitrate reduction and environment. He regards temperate perennial species as predominantly root assimilators whereas annual species and tropical perennial species are considered to be predominantly leaf assimilators. The trouble with this, as with many other generalisations, is that many species are exceptions to the rule (Table 2). It is evident that many tropical perennials, both woody and herbaceous, exhibit only small rates of leaf nitrate reduction whereas, from the limited data available (Table 2), it would seem that many temperate perennial species are leaf assimilators, having large activity of leaf nitrate reductase activities compared to root activity (Stewart & Orebamjo, 1983; Stewart, Sumar & Patel, 1987). The evidence of Table 2 suggests, contrary to the conclusion reached by Andrews (1986), that there is little difference between annual species, either tropical or temperate, or perennial tropical species all of which are mainly leaf assimilators. However, annual species tend to favour root assimilation when compared with perennials. Smirnoff & Stewart (1985*b*) have attempted to relate the site of nitrogen assimilation in different species to their ecological preferences. They suggest that shade tolerant species tend towards root nitrate assimilation. This suggestion is supported by comparative studies on ferns in which root nitrate assimilation is common; root assimilation is most pronounced in those tolerant of shade (Stewart *et al.*, 1987). Similarly a study of Australian rain forest trees and vines showed that pioneer species were predominantly leaf nitrate assimilators while those of closed mature forest reduced little nitrate in their leaves (Stewart *et al.*, 1988). The latter are not normally nitrate assimilators and their apparent preference for root nitrate assimilation may simply reflect the fact that the root is the site for the assimilation of forms of nitrogen other than nitrate. These species of the closed forest have the necessary biochemical components in their roots for nitrogen assimilation; they actively transfer nitrogen from root to shoot and can catabolise transported nitrogenous compounds, such as amides or ureides, in their leaf cells. Thus, many so-called root nitrate assimilators may be species which in their natural environment utilise dinitrogen, ammonium or even organic nitrogen sources.

Nitrate utilisation is widespread among plants even though it is a less cost effective nitrogen source when compared with ammonium. The prevalence

Table 2. *Distribution of nitrate reductase between the roots and shoots of annual and perennial plants.*

Species	Distribution	Nitrate reductase (pkat g fw^{-1})	
		Shoot	Root
Annual			
Amaranthus viridis	Tropical	1300	278
Atriplex hastata	Temperate	2346	231
Celosia argentea	Tropical	888	169
Chenopodium album	Temperate	1668	139
Glaucium flavum	Temperate	1184	516
Euphorbia glomerifera	Tropical	1478	121
Lathyrus sylvestris	Temperate	306	87
Plantago lagopus	Temperate	286	93
Spigelia anthelmia	Tropical	545	134
Urtica dioica	Temperate	4059	400
Perennial			
Albizzia harveyi	Tropical	202	44
Avicennia nitida	Tropical	175	53
Calophyllum brasiliense	Tropical	211	40
Cecropia glaziovii	Tropical	500	200
Colophospermum mopane	Tropical	447	76
Crataegus monogyna	Temperate	556	208
Dimorphandra mollis	Tropical	125	28
Juglans nigra	Temperate	1112	150
Kielmeyera coriacea	Tropical	133	10
Malaisia scandens	Tropical	150	62
Neolitsea dealbata	Tropical	37	32
Polyscias elegans	Tropical	120	12
Sambucus nigra	Temperate	1946	100
Trema aspera	Tropical	920	25

of nitrate reduction among plants may reflect the use of available energy from both photosynthesis and from respiration, so that unlike ammonia or dinitrogen, nitrate can be reduced in shoot as well as root tissues. Also, nitrate ions, unlike ammonium ions, can be stored in plant cells. However, the assimilatory capacity of the root may be insufficient to meet the demands of the whole plant as it relies on translocated assimilates and does not, therefore, exploit the availability of rapidly changing solar energy. The morphology of a plant, in particular the allocation of biomass above and below ground, may also limit the extent to which ammonium ions can meet all of a plant's nitrogen requirements.

The product of nitrate reduction is ammonia, which is also the first product of dinitrogen reduction and, in some soils, is the available form of nitrogen in the soil solution. Prior to the mid-1970s it was generally thought that the entry of ammonia into organic combination was catalysed by the enzyme glutamate dehydrogenase. However, most workers now accept that the only major pathway of ammonia assimilation is the glutamate synthase route (Fig. 1), in which glutamine synthetase and glutamate synthase catalyse an ATP driven synthesis of glutamate (Miflin & Lee, 1980). There is little, if any, experimental evidence which would support a role for glutamate dehydrogenase in ammonia assimilation. Labelling studies with ^{15}N, used in combination with specific enzyme inhibitors, indicate that when glutamine synthetase is inhibited incorporation of ammonia ceases after a short time and when glutamate synthase is inhibited no further labelling of glutamate can be detected (Fentem, Lea & Stewart, 1983). The glutamate synthase route would appear to be the only pathway for primary ammonia assimilation in species as diverse as fast growing nitrophilous plants such as *Datura stramonium* (Lewis & Probyn, 1978) and slow growing mycorrhizal species such as *Fagus sylvatica* (Martin *et al*., 1986). These studies suggest that ammonium assimilation occurs exclusively through the glutamate synthase route even in plants from habitats differing in the form of available nitrogen.

Biochemical and molecular studies of glutamine synthetase indicate that the enzyme exists in multiple molecular forms which are coded for by a small multigene family (Wallsgrove, 1987). Leaf tissue of many plants contains two forms of the enzyme, one localised in the chloroplasts the other in the cytosol and both forms are also present in some roots (McNally & Hirel, 1983). There is also evidence of an isoform specific to nodules in some nitrogen fixing plants (Cullimore *et al*., 1983).

The role of chloroplastic and cytosolic forms of glutamine synthetase is not established. Keys *et al*. (1978) suggested that the cytosolic isoform functions in the reassimilation of ammonia released in photorespiration. However, the absence of the cytosolic isoform from the leaves of many species exhibiting high rates of photorespiration (McNally *et al*., 1983) argues against this role, and more recent studies with mutants lacking chloroplastic glutamine synthetase have shown that they accumulate much ammonia under photorespiratory conditions. These observations suggest that it is the chloroplastic isoform of glutamine synthetase which is active in photorespiration (Wallsgrove, 1987). An alternative role proposed for the cytosolic glutamine synthetase is in the dark assimilation of ammonia (Hirel & Gadal, 1982). In relation to this suggestion, it is interesting that leaves of trees characteristic of closed forests exhibit relatively higher activity of cytosolic glutamine synthetase than those found in the leaves of pioneer species (Stewart *et al.*, 1988). In ferns, levels of chloroplastic glutamine synthetase correlated with photosynthetic capacity which was interpreted as indicating a relation between this enzyme and photorespiration (Stewart *et al.*, 1986). However, it could also relate to the well established link between nitrogen supply and photosynthetic capacity (Sharkey, 1985). In woody plants, amounts of chloroplastic glutamine synthetase correlated with those of leaf nitrate reductase (Stewart *et al.*, 1988) and in ferns the activities of nitrate reductase and glutamine synthetase in roots were closely related (Stewart *et al.*, 1986). These results suggest that there is a relation between the characteristics of nitrogen assimilation exhibited by a species and the environment in which it grows. Typically, pioneer species are leaf nitrate assimilators and have high activities of leaf nitrate reductase and chloroplastic glutamine synthetase. In contrast, species from climax vegetation tend to have a small capacity for nitrate reduction and exhibit small amounts of chloroplastic glutamine synthetase.

Nitrogen allocation

Differences between species in the site of nitrogen assimilation and in the source of nitrogen utilised are reflected in variation in the forms of nitrogen transported. Pate and colleagues have described the types of nitrogenous compounds transported by higher plants. Most common are the amides, asparagine and glutamine; less common are arginine and citrulline (either together or separately), and in some species the ureides, allantoin and allantoic acid (Pate, 1983). Sprent (1980) has made an interesting comparison of the forms of nitrogen transported in tropical and temperate legumes; she concluded that while amides are typical of temperate species, ureides are characteristic of tropical species. She explains this difference in terms of the solubility of ureides with temperature, their high C:N ratio compared with amides, and their cost effectiveness with respect to water loss.

Species also differ with respect to the forms of nitrogen stored when a large external supply is available. Frequently the form of nitrogen stored is also that which is transported. There are many nitrate accumulating plants and these are frequently also ruderal species. It is often suggested that nitrate accumulation simply reflects an excess of uptake over the capacity for reduction. However, in some species nitrate accumulation persists even under conditions where nitrate supply limits growth. An explanation is that nitrate accumulation has a specific physiological role. In some species nitrate concentrations can exceed 100 mM (Stewart *et al.*, 1979) and a role as an osmotic solute seems likely (Smirnoff & Stewart, 1985*b*). Nitrate accumulation in plants of desert and semi-arid regions has been suggested to play a protective role for photosynthesis during water stress when stomata are closed, in that nitrate reduction could lead to the dissipation of excess photochemical energy and thereby minimise photoinhibition (Smirnoff & Stewart, 1985*b*).

Nitrogenous compounds, besides being allocated for growth and storage, may also function in the adaptation of plants to environmental stresses, both chemical and physical, and also in protection against biotic stresses. Plants characteristic of saline and arid environments often contain very large amounts of nitrogen in the form of proline or glycine betaine (Stewart *et al.*, 1979; Wyn Jones & Storey, 1981; Poljakoff-Mayber *et al.*, 1987). Under some conditions as much as 40% of nitrogen in plants may be present in these and related compounds, which may function in osmotic adjustment (Wyn Jones, 1984; Stewart & Larher, 1980). They may also have a role in protecting cellular enzymes and membrane proteins against heat denaturation, dehydration and chemical denaturation (Paleg *et al.*, 1981; Jolivet, Hamelin & Larher, 1983; Smirnoff & Stewart, 1985*b*; Paleg, Stewart & Starr, 1985). Generally, the accumulation of proline is inducible and occurs only when plants are exposed to salinity or to water stress. In contrast, many of the species accumulating glycine betaine do so constitutively; thus a halophyte such as *Suaeda maritima* contains large concentrations of glycine betaine even when grown under non-saline conditions (Stewart *et al.*, 1979). Glycine betaine accumulating plants would appear to have a continuous demand for 'stress nitrogen' while for proline accumulators the demand is temporary and restricted to periods of stress. Moreover, proline accumulated under stress conditions is rapidly metabolised following the removal of the stress, whilst there is little evidence for the catabolism or reutilisation of nitrogen allocated to glycine betaine (Rhodes, 1987).

Many species of plants allocate large quantities of nitrogen to secondary compounds which may have a protective role against herbivores. Most studied in this connection are the alkaloid accumulating plants which comprise nearly 20% of all species. Other, potentially protective nitrogenous compounds include some of the non-protein amino acids, amines, cyanogenic glycosides, peptides and even proteins such as abrin and ricin (Harborne, 1982). It is curious that so many of the secondary compounds which are accumulated as a defence against damage by herbivores are nitrogenous compounds when it is the nitrogen content of vegetation which is the limiting nutritional factor for many herbivores (Mattson, 1980).

Conclusions

Much comparative ecophysiology has been exclusively concerned with photosynthesis and the carbon economy of plants. Yet, given the importance of the nitrogen status of tissues in determining the photosynthetic rates and plant size, it is surprising that there are relatively few ecophysiological studies of plant nitrogen assimilation. The work discussed here indicates that plants are adapted in various ways to the nitrogen supply, both the absolute amount and the fluctuations in supply from the environment. Plants exhibit great differences in their requirement for nitrogen, i.e. the nitrogen resource demand, and in the efficiency with which they acquire it from the environment. Differences in the capacity of species to utilise nitrate and in the site of its assimilation into organic compounds relate both to environmental and plant characteristics. Under natural conditions many plants allocate substantial amounts of nitrogen to compounds which may play a role in protection against stresses such as salinity, water deficits and attack by herbivores, but little is known of how this allocation to different physiological functions is regulated or of the extent to which these functions are competitive or indeed how they are balanced with demands for growth. The obvious and fundamental link between leaf nitrogen and photosynthesis suggests that there is competition between the assimilatory and protective functions of nitrogen and that this may have implications with respect to growth and survival of plants, particularly in extreme environments. This points to the need to integrate ecophysiological studies of the carbon and nitrogen economy of plants.

References

Abuzinadah, R.A., Finlay, R.D. & Read, D.J. (1986). The role of proteins in the nitrogen nutrition of ectomycorrhizal plants. 2. Utilization of protein by mycorrhizal plants of *Pinus contorta*. *New Phytologist*, **103**, 495–506.

Abuzinadah, R.A. & Read, D.J. (1986*a*). The role of proteins in the nitrogen nutrition of ectomycorrhizal plants. 1. Utilization of peptides and proteins by ectomycorrhizal fungi. *New Phytologist*, **103**, 481–93.

Abuzinadah, R.A. & Read, D.J. (1986*b*). The role of proteins in the nitrogen nutrition of ectomycorrhizal plants. III. Protein utilization by *Betula*, *Picea* and *Pinus* in mycorrhizal association with *Hebeloma crustuliniforme*. *New Phytologist* **103**, 507–14.

Alexander, I.J. (1985). The significance of ectomycorrhizas in the nitrogen cycle. In *Nitrogen as an Ecological Factor*, eds. J.A. Lee, S. McNeill & I.H. Rorison, pp. 69–94. Oxford: Blackwell.

Al Gharbi, A. & Hipkin, C.R. (1984). Studies on nitrate reductase in British angiosperms. 1. A comparison of nitrate reductase activity in ruderal, woodland edge and woody species. *New Phytologist*, **97**, 629–39.

Andrews, M. (1986). The partitioning of nitrate assimilation between root and shoot of higher plants. *Plant, Cell and Environment*, **9**, 511–19.

Bray, C.M. (1983). *Nitrogen Metabolism in Plants*, p. 214. London: Longman.

Chandler, G.E. & Anderson, J.W. (1976). Studies on the nutrition and growth of *Drosera* species with special reference to the carnivorous habit. *New Phytologist*, **76**, 129–41.

Cole, M.M. (1986). *The Savannas, Biogeography and Geobotany*, pp. 1–438. London: Academic Press.

Cullimore, J.V., Lara, M., Lea, P.J. & Miflin, B.J. (1983). Purification and properties of two forms of glutamine synthetase from the plant fraction of *Phaseolus* root nodules. *Planta*, **157**, 245–53.

Ehleringer, J.R., Schulze, E.D., Ziegler, O.L., Lange, O.L., Farquhar, G.D. & Cowan, I.R. (1985). Xylem tapping mistletoes: water or nutrient parasites? *Science*, **227**, 1479–81.

Fentem, P.A., Lea, P.J. & Stewart, G.R. (1983). Action of inhibitors of ammonia assimilation on amino acid metabolism in *Hordeum vulgare* L. (cv Golden Promise). *Plant Physiology*, **71**, 502–6.

Ferrar, P.J. & Osmond, C.B. (1986). Nitrogen supply as a factor influencing photoinhibition and photosynthetic acclimation after transfer of shade grown *Solanum dulcamara* to bright light. *Planta*, **168**, 563–70.

Field, C. & Mooney, H.A. (1986). The photosynthesis–nitrogen relationship in wild plants. In *On the Economy of Plant Form and Function*, ed. T.J. Givnish, pp. 25–55. Cambridge University Press.

Fitter, A.H. (1987). An architectural approach to the comparative ecology of plant root systems. In *Frontiers of Comparative Plant Ecology*, eds. I.H. Rorison, J.P. Grime, R. Hunt, G.A.F. Hendry & D.H. Lewis, pp. 61–78. London: Academic Press.

Glatzel, G. & Balasubramaniam, S. (1987). Mineral nutrition of mistletoes: general concepts. In *Parasitic Flowering Plants*, eds. H. Chr. Weber & W. Forstreuter, pp. 263–76. FRG: Marburg.

Greenwood, D.J. (1986). Prediction of nitrogen fertilizer needs of arable crops. *Advances in Plant Nutrition*, **2**, 1–62.

Harborne, J.B. (1982). *Introduction to Ecological Biochemistry*. London: Academic Press.

Hirel, B. & Gadal, P. (1982). Glutamine synthetase in rice. A comparative study of the enzymes from roots and leaves. *Plant Physiology*, **66**, 619–23.

Jeffrey, D.W. (1987). *Soil–Plant Relationships an Ecological Approach*, pp. 1–295. Kent: Croom-Helm Ltd.

Jolivet, Y., Hamelin, J. & Larher, F. (1983). Osmoregulation in halophytic higher plants: the protective effects of glycine betaine and other related solutes against oxalate destabilization of membranes in beetroot cells. *Zeitschrift für Planzenphysiologia*, **109**, 171–80.

Keys, A.J., Bird, I.F., Cornelius, M.F., Lea, P.J., Wallsgrove, R.M. & Miflin, B.J. (1978). Photorespiratory nitrogen cycle. *Nature*, **275**, 741–2.

Lamont, B. (1982). Mechanisms for enhancing nutrient uptake in plants, with particular reference to mediterranean South Africa and Western Australia. *Botanical Review*, **48**, 597–689.

Lawlor, D.W. (1987). *Photosynthesis: Metabolism, Control and Physiology*. Harlow: Longman Scientific & Technical.

Lawlor, D.W. (1991). Concepts of nutrition in relation to cellular processes and environment. In *Plant Growth: Interactions with Nutrition and Environment*, eds. J.R. Porter & D.W. Lawlor, Society for Experimental Biology, Seminar Series 43, pp. 1–32. Cambridge University Press.

Lee, J.A. & Stewart, G.R. (1978). Ecological aspects of nitrogen assimilation. *Advances in Botanical Research*, **6**, 1–43.

Lewis, O.A.M. & Probyn, T.A. (1978). ^{15}N incorporation and glutamine synthetase inhibition studies of nitrogen assimilation in leaves of the nitrophile *Datura stramonium* L. *New Phytologist*, **81**, 519–26.

Martin, F., Stewart, G.R., Genetet, I. & Le Tacon, F. (1986). Assimilation of $^{15}NH_4$ by beech (*Fagus sylvatica* L.) ectomycorrhizas. *New Phytologist*, **102**, 85–94.

Mattson, W.J. (1980). Herbivory in relation to plant nitrogen content. *Annual Review of Ecology and Systematics*, **11**, 119–61.

McNally, S.F. & Hirel, B. (1983). Glutamine synthetase isoforms in higher plants. *Physiologie Vegetale*, **21**, 761–74.

McNally, S.F., Hirel, B., Gadal, P., Mann, A.F. & Stewart, G.R. (1983). Glutamine synthetases of higher plants. Evidence for a specific isoform content are related to their possible physiological role and compartmentation within the leaf. *Plant Physiology*, **72**, 22–5.

McNally, S.F. & Stewart, G.R. (1987). Inorganic nitrogen assimilation by parasitic angiosperms. In *Parasitic Flowering Plants*, eds. H. Chr. Weber & W. Forstreuter, pp. 539–46. FRG: Marburg.

Miflin, B.J. & Lee, P.J. (1980). Ammonia assimilation. In *Biochemistry of Plants Vol. 5*, ed. B.J. Miflin, pp. 169–202. London: Academic Press.

Paleg, L.G., Douglas, T.J., Van Daal, A. & Keech, D.B. (1981). Proline, betaine and other organic solutes protect enzymes against heat inactivation. *Australian Journal of Plant Physiology*, **8**, 107–14.

Paleg, L.G., Stewart, G.R. & Starr, R. (1985). The effect of compatible solutes on proteins. *Plant and Soil*, **89**, 83–96.

Pate, J.S. (1983). Patterns of nitrate metabolism in higher plants and their ecological significance. In *Nitrogen as an Ecological Factor*, eds. J.A. Lee, S. McNeill & I.H. Rorison, pp. 225–56. Oxford: Blackwell.

Pate, J.S. & Dixon, K.W. (1978). Mineral nutrition of *Drosera erythrorhiza* Lindl. with special reference to its tuberous habit. *Australian Journal of Botany*, **26**, 455–64.

Poljakoff-Mayber, A., Symon, D.E., Jones, G.P., Naidu, N.P., & Paleg, L.G. (1987). Nitrogenous compounds in Native South Australian Plants. *Australian Journal of Plant Physiology*, **14**, 341–50.

Press, M.C., Shah, N., Tuohy, J.M. & Stewart, G.R. (1987). Carbon isotope ratios demonstrate carbon flux from C4 host to C3 parasite. *Plant Physiology*, **85**, 1143–5.

Rabinovitch-Vin, A. (1983). Influence of nutrients on the composition and distribution of plant communities in Mediterranean-type ecosystems of Israel. In *Mediterranean-Type Ecosystems; The Role of Nutrients*, eds. F.J. Kruger, D.T. Mitchell & J.U.M. Jarvis, pp. 74–85. Berlin: Springer-Verlag.

Raven, J.A. (1985). Regulation of pH and generation of osmolarity in vascular land plants: costs and benefits in relation to efficiency of use of water, energy and nitrogen. *New Phytologist*, **101**, 25–77.

Raven, J.A. & Smith, F.A. (1976). Nitrogen assimilation and transport in vascular land plants in relation to intracellular pH regulation. *New Phytologist*, **76**, 415–31.

Rhodes, D. (1987). Metabolic responses to stress. In *The Biochemistry of*

Plants, Vol. 12, Physiology of Metabolism, ed. D.D. Davies, pp. 202–42. New York: Academic Press.

Robertson, G.P. (1982). Nitrification in forested ecosystems. *Philosophical Transactions of the Royal Society of London*, **B296**, 445–57.

Robinson, D. & Rorison, I.H. (1983). Relationships between root morphology and nitrogen availability in a recent theoretical model describing nitrogen uptake from soil. *Plant, Cell and Environment*, **6**, 641–7.

Schulze, E.-D. & Ehleringer, J.R. (1984). The effect of nitrogen supply on the growth and water use efficiency of xylem-tapping mistletoes. *Planta*, **162**, 268–75.

Schulze, E.-D., Turner, N.C. & Glatzel, G. (1984). Carbon, water and nutrient relations of two mistletoes and their hosts. *Plant, Cell and Environment*, **7**, 293–9.

Shah, N., Smirnoff, N. & Stewart, G.R. (1987). Photosynthesis and stomatal characteristics of *Striga hermonthica* in relation to its parasitic habit. *Physiologia Plantarum* **69**, 699–703.

Sharkey, T.D. (1985). Photosynthesis in intact leaves of C3 plants: physics, physiology and rate limitations. *Botanical Review*, **51**, 53–105.

Silvester, W.B. (1977). Dinitrogen fixation by plant associations excluding legumes. In *A Treatise on Dinitrogen Fixation, Section IV, Agronomy and Ecology*, eds. R.W.F. Hardy & A.H. Gibson, pp. 141–90. New York: John Wiley & Sons.

Smirnoff, N. & Stewart, G.R. (1985*a*). Nitrate assimilation and translocation by higher plants; comparative physiology and ecological consequences. *Physiologia Plantarum*, **64**, 133–40.

Smirnoff, N. & Stewart, G.R. (1985*b*). Stress metabolites and their role in coastal plants. *Vegetatio*, **62**, 273–8.

Smirnoff, N., Todd, P. & Stewart, G.R. (1984). The occurrence of nitrate reduction in the leaves of woody plants. *Annals of Botany*, **54**, 363–74.

Sprent, J.I. (1980). Root nodule anatomy, type of export product and evolutionary origin in some leguminosae. *Plant, Cell and Environment*, **3**, 35–43.

Stewart, G.R., Hegarty, E.E. & Specht, R.L. (1988). Inorganic nitrogen metabolism in plants of Australian Rainforest communities. *Physiologia Plantarum*, **74**, 26–33.

Stewart, G.R. & Larher, F. (1980). The accumulation of amino acids and related compounds in relation to environmental stresses. In *The Biochemistry of Plants, Vol. 3, Amino Acid and Derivatives*, ed. B.J. Miflin, pp. 609–35. London: Academic Press.

Stewart, G.R., Larher, F., Ahmad, I. & Lee, J.A. (1979). Nitrogen metabolism and salt tolerance in higher plant halophytes. In *Ecological Processes in Coastal Environments*, eds. R.L. Jeffries & A.J. Davy, pp. 211–27. Oxford: Blackwell Scientific Publications.

Stewart, G.R. & Orebamjo, T.O. (1979). Some unusual characteristics of nitrate reduction in *Erythrina senegalensis*. *New Phytologist*, **83**, 311–19.

Stewart, G.R. & Orebamjo, T.O. (1983). Studies of nitrate utilization by dominant species of regrowth vegetation of tropical West Africa; a Nigerian example. In *Nitrogen as an Ecological Factor*, eds. J.A. Lee, S. McNeill & I.H. Rorison, pp. 167–88. Oxford: Blackwell Scientific Publications.

Stewart, G.R., Popp, M., Holzapfel, I., Stewart, J.A. & Dickie-Eskew, A. (1986). Localization of nitrate reduction in ferns and its relationship to environment and physiological characteristics. *New Phytologist*, **104**, 373–84.

Stewart, G.R., Sumar, N. & Patel, M. (1987). Comparative aspects of inorganic nitrogen assimilation in higher plants. In *Inorganic Nitrogen Metabolism*, eds. W.R. Ulrich, P.J. Aparicio, P. J. Syrett & F. Castilla, pp. 137–41. Berlin: Springer-Verlag.

Terry, N. & Rao, I.M. (1991). Nutrients and photosynthesis: iron and phosphorus as case studies. In *Plant Growth: Interactions with Nutrition and Environment*, eds. J.R. Porter & D.W. Lawlor, Society for Experimental Biology, Seminar Series 43, pp. 55–79. Cambridge University Press.

Tinker, P.B. (1980). The role of rhizosphere microorganisms in phosphorus uptake by plants. In *The Role of Phosphorus in Agriculture*, eds. F. Kwasaneh & E. Sample, pp. 617–54. Madison: American Society of Agronomy.

Van der Dijk, S.J., Lanting, L., Lambers, H., Posthumus, F., Stulen, I. & Hofstra, R. (1982). Kinetics of nitrate uptake by different species from nutrient-rich and nutrient poor habitats as affected by nutrient supply. *Physiologia Plantarum*, **55**, 103–10.

Wallsgrove, R.M. (1987). The role of glutamine synthetase and glutamate synthase in nitrogen metabolism of higher plants. In *Inorganic Nitrogen Metabolism*, eds. W.R. Ulrich, P.J. Aparico, P.J. Syrett & F. Castilla, pp. 137–41. Berlin: Springer-Verlag.

Watson, A.P., Matthiessen, J.N. & Springett, B.P. (1982). Arthropod associates and macronutrient status of the red-ink sundew (*Drosera erythrorhiza* Lindl.). *Australian Journal of Ecology*, **7**, 13–22.

Willis, A.J. (1961). Braunton Burrows; the effect of the addition of mineral nutrients to the dune soils. *Journal of Ecology*, **51**, 353–74.

Wyn Jones, R.G. (1984). Phytochemical aspects of osmotic adaptation. *Recent Advances in Phytochemistry*, **18**, 55–78.

Wyn Jones, R.G. & Storey, R. (1981). Betaines. In *The Physiology and Biochemistry of Drought Resistance in Plants*, eds. L.G. Paleg & D. Aspinall, pp. 171–204. Sydney: Academic Press.

B. MARSHALL AND J.R. PORTER

Concepts of nutritional and environmental interactions determining plant productivity

Interactions

Interactions within plants are perceived as a balance between the supply and demand of carbon and nutrients needed for growth. The environment is a dominant factor in determining this balance. The genetic codes of the plants determine the range of phenotypic responses to the environment, and, in particular, the nature and scale of the interactions between nutrients and the carbon within them (Lawlor, 1991, this volume). Earlier chapters of this book deal with interactions at the cellular and tissue level and, to a lesser extent, at lower levels of organisation, i.e. organelle, molecular and atomic. This chapter is concerned with interactions within the individual plant and populations of them (monocultures, crops). Later chapters consider interactions at the community and ecosystem level (Fig. 1).

What is meant by interaction? Those familiar with statistics, and in particular the *analysis of variance*, know that interaction is detected with a certain confidence when, for example, the yield of a crop is influenced by several factors and the response to one is dependent on the value of one or more of the remaining factors. Variation in yield may be correlated with variation in a single factor (e.g. nitrogen level) or a plurality of factors (e.g. nitrogen level and fungicide use or non-use). However, these analyses of the effects of single and/or interacting factors on plant/crop performance ignore several important features about how interactions actually *operate* in biological systems. For example, interactions occur in space, in time, or in both. A simple example of an interaction in space is between neighbouring plants competing for solar radiation above- and/or water and nutrients below-ground. An example of a temporal interaction is when conditions early in the development of a plant favour vegetative growth and rapid exploitation of soil resources. If these resources are limited at this time then subsequent reproductive growth may be reduced or fail completely.

A statistical analysis describes the operation of a system at an *empirical* (model) level and is useful in predicting the response of the system in

Ecosystem
Community

Population
Individual
Organ

Tissues
Cell
Organelle
Molecule
Atom

Fig. 1. Levels of biological organisation. This chapter is concerned with functioning at the levels of organ up to population.

situations similar to those within which the model was developed. While it may point to the major factors that influence the system it does not give one insight into the underlying mechanisms that operate within the system. It is not always necessary to understand mechanisms. Driving a car requires only a basic knowledge of the interface between car and driver (ignition, accelerator, brake, steering and gear system). However, if the car breaks down, previously configured empirical relations no longer hold. A knowledge of the component parts, their function and their relation to each other is required in order to diagnose and repair the fault. Such knowledge is also required when designing a new car. The latter knowledge is called *mechanistic*. The performance of the overall system is predicted by integrating the relations from lower levels of organisation. Clearly, it is wise to test the validity of the predictions before production starts! This is where models can be of great assistance.

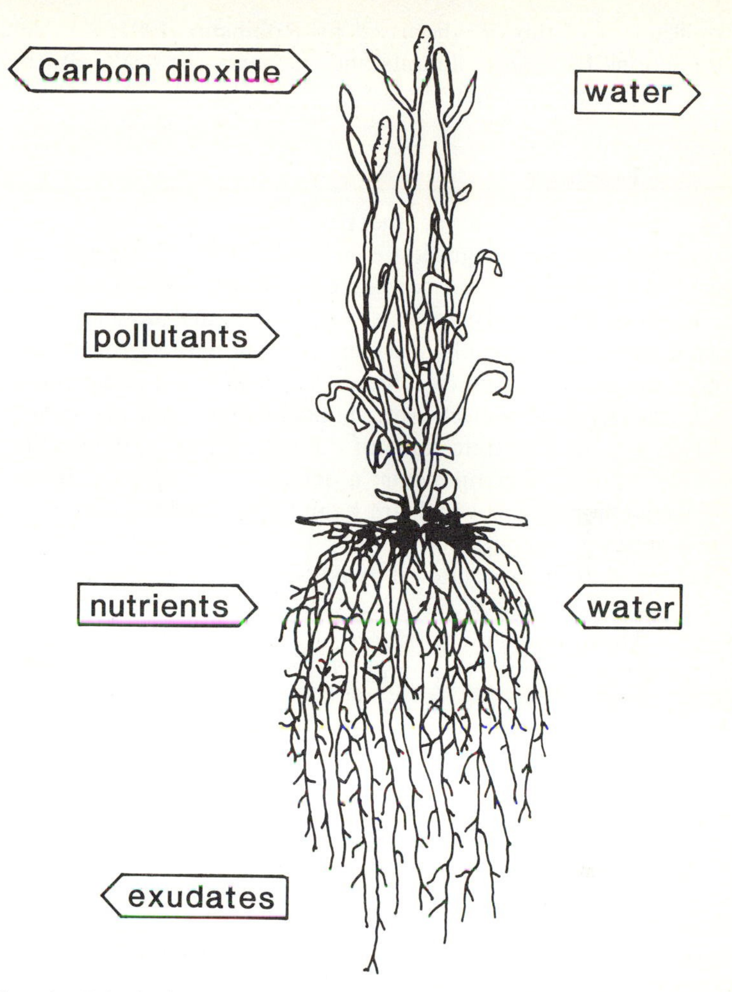

Fig. 2. Principal chemical exchanges between the plant and its environment.

In the present context we are concerned with the plant, that is both its shoot and root. The major exchanges between the plant and its environment via the roots are the uptake of nutrients, water, and oxygen (nitrogen fixation in some species) and the loss of organic and inorganic molecules through exudation, respiration and sloughing of root cap cells. Exchanges via the shoot are the loss of water through transpiration, the uptake of carbon through photosynthesis and uptake of pollutants (Fig. 2). The plant is able to adjust, to some extent, the growth rate of the root system relative to its shoot in response to imbalances in nutrient and carbon supply – the implications of this are discussed by Robinson (1991, this volume). The interactions between pollutants and nutrients are dealt with by Huettl & Fink (1991, this volume).

Feedback

The prospect of predicting the performance of a plant soon becomes daunting when the integration of mechanisms of fine detail at the cellular and sub-cellular level are considered. Nevertheless, plants do grow in a highly coordinated manner due, in part, to the presence of *networks* of mechanisms running both in sequence and in parallel, and to the feedbacks that exist between them (see Trewavas, 1986, for a detailed discussion).

A relatively simple electro-mechanical system which controls a flow rate will serve to demonstrate some of the features of feedback (Fig. 3). There are three principal components: a device for comparing the levels of two inputs (a comparator), which are V_C and V_I, and whose output, V_O, becomes the input to a pump, the output of which is a flow of material at a rate F. The flow, in turn, is monitored by a flow meter with an output, V_F. Typical relations between the output of each component and its input are shown in Fig. 4. The comparator produces an output that is proportional to the imbalance between the two inputs (Fig. 4*a*). V_C is usually preset and known as the control level voltage and V_I is variable. When the two voltages are equal there is zero output. As V_I decreases the imbalance between V_C and V_I increases as does V_O until, for this example, the output reaches a maximum.

The principal features of the pump are that it requires a minimum voltage before starting and that the flow rate increases steadily thereafter without an upper limit; in reality there will always be an upper limit but it may be several times higher than the intended operational rate (Fig. 4*b*). Typical characteristics of the flow meter are a minimum flow rate, below which it stalls, and an upper limit, beyond which V_F ceases to increase with flow rate (Fig. 4*c*). The presence of thresholds and limits are included for generality but are not critical in this example.

Without feedback the overall transfer function for the system, i.e. the relation between output V_F and input V_I is shown in Fig. 4*d* (line T). When V_I is small the flow rate is at a maximum determined by the maximum output voltage of the comparator. As V_I increases there comes a point when the flow rate is no longer fast enough to saturate the flow meter and V_F starts to decline, reaching zero when the flow rate has fallen to the stall point of the meter. With feedback present, that is the output from the flow meter is fed directly to the input of the comparator (dashed line, Fig. 3), the system is held at a fixed point. Since V_I now equals V_F the point must lie on the intersection between the 1:1 line and the overall transfer function (Fig.

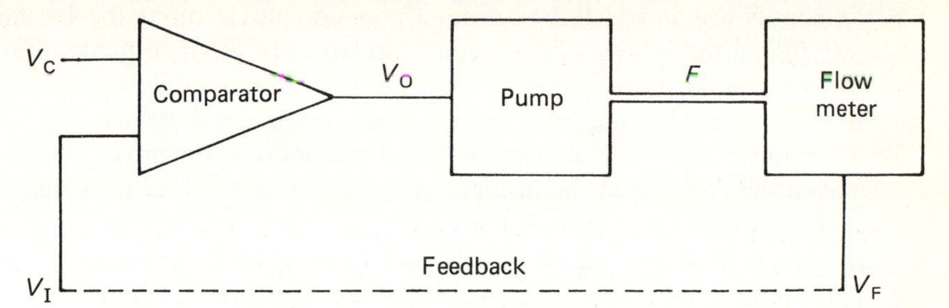

Fig. 3. A diagram of a simple electrical controller for gas flow rate, F. V_C, V_I, V_F and V_O are the control, input, flow meter and the comparator output voltages respectively. Feedback is present when V_F is connected back to V_I (dashed line).

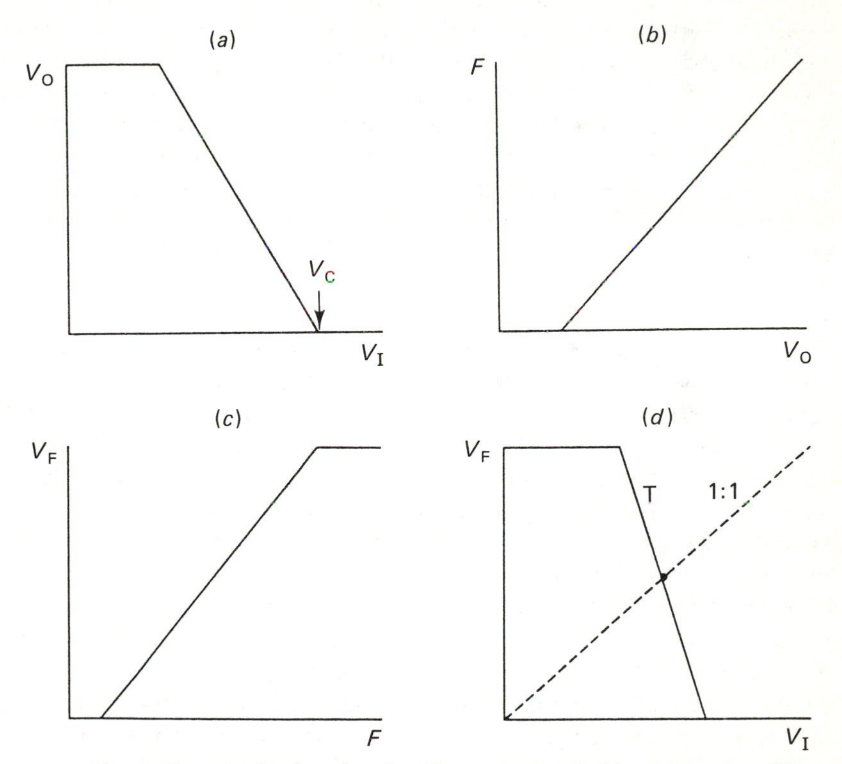

Fig. 4. The transfer function for the components, (*a*) comparator, (*b*) gas pump and (*c*) electrical gas flow meter, of a simple electrical flow controller (see Fig. 3.) and (*d*) the overall transfer function with no feedback present (see text for an explanation of the 1:1 line and its intersection with line T, the overall transfer function).

4*d*). The only way in which the fixed point can be moved along the 1:1 line is by changing the value of V_C. It is in the relation between control voltage and flow rate that we are actually interested (uppermost line, Fig. 5).

Now consider what happens to the system when a leak develops in the flow line and only half the flowing material reaches the flow meter. Assume the system was resting at the point A (Fig. 5) before the leak occurred. If there is no feedback between the flow meter and the comparator, the flow rate reaching the meter would simply be halved and the new state for the system would be at point C. However, if the feedback is connected, the lowered output from the flow meter would induce a higher level, V_O, to the pump. The flow rate reaching the meter would still fall but, in this case, would decline by only 18 per cent as opposed to the 50 per cent decline without feedback. Thus, feedback can both simplify the operation of a complex system and reduce the effect of damage on its overall functioning. The overall structure of a system may well be as important as the separate effects of individual elements (Ferrari, 1982).

The regulation of uptake of carbon dioxide from the atmosphere and the loss of water through stomata provide examples of biological systems which have similar feedback mechanisms to the electro-mechanical analogue. It is generally accepted that stomata are sensitive to carbon dioxide concentration (Heath, 1948; Raschke, 1975). Goudriaan & van Laar (1977) have shown, for bean and maize, that the concentration of carbon dioxide in the sub-stomatal cavity is kept constant, by adjustment of stomatal aperture, when the net photosynthetic rate is altered. The pump, in this case, is a combination of the diffusion gradient for CO_2 and the resistance to its flow through the stomatal aperture. The desired flow or photosynthetic rate (equivalent to V_C) is determined by the incident irradiance. By reducing stomatal aperture, when photosynthetic rate declines, the rate of water loss is also reduced and so the efficiency with which water is used is improved.

Feedback is also present in the regulation of nutrient uptake. The net uptake rate of sulphate or nitrate ions by roots is the result of a balance between the rates of influx and efflux. In both cases it appears that when supply is not meeting plant demand then net uptake, in the short term, is increased by reducing the rate of efflux not by increasing influx (Cram, 1988). Although influx and efflux are active processes the energy cost incurred, as a proportion of the total energy turnover in the plant, is small and is far outweighed by the benefits of possessing a control mechanism (Clarkson, 1985). On a longer time-scale, when nutrients are limiting, the balance between shoot and root growth can be modified in favour of the root and, as a result, improve the efficiency with which resources are used (Robinson, 1991, this volume). Feedback may also be positive as when

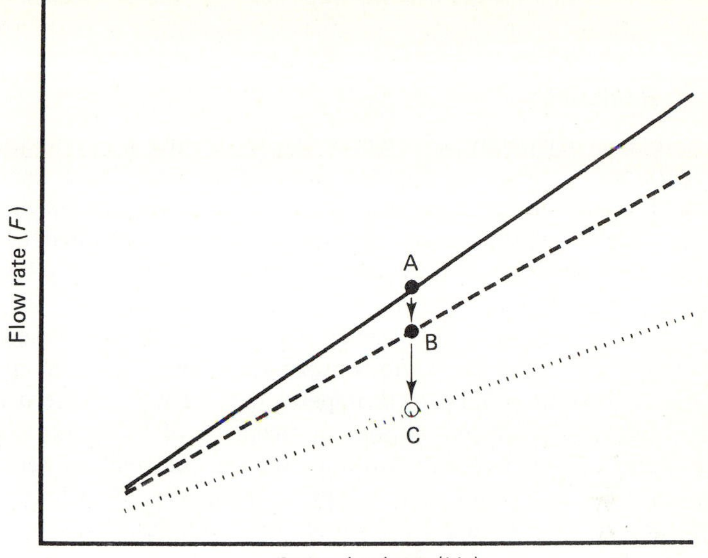

Fig. 5. The effect of feedback on the sensitivity of the electrical flow controller (see Figs 3 and 4) to partial damage. The uppermost line represents the response function of the undamaged system which is assumed to be at point A. When the pump develops a 50 per cent leak the new resting position is point B with feedback present and point C when there is no feedback.

plant weight increases exponentially as a result of positive feedback between increasing leaf area, which increases the amount of light intercepted, and plant growth rate which increases leaf area etc. Feed forward may also be present, as when a plant is infected by a pathogen. The initial reactions can cause specific genes to be activated, which in turn produce further reactions, such as cell necrosis, which effectively prevent the further spread of the pathogen.

Potential productivity

Potential productivity can be defined as the growth rate of a plant or crop when water, nutrients and disease are not limiting growth. In agricultural crops these factors are under the control of the grower. The two major climatic factors which, in the field, cannot be controlled are solar radiation and temperature. A third factor, the concentration of carbon dioxide in the atmosphere, is relatively constant during the life span of an annual plant's life cycle. This may not be true for long-lived shrubs and trees in view of current predictions of climate change (Houghton & Woodwell, 1989).

Radiation

The increase in dry weight of a plant or crop results from a balance between photosynthetic gain and respiratory loss of carbon. The latter releases energy for maintaining the integrity of cellular function and for driving the processes which synthesise and transport molecules required in various parts of the plant (Penning de Vries, 1971; Lawlor, 1991, this volume). The production of dry matter by a plant or crop is determined by the amount of light intercepted and the efficiency with which the intercepted light is used (Monteith, 1977). The various processes are not independent of each other: excepting an initial growth impetus derived from the seed reserves, no dry weight is produced without interception of light and vice versa. When water and/or nutrients are not limiting, the relation between dry weight and the amount of solar energy intercepted is often linear. This is not surprising as they are two cumulative variables (Russell, Jarvis & Monteith, 1989). What is more significant is that, for C_3 species, the slope of the line, the light conversion coefficient, is a conservative figure of about 1.8 g *dry weight* MJ^{-1} *total solar radiation*. For C_4 species the coefficient has a slightly greater value, reflecting the different photosynthetic pathways (Monteith, 1978). Fig. 6 shows a typical relation between the total dry weight (excluding roots) of potato (C_3 species) and millet (C_4 species) crops and the amounts of total solar radiation intercepted as they progress through the growing season. This approach has been used, by several groups, to analyse the effects of treatments – husbandry (Green, 1984; Allen & Scott, 1979), intercropping (Marshall & Willey, 1983) and disease (Marshall, Barker & Verrall, 1988) – on the productivity of crops. The true value of the light conversion coefficients will be slightly higher than current estimates based on most field observations, which frequently ignore the dry weight of the roots and always omit carbon lost by root exudation and root turnover.

Temperature

Temperature is a major determinant of the rate at which a plant, or organ within a plant, develops towards maturity. Often, the rate of organ initiation and development can be approximated by its linear relation with the temperature of the tissue under investigation. This is true for germination (Garcia Huidobro, Monteith & Squire, 1982), leaf primordium initiation (Kirk, Davies & Marshall, 1985; Baker & Gallagher, 1983), leaf expansion (Gallagher, 1979; Kirk, 1986; Milford, Pocock & Riley, 1985*a*,

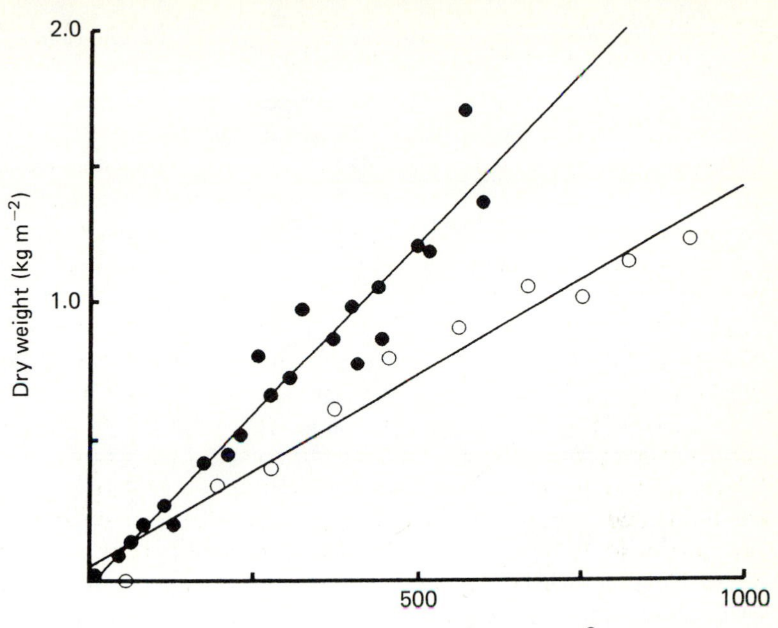

Fig. 6. Typical relation between total plant dry weight (excluding roots) and accumulated intercepted radiation when water and nutrients are not limiting growth rate. Data is from sequential harvests taken from (●) millet (ODA, 1987) and (○) potato (MacKerron, unpublished) crops.

b, *c*) and root development and extension (Gregory, 1983; 1986). There is a minimum or base temperature below which development is suspended. As temperature increases above the base the rate of development steadily increases to a maximum and then declines rapidly to zero. This has led to the concept of thermal time to replace chronological time as an appropriate scale for measuring plant development (Ong, 1983*a*, *b*). Angus *et al.* (1981) have suggested that the relation is non-linear near the base but the rate is, nevertheless, still determined by temperature. Exposure to more extreme temperatures may result in cold or heat stress and permanent damage to the exposed tissue. Daylength can modify the dependency of development on temperature (Kirby *et al.*, 1987). Effects of nutrient deficiency on development are less clear and may be mediated through effects on tissue temperature (van Keulen & Seligman, 1987).

Table 1. *The three crop processes leading to the dry weight yield of an organ.*

Accumulated intercepted radiation
(MJ m^{-2})
×
Light conversion coefficient
(g MJ^{-1})
×
Partitioning coefficient
(g organ g $plant^{-1}$)
=
Yield
(g organ m^{-2})

Crop processes

The dry matter yield of a crop is the product of three processes: the quantity of solar radiation intercepted over the life of a crop, the amount of dry weight produced per unit of solar energy intercepted, and the proportion of that dry weight which is partitioned to the harvested organ (Table 1) Marshall *et al.* (1988) provide an example of this approach to the analysis of the growth of potato.

Temperature determines the maximum rate of canopy expansion and, hence, the potential for interception of radiation by the leaves and crop water loss via transpiration. Fig. 7 demonstrates this clearly in an experiment investigating the effect of temperature on the development and growth of groundnut. As the mean daily temperature is increased, from 22 °C to 28 °C, so the rate of canopy expansion increases (Fig. 7*a*). Studies of leaf appearance and expansion show that below 10 °C groundnut ceases to develop. Using this value as the base temperature, the data were replotted on a thermal time-scale and now there is a single unique relation (Fig. 7*b*).

Temperature and solar radiation determine the potential growth of a plant or crop. However, a plant requires more than carbon to build the molecules necessary for proper functioning.

Nutrient limited productivity

In order to function properly a plant requires a wide range of nutrients, macronutrients in relatively large amounts, and micronutrients in small amounts (Marschner, 1986). Nitrogen, phosphorus and potassium constitute the macronutrients. Potassium plays a predominantly osmotic role;

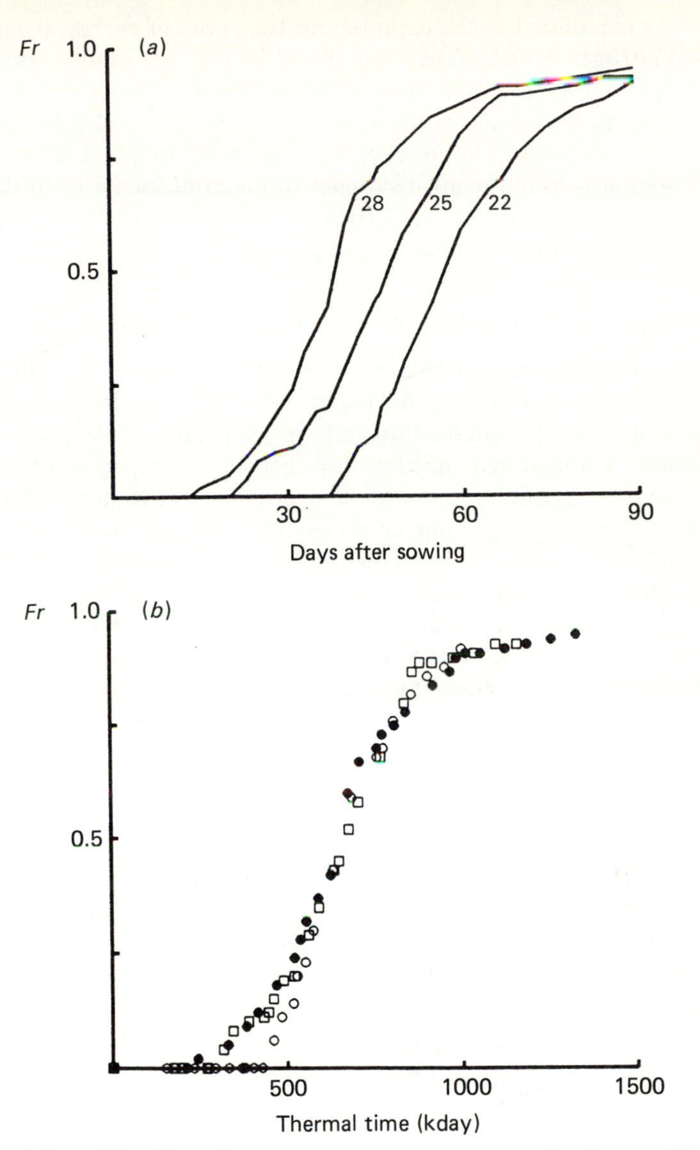

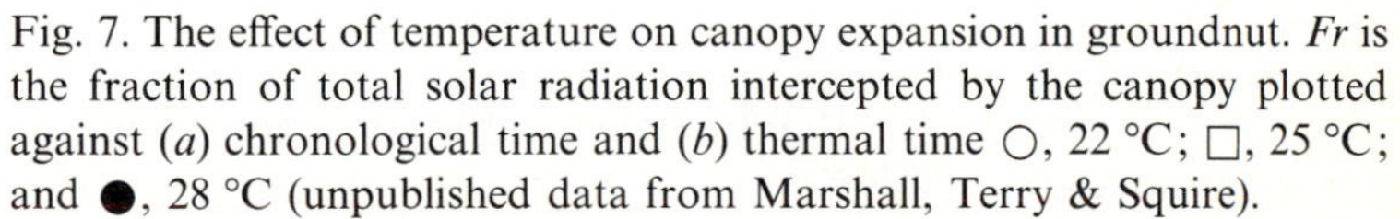

Fig. 7. The effect of temperature on canopy expansion in groundnut. *Fr* is the fraction of total solar radiation intercepted by the canopy plotted against (*a*) chronological time and (*b*) thermal time ○, 22 °C; □, 25 °C; and ●, 28 °C (unpublished data from Marshall, Terry & Squire).

phosphorus is required in the capture and transport of energy around the plant and nitrogen, which is the most abundant nutrient in the plant, plays a central role in the operation of the photosynthetic system (Lawlor, 1991, this volume). Yields of field crops in western agriculture show the greatest sensitivity to nitrogen. This is due, in large part, to plant demand for nitrogen and also to the relative supplies of macronutrients from the soil. On a world scale, phosphorus is likely to become the major macronutrient limiting growth as mineral reserves are depleted.

Leaf nitrogen concentration and photosynthesis

Chloroplasts, where the carbon dioxide is fixed, contain 75 per cent of the nitrogen in leaf tissue (Stocking & Ongun, 1962). Thus, there is an intimate connection between carbon fixation and nitrogen status of the plant. Many studies have demonstrated the close link between nitrogen concentration and potential or maximum rate of photosynthesis. Van Keulen & Seligman (1987) have collated a large number of examples, covering a range of annual species, from the literature and show that the relation between the rate of leaf photosynthesis in bright light and the nitrogen concentration of the leaf can be approximated by a straight line. The same is true for the perennial herb *Solidago altissima* L. (Compositae – Hirose & Werger, 1987*a*, *b*) and for potato (Marshall & Vos, 1990).

The rate of photosynthesis of a leaf is dependent on the irradiance incident on the leaf (Fig. 8*a*). Essentially, in dull light the rate increases linearly with irradiance, the slope (α) being a measure of the photochemical efficiency of the leaf. In bright light the photosynthetic system reaches a maximum rate, P_{MAX}, which is dependent on the rate of supply of carbon dioxide. The shape of the transition between these two limits varies between species (Marshall & Biscoe, 1980*a*, *b*) and may vary with environment (Marshall & Biscoe, 1977), but this is not important for the arguments that follow.

What is the optimal distribution of nitrogen in the canopy?

The relation between photosynthetic rate of a leaf, P_L, the irradiance, I, can be written as

$$P_L = f(I) \tag{1}$$

where $f(\,)$ means *function of*. It is convenient to use a non-dimensional form for the irradiance

$$\hat{I} = \frac{\alpha I}{P_{MAX}}$$

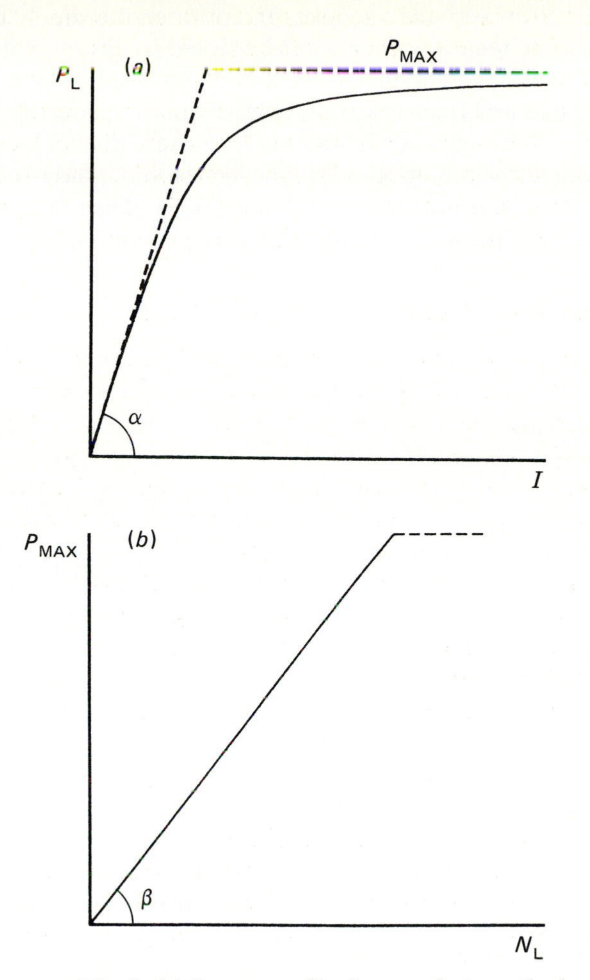

Fig. 8. (*a*) Response of leaf gross photosynthesis, P_L, to irradiance, I, and (*b*) the dependency of the maximum rate of gross photosynthesis, P_{MAX}, on leaf nitrogen concentration, N_L.

and separate P_{MAX} from the function of irradiance,

$$P_L = P_{MAX} f(\hat{I}) \tag{2}$$

The relation between the maximum rate of photosynthesis and nitrogen can be represented by a straight line (Fig. 8*b*). In practice most, but not all, of the nitrogen in a leaf is involved directly with photosynthesis and resides within the chloroplasts. The remaining nitrogen is in the form of proteins and amino-acids in the cytoplasm and vacuoles, except when the supply of nitrogen is in excess of requirements, when it can be stored as nitrate in the vacuoles. Because of this, the relation in Fig. 8*b* is shifted to the right when total nitrogen is considered and there is a positive intercept on the nitrogen axis. For the purpose of this analysis, the essential feature is the slope of the line, β, and the N-intercept will be ignored. The inclusion of a non-zero intercept would alter detail but not the general conclusion. Thus P_{MAX} in Eqn 2 can be replaced by a linear function of leaf nitrogen concentration, $N_L(l)$,

$$P_L = \beta N_L(l) f(\hat{I}) \tag{3}$$

The photosynthesis of the canopy,

$$P_C = \int_0^l \beta N_L(l) f(\hat{I}) \mathrm{d}l \tag{4}$$

is the integral of the photosynthesis of all leaves (l) in the canopy.

To find the optimum distribution of nitrogen in the canopy that maximises canopy photosynthesis one needs to find the stationary point,

$$\Delta P_C = \int_0^l \beta \Delta N_L(l) f(\hat{I}) \mathrm{d}l = 0$$

subject to the constraint,

$$\int_o^l \Delta N_L(l) \mathrm{d}l = 0$$

The constraint means that there is no net change in the total amount of nitrogen in the canopy, although nitrogen can be relocated within the canopy in order to maximise the rate of canopy photosynthesis. The general solution to this problem is

$$\int_0^l \Delta N_L(l) [\beta f(\hat{I}) - \lambda] \mathrm{d}l = 0 \tag{5}$$

where λ is the Lagrange multiplier. Since the adjustments to individual leaf nitrogen concentrations, ΔN_L, are considered to be arbitrary then, for Eqn 5 to be true, the term in brackets must equal zero. Because λ is a constant $f(\hat{I})$, and hence $\hat{I}$, must be constant down the canopy. Referring back to the definition of the non-dimensional irradiance, $\hat{I}$, this means that

$$P_{MAX} \propto I$$

In other words, under optimal conditions the nitrogen is distributed down the canopy such that the maximum rate of photosynthesis (determined by the nitrogen status of the leaf) mirrors the profile of irradiance within the canopy. This approach has been used by Field (1983) and Hirose & Werger (1987*b*) in a more detailed investigation of nitrogen distribution within a canopy, which includes non-photosynthetic nitrogen and the costs of relocation. Their conclusions are essentially the same; namely, when nitrogen is present in just sufficient quantity to sustain maximum growth rates set by the radiative and thermal environments then the amount of nitrogen required per unit dry weight declines as the crop increases in size.

Plant nitrogen concentration and processes

What is the evidence that plants optimise their nitrogen distribution? Greenwood (1982) developed a simple model for crop response to nitrogen fertiliser. At the core of the model was the observation that, in crops that have received optimal amounts of nitrogen fertiliser – any less nitrogen reduced growth whereas extra nitrogen applied had no effect – there is a unique relation between the nitrogen concentration of the whole plant (excluding fibrous roots) and its total dry weight (see later, Fig. 11). The nitrogen concentration does indeed fall with increasing size of the plant, consistent with a reducing demand for nitrogen per unit of carbon fixed, which is a result of self-shading (as discussed in the previous section) and also an increase in the proportion of structural carbon in the plant (van Dobben, 1962). This work was based on results from 22 different species. There is evidence that when super-optimal amounts of nitrogen are applied to potato crops they do accumulate extra nitrogen not associated with an increase in carbon (Greenwood, Neeteson & Draycott, 1985*a*; Millard & Marshall, 1986). Some of this extra nitrogen accumulates in the leaf, which would be consistent with a shift to the right of the relation between maximum photosynthetic rate and nitrogen concentration referred to earlier. Even when nitrogen supply exceeds that required for photosynthesis the nitrogen concentration in all plant parts falls with time (Seligman *et al.*, 1976; Vos, 1981).

Greenwood, Neeteson & Draycott (1985*b*) proposed that, when nitrogen uptake was sub-optimal, then growth rate was reduced by a factor, G_r;

$$G_r = \frac{(P_W - P_O)}{(P_M - P_O)}$$

where P_W, P_M and P_O are, respectively, the nitrogen concentration of the whole plant, the minimum concentration to sustain maximum growth rate and the concentration at which growth ceases. P_M is read directly from the optimal curve (solid line, Fig. 11). In the early growth stages of a potato

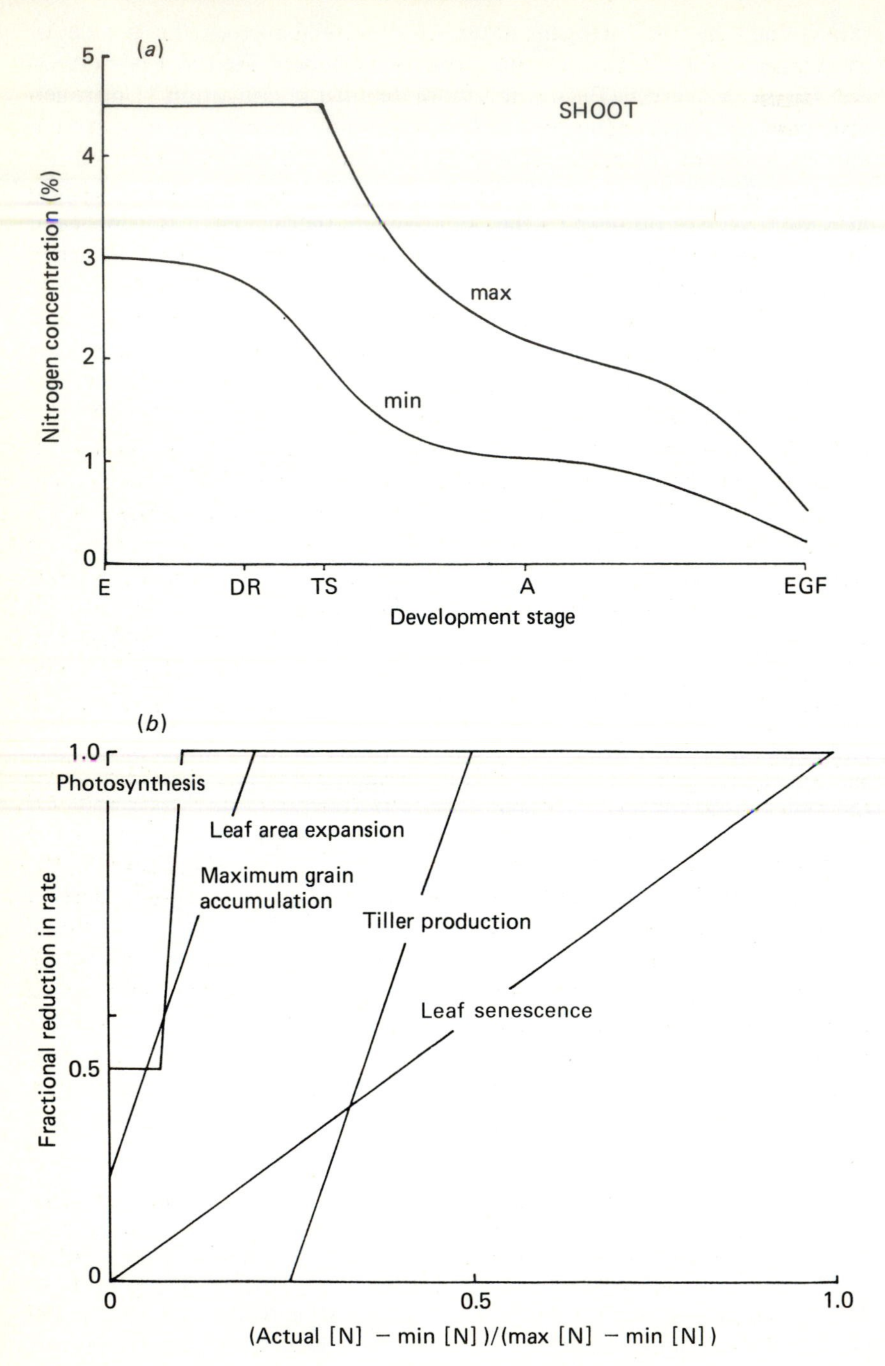
(a)
SHOOT
Nitrogen concentration (%)
5
4
3
2
1
0
max
min
E
DR
TS
A
EGF
Development stage
(b)
1.0
0.5
0
Photosynthesis
Leaf area expansion
Maximum grain
accumulation
Tiller production
Leaf senescence
Fractional reduction in rate
0
0.5
1.0
(Actual [N] − min [N])/(max [N] − min [N])

plant, P_O was found to be zero but, as the tubers became the major proportion of total plant weight, rose to 0.8 per cent. Van Keulen & Seligman (1987) and Porter *et al.* (unpublished) have used a similar approach to model the effects of nitrogen shortage on plant processes in spring and winter wheats respectively. In both cases they refer to the nitrogen concentration of the shoot as opposed to the whole plant or individual leaves. Van Keulen & Seligman (1987), after an extensive literature review, decided on a constant value of 1 per cent for P_O. In their model P_M declines with development age of the plant rather than with plant size. Porter *et al.*, as well as defining the decline in P_M with development age, also showed that P_O declines with development age, starting in a young crop at 3 per cent and falling to less than 0.5 per cent in a mature crop (Fig. 9*a*).

There is a range of processes that are affected by shortages of nitrogen and which vary in their thresholds and sensitivities. Fig. 9*b* shows the reduction factors used in the latest version of the AFRC winter wheat model (AFRCWHT2, Porter *et al.*, unpublished). Leaf senescence is the first process to be affected, leaves being a major source of nitrogen for relocation. It would appear that photosynthesis is least affected, but one must remember that this is referring to the nitrogen concentration of the shoot and, with relocation of nitrogen to the younger leaves, the effects of nitrogen shortage on photosynthesis are reduced. Nitrogen has a much greater effect on leaf area expansion than on photosynthesis (Gregory, Marshall & Biscoe, 1981).

Ingestadt (1982) has demonstrated the existence of a unique relation between the nutrient concentration of a plant and its relative growth rate (Fig. 10). Between levels of about 0.1 and 0.5 per cent nitrogen (on a fresh weight basis) the relative growth increases linearly. Beyond 0.5 per cent the increased uptake is not matched by a corresponding increase in carbon growth, relative growth falls rapidly and nitrogen concentrations ultimately become lethal. It is possible, using a fine control system for delivering nutrients to the root system, to maintain plants at a range of internal nutrient concentrations by matching the relative addition of nutrients to the intended relative growth rate of the plant (the internal nutrient concentration can be read off the line, Fig. 10). Providing the relative addition rate is maintained constant, the plants show no deficiency symptoms. It is

---

Fig. 9. Effects of nitrogen shortage on plant and crop processes included in the AFRC mathematical model of winter wheat growth; (*a*) critical nitrogen concentrations in the shoot and (*b*) fractional reductions in the rates of different processes. The development stages indicated are E, emergence; DR, double ridge; TS, terminal spikelet; A, anthesis; and EGF, end of grain fill.

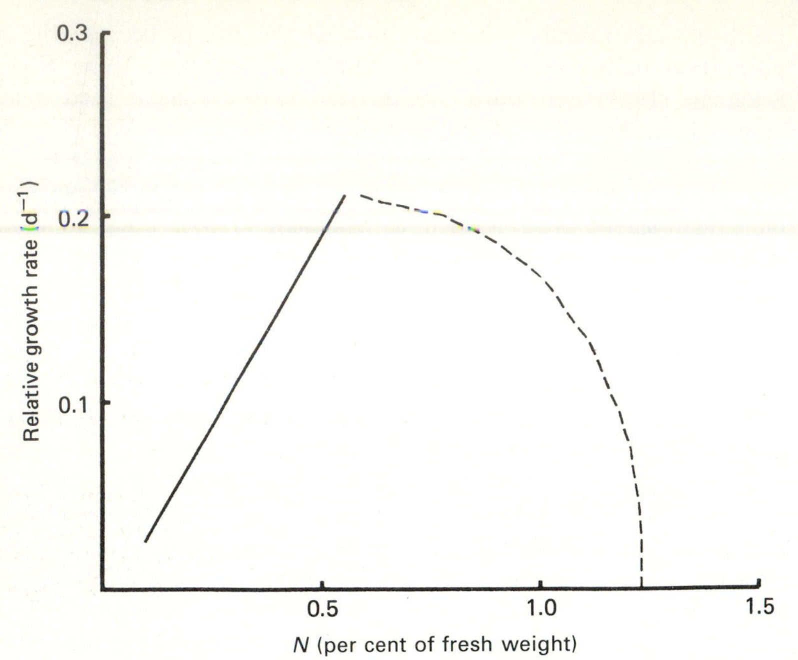

Fig. 10. Relation between the relative growth rate and nitrogen concentration of an entire plant (*N*, fresh weight; Ingestadt, 1982).

only when the relative rate is changed abruptly that symptoms appear. It could be argued that the 1 : 1 relation between relative addition rate of nutrients and relative growth rate is simply a consequence of the very low concentrations used in the nutrient solution, i.e. the plants rapidly assimilate a new burst of nutrients and are then starved for the remainder of the period so that nutrients are effectively being supplied on a pulsed on/off basis. However, the lack of deficiency symptoms and the close relation between the plant's internal nutrient concentration and its relative growth rate suggest that this is more than a casual relation.

Absolute and relative growth rates

Referring to the linear part of the response shown in Fig. 10, the relative growth rate can be defined as

$$\frac{1}{W}\frac{\mathrm{d}W}{\mathrm{d}t} = \beta(N_{\mathrm{C}} - N_{\mathrm{MIN}}) \qquad (6)$$

where N_{C} is the nitrogen concentration of the plant, N_{MIN} is the nitrogen concentration at which growth ceases and β is the slope of the response.

The experiments of Ingestadt and colleagues were carried out on young seedlings of relatively slow growing species (birch, alder) without competition from neighbouring plants. In reality, plants frequently grow in stands. In the case of a single species community, the absolute growth rate of the stand – when nutrients, water, pest and disease, are not limiting growth – is determined by the quantity of solar radiation intercepted by the canopy. This can be estimated assuming an exponential attenuation of radiation down the canopy,

$$\Delta I = I_0(1 - e^{-K\,LAI}) \tag{7}$$

where I_0 is the solar radiation incident at the top of the canopy, K is the extinction coefficient and LAI the cumulative leaf area index from the top of the canopy (Szeicz, 1974). LAI is simply related to the weight of the crop, W, by the specific leaf area, Φ,

$$LAI = \Phi W \tag{8}$$

The absolute growth rate of the crop is then given by

$$\frac{dW}{dt} = \varepsilon \Delta I = \varepsilon I_0(-e^{-K\Phi W}) \tag{9}$$

where ε is the light conversion coefficient (e.g. slope of relation similar to those in Fig. 6). When plants first emerge, leaf area is small and ΔI is proportional to LAI. As they grow they intercept more radiation each day and the potential growth rate increases, initially in an exponential manner. However, as the canopy continues to expand the mutual shading of leaves both within and between plants increases. Eventually, when LAI has reached a value of about 3, there is no change in the proportion of radiation intercepted, despite further expansion of the canopy, and the upper limit to absolute growth rate is determined entirely by the incident solar radiation, I_0.

Combining the equations for relative (6) and absolute (9) growth rates gives the nitrogen concentration of the plant as a function of the dry weight of the crop,

$$N_C = \frac{\varepsilon I_0}{\beta}\frac{(1 - e^{-K\Phi W})}{W} + N_{MIN} \tag{10}$$

Taking typical values for the parameters (see Table 2) in Eqn 10, the predicted relation (dashed line, Fig. 11) compares very favourably with the observations made by Greenwood (1982). We assumed an average receipt of 14 MJ m^{-2} for the solar radiation incident each day. We converted the parameters β and N_{MIN} from the Ingestadt relation (Eqn 6 and Fig. 10) to

Table 2. *Values for the parameters in Eqn 10 used to derive the second relation in Fig. 11 (dashed line).*

$\beta = 3.15\ \mathrm{d}^{-1}$
$N_{MIN} = 7.6\ \mathrm{mg\ g}^{-1}$
$\varepsilon = 1.8\ \mathrm{g(DW)\ MJ}^{-1}$
$I_o = 14\ \mathrm{MJ\ m}^{-2}\ \mathrm{d}^{-1}$
$\mathrm{K} = 0.5$
$\Phi = 0.016\ \mathrm{m}^2\ \mathrm{g}^{-1}$

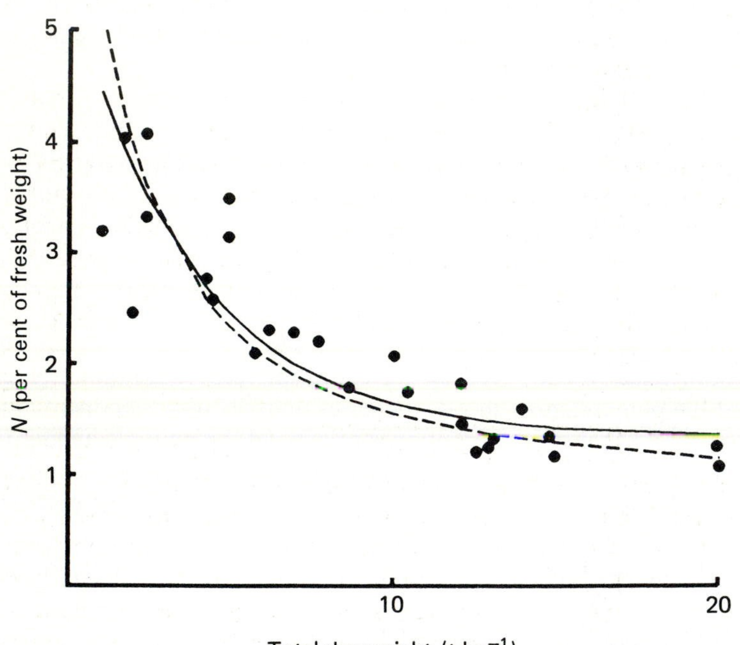

Fig. 11. Relation between average nitrogen concentration of plants (*N*, dry weight), grown with optimum levels of fertiliser, and the total dry weight (excluding roots) at harvest of 22 different species (Greenwood, 1982, and Greenwood *et al.*, 1985*a*). The solid line is the equation fitted by Greenwood (see text for an explanation of the dashed line).

a dry weight basis assuming 7.5 per cent dry matter concentration in plant tissue.

Demand and supply

Given unlimited resources, the maximum growth rate of a plant is determined by the genetic information contained within each cell of the plant (Lawlor, 1991, this volume). Plant weight increases exponentially with time and thus its relative growth rate is constant. For a given genetic makeup, temperature and the intensity of radiation incident on the plant determine the precise value of relative growth rate attainable. In order to sustain these relative growth rates, minimum or critical concentrations of nutrients within the plant's tissues must be achieved and, hence, a demand for nutrients is defined.

The precise mechanisms which determine these critical concentrations for the various nutrients within the different plant tissues are not well understood. The best known is that for nitrogen, because of its major role in photosynthesis. There are distinct differences between C_3 and C_4 species in the critical nitrogen concentrations which reflect the different photosynthetic systems. Even within C_3 or C_4 species there are small differences which probably reflect differences in partitioning between nitrogen used for photosynthetic and non-photosynthetic purposes. For example, potatoes have a large non-photosynthetic nitrogen store in the tubers whereas in grain crops a large non-photosynthetic nitrogen pool develops during the phase of rapid reproductive growth.

When supply satisfies demand, the genetically determined rate of exponential growth is maintained. However, even with isolated plants, there is inevitably increasing overlap of foliage with time. As the plant increases in size, the amount of light intercepted per unit of foliage declines with a concomitant decline in the relative growth rate. Plants growing in a single species community (crop) will often reach the state when the absolute, as opposed to relative, rate of growth is constant and any further increase in foliage area produces no significant increase in the amount of radiant energy captured. Hence, the demand for further nutrient uptake, compared to that of a plant growing at its unrestrained genetic potential, is reduced because both the relative growth rate is declining and the required nutrient concentration to sustain the slower relative growth is less.

The relative constancy of the relation presented in Eqn 10 within C_3 or C_4 species implies that there must be a high degree of correlation between the parameters in the equation and the environment as the latter changes. The value of β depends on the environmental conditions and, presumably, must cancel out changes in the quantity εI_0; equally there must be compensation between K and Φ. The constancy of the relation noted by Greenwood (1982) was observed on arable crop species. Whether precisely the same relation applies to non-cultivated species, and in particular those growing in more extreme climates (shade, high altitude etc.), remains to be seen.

When nutrient supply cannot meet demand then both the relative growth and the average nutrient concentration in the plant are reduced with the net effect that the absolute amount of growth per unit of nutrient taken up is increased. This must involve a change in the relative partitioning of nutrients and carbon between the organs which comprise the plant and the biochemical processes within the organs. A frequent observation when nutrients limit growth is that a greater proportion of the carbon fixed is partitioned to root growth. Various strategies of resource partitioning between root and shoot growth, in relation to nutrient supply and the implications for plant growth, are considered by Robinson (1991, this volume).

Mathematical models dealing with nutrient-limited productivity have tended to use the nitrogen concentration averaged over the entire plant, or at least over the shoot. As a result of this approach it would appear that, as a plant increases in weight, it requires less and less nitrogen to produce the same amount of extra tissue, at least when the plant is in a stand. When nitrogen is non-limiting analysis has shown that new tissue is produced with an initial nutrient concentration which is independent of plant size. Equally, when nutrients are limiting a gradient of nutrient concentration develops down the canopy which tends to optimise the photosynthetic efficiency of the canopy. Clearly, models which deal only with average shoot or plant concentration are really only empirical descriptions of the relations between nutrients and productivity. Should one place more emphasis, in future, on considering the production of new tissue, particularly by identifying what determines the tissue's demand for resources and what are the consequences if one or more of these demands cannot be met? How does a cell, organ or plant sense nutrient status and respond?

Nitrogen, not surprisingly, has received much attention and, for much of the temperate regions, this is the major nutrient limiting productivity. But, on a world scale, phosphorus may become the dominant nutrient limitation. How do we define a demand for this and other nutrients, and what mechanisms should be invoked to link the effects of shortfall in supply of phosphorus to productivity? What determines partitioning? How do we approach it and at what level of organisation (shoot to root, organ to organ etc.)? These are still crucial questions that remain to be answered with respect to both carbon and nutrients. In any event, it is essential that one does not lose sight of the whole and become lost in the complexities within.

References

Allen, E.J. & Scott, R.K. (1979). An analysis of growth of the potato crop. *Journal of Agricultural Science, Cambridge*, **94**, 583–606.

Angus, J.F., MacKenzie, D.H., Morton, R. & Schafer, C.A. (1981). Phasic development in field crops. II. Thermal and photoperiodic responses of spring wheat. *Field Crops Research*, **4**, 269–83.

Baker, C.K. & Gallagher, J.N. (1983). The development of winter wheat in the field. I. Relation between apical development and plant morphology within and between species. *Journal of Agricultural Science, Cambridge*, **101**, 327–35.

Clarkson, D.T. (1985). Factors affecting mineral nutrient acquisition by plants. *Annual Review of Plant Physiology*, **36**, 77–115.

Cram, W.J. (1988). Transport of nutrient ions across cell membranes *in vivo*. In *Advances in Plant Nutrition*, Vol. 3, eds. B. Tinker & A. Lauchli, pp. 1–53. New York: Praeger.

Dobben, W.H., van (1962). Nitrogen uptake of spring wheat and poppies in relation to growth and development. *Jaarboek Instituut voor Biologisch en Scheikundig Onderzoek van Landbouwgewassen*, **1959,** 93–105. (Dutch with English summary.)

Ferrari, Th. J. (1982). Introduction to dynamic simulation. In *Simulation of Plant Growth and Crop Production*, eds. F.W.T. Penning de Vries & H.H. van Laar, pp. 35–49. Wageningen: Pudoc.

Field, C. (1983). Allocating leaf nitrogen for the maximisation of carbon gain: Leaf age as a control on the allocation program. *Oecologia (Berlin)*, **56**, 341–7.

Gallagher, J.N. (1979). Field studies of cereal leaf growth. I. Initiation and expansion in relation to temperature and ontogeny. *Journal of Experimental Botany*, **30**, 625–36.

Garcia Huidobro, J., Monteith, J.L. & Squire, G.R. (1982). Time, temperature and germination of pearl millet (*Pennisetum typhoides* S.& H.). 1. Constant temperature. *Journal of Experimental Botany*, **33**, 288–96.

Goudriaan, J. & van Laar, H.H. (1977). Relations between leaf resistance, CO_2-concentration and CO_2-assimilation in maize, beans, lalang grass and sunflower. *Photosynthetica*, **12**, 241–9.

Green, C.F. (1984). Discriminants of productivity in small grain cereals: A review. *Journal of the National Institute of Agricultural Botany*, **16**, 453–63.

Greenwood, D.J. (1982). Modelling of crop response to nitrogen fertilizer. *Philosophical Transactions of the Royal Society of London*. B**296**, 351–62.

Greenwood, D.J., Neeteson, J.J. & Draycott, A. (1985*a*). Response of potatoes to N fertilizer: Quantitative relations for components of growth. *Plant and Soil*, **85**, 163–83.

Greenwood, D.J., Neeteson, J.J. & Draycott, A. (1985*b*). Response of potatoes to N fertilizer: Dynamic model. *Plant and Soil*, **85**, 185–203.

Gregory, P.J. (1983). Response to temperature in a stand of pearl millet (*Pennisetum typhoides* S. & H.). 3. Root development. *Journal of Experimental Botany*, **34**, 744–56.

Gregory, P.J. (1986). Response to temperature in a stand of pearl millet (*Pennisetum typhoides* S. & H.). 8. Root growth. *Journal of Experimental Botany*, **37**, 379–88.

Gregory, P.J., Marshall, B. & Biscoe, P.V. (1981). Nutrient relations of winter wheat. 3. Nitrogen uptake, photosynthesis of flag leaves and translocation of nitrogen to grain. *Journal of Agricultural Science, Cambridge*, **96**, 539–47.

Heath, O.V.S. (1948). Control of stomatal movement by reduction in the normal carbon dioxide content of the air. *Nature*, **161**, 179–81.

Hirose, T. & Werger, M.J.A. (1987*a*). Nitrogen use efficiency in instantaneous and daily photosynthesis of leaves in the canopy of a *Solidago altissima* stand. *Physiologia Plantarum*, **70**, 215–22.

Hirose, T. & Werger, M.J.A. (1987*b*). Maximising daily canopy photosynthesis with respect to the leaf nitrogen allocation pattern in the canopy. *Oecologia (Berlin)*, **72**, 520–6.

Houghton, R.A. & Woodwell, G.M. (1989). Global climatic change. *Scientific American*, **260**, 18–26.

Huettl, R.F. & Fink, S. (1991). Pollution, nutrition and plant function. In *Plant Growth: Interactions with Nutrition and Environment*, eds. J.R. Porter & D.W. Lawlor, Society for Experimental Biology, Seminar Series 43, pp. 207–26. Cambridge University Press.

Ingestadt, T. (1982). Relative addition rate and external concentration; Driving variables used in plant nutrition research. *Plant, Cell and Environment*, **5**, 443–53.

Keulen, H., van & Seligman, N.G. (1987). *Simulation of Water Use, Nitrogen Nutrition and Growth of a Spring Wheat Crop*, 310 pp. Wageningen: Pudoc.

Kirby, E.J.M., Porter, J.R. & 16 other authors (1987). An analysis of primordium initiation in Avalon winter wheat crops with different sowing dates and at nine sites in England and Scotland. *Journal of Agricultural Science, Cambridge*, **109**, 123–34.

Kirk, W.W. (1986). *Leaf and Canopy Development in the Potato*. Ph.D. Thesis, University of Dundee, Scotland, 135 pp.

Kirk, W.W., Davies, H.V. & Marshall, B. (1985). The effect of temperature on the initiation of leaf primordia in developing potato sprouts. *Journal of Experimental Botany*, **36**, 1634–1643.

Lawlor, D.W. (1991). Concepts of nutrition in relation to cellular processes and environment. In *Plant Growth: Interactions with Nutrition and Environment*, eds. J.R. Porter & D.W. Lawlor, Society for Experimental Biology, Seminar Series 43, pp. 1–32. Cambridge University Press.

Marschner, H. (1986). *Mineral Nutrition of Higher Plants*, 674 pp. London: Academic Press.

Marshall, B., Barker, H. & Verrall, S.R. (1988). Effects of potato leafroll virus on the crop processes loading to tuber yield in potato cultivars which differ in tolerance of infection. *Annals of Applied Biology*, **113**, 297–305.

Marshall, B. & Biscoe, P.V. (1977). A mobile apparatus for measuring leaf photosynthesis in the field. *Journal of Experimental Botany*, **28**, 1008–17.

Marshall, B. & Biscoe, P.V. (1980*a*). A model of C_3 leaves describing the dependence of net photosynthesis on irradiance. I. Derivation. *Journal of Experimental Botany*, **31**, 29–39.

Marshall, B. & Biscoe, P.V. (1980*b*). A model of C_3 leaves describing the dependence of net photosynthesis on irradiance. II. Application to the analysis of flag leaf photosynthesis. *Journal of Experimental Botany*, **31**, 41–8.

Marshall, B. & Vos, J. (1990). Influence of leaf nitrogen concentration on the light saturated rate of photosynthesis. *Abstracts, 11th Triennial Conference of the European Association for Potato Research, 1990 E*, pp. 197–8.

Marshall, B. & Willey, R.W. (1983). Radiation interception and growth in an intercrop of pearl millet/groundnut. *Field Crops Research*, **7**, 141–60.

Milford, G.F.J., Pocock, T.O. & Riley, J. (1985*a*). An analysis of leaf growth in sugar beet. I. Leaf appearance and expansion in relation to temperature under controlled conditions. *Annals of Applied Biology*, **106**, 163–72.

Milford, G.F.J., Pocock, T.O. & Riley, J. (1985*b*). An analysis of leaf growth in sugar beet. II. Leaf appearance in field crops. *Annals of Applied Biology*, **106**, 173–85.

Milford, G.F.J., Pocock, T.O., Riley, J. & Messem, A.B. (1985*c*). An analysis of leaf growth in sugar beet. III. Leaf expansion in field crops. *Annals of Applied Biology*, **106**, 187–203.

Millard, P. & Marshall, B. (1986). Growth, nitrogen uptake and partitioning within the potato (*Solanum tuberosum* L.) crop, in relation to nitrogen application. *Journal of Agricultural Science, Cambridge*, **107**, 421–9.

Monteith, J.L. (1977). Climate and the efficiency of crop production in Britain. *Philosophical Transactions of the Royal Society of London*, **B281**, 277–94.

Monteith, J.L. (1978). Reassessment of maximum growth rates for C_3 and C_4 crops. *Experimental Agriculture*, **14**, 1–5.

ODA (1987). *Microlimatology in Tropical Agriculture, Vol. 1, Introduction, Methods and Principles*. ODA, London, Research Schemes R3208 and R3819.

Ong, C.K. (1983*a*). Response to temperature in a stand of pearl millet (*Pennisetum typhoides* S. & H.). I. Vegetative development. *Journal of Experimental Botany*, **34**, 322–36.

Ong, C.K. (1983*b*). Response to temperature in a stand of pearl millet (*Pennisetum typhoides* S. & H.). II. Reproductive development. *Journal of Experimental Botany*, **34**, 337–48.

Penning de Vries, P.W.T. (1971). Respiration and growth. In *Crop Processes in Controlled Environments*, Proceedings of an International Symposium held at Glasshouse Crops Research Institute, Littlehampton, eds. A.R. Rees, K.E. Cockshull, D.W. Hand & R.G. Hurd, pp. 327–470. London and New York: Academic Press.

Raschke, K. (1975). Stomatal action. *Annual Review of Plant Physiology*, **26**, 309–40.

Robinson, D. (1991). Strategies for optimising growth in response to nutrient supply. In *Plant Growth: Interactions with Nutrition and Environment*, eds. J.R. Porter & D.W. Lawlor, Society for Experimental Biology, Seminar Series 43, pp. 177–205. Cambridge University Press.

Russell, G., Jarvis, P.G. & Monteith, J.L. (1989). Absorption of radiation by canopies and stand growth. In *Plant Canopies: Their Growth, Form and Function*, Society for Experimental Biology, Seminar Series 31, pp. 21–39. Cambridge University Press.

Seligman, N.G., Keulen, H. van, Yulzari, A., Yonathan, R. & Benjamin, R.W. (1976). *The effect of abundant nitrogen fertilizer application on the seasonal change in mineral concentration in annual Mediterranean pasture species*. Preliminary Report no. 754, Division of Scientific Publications, Bet Dagan, Israel, 12 pp.

Stocking, C.R. & Ongun, A. (1962). The intracellular distribution of some metallic elements in leaves. *American Journal of Botany*, **49**, 284–9.

Szeicz, G. (1974). Solar radiation in crop canopies. *The Journal of Applied Ecology*, **11**, 1117–56.

Trewavas, A. (1986). Understanding the control of plant development and the role of growth substances. *Australian Journal of Plant Physiology*, **13**, 447–57.

Vos, J. (1981). *Effects of temperature and nitrogen supply on post-floral growth of wheat; measurements and simulations*. Verslagen van Landbouwkundige Onderzoekingen (Agricultural Research Reports) 911, 164 pp. Wageningen: Pudoc.

H. MARSCHNER

Plant–soil relationships: acquisition of mineral nutrients by roots from soils

Introduction

For the mineral nutrition of plants the amount of so-called available nutrients in the soil is of crucial importance. However, nutrient availability, as characterised by chemical extraction methods (soil testing), is often poorly correlated with the actual nutrient uptake by plants. This is because factors, not taken into account in the extraction procedure, are also important in the acquisition of mineral nutrients by plants from soils. These are particularly:

- spatial availability, determined by root growth and its surface area and by the mobility of mineral nutrients in the soil solution via mass flow and diffusion,
- mobilisation and immobilisation of mineral nutrients in the rhizosphere by root-induced changes (pH, exudates, microbial activity).

This subject matter has been discussed recently in a general review (Marschner, 1986), as well as in a number of more specialised reviews emphasising natural vegetation (Chapin, 1988), spatial availability (Jungk & Claassen, 1989), root metabolic activity (Schubert & Mengel, 1989) and root exudates (Uren & Reisenauer, 1988).

Root growth

For the acquisition of mineral nutrients with low mobility in the soil (e.g. phosphorus) root growth and surface area are of particular importance. Although there are distinct genotypic differences in root surface area, within a given genotype the surface area may be modified in response to both external and internal factors. For example, when a nutrient is inadequately supplied, root growth and surface area are generally increased at the expense of shoot growth as shown for maize in Table 1.

Table 1. *Effects of phosphorus starvation on parameters of shoot and root growth of 12-day-old maize plants (Anghinoni & Barber, 1980).*

	Shoot		Root		
No. of days without P	Dry weight (g pot^{-1})	P (%)	Dry weight (g pot^{-1})	Length (m pot^{-1})	Radius ($\times 10^2$ cm^{-1})
1	2.10	0.95	0.27	4.64	2.27
2	2.34	0.65	0.31	5.77	2.23
4	1.93	0.32	0.40	7.57	1.99
6	1.65	0.27	0.43	9.08	1.84

Table 2. *Effect of P nutritional status of white lupin (*Lupinus albus *L.) on number of proteoid roots per plant (Marschner, Römheld & Cakmak, 1987a).*

Treatment	Shoot P concentration	Root (mg g^{-1} dry wt)	No. proteoid roots
Control	2.2	1.8	53
Foliar spray with P	3.6	3.4	15

One of the factors regulating the change in shoot and root growth is the stronger sink competition by the roots for photosynthates when the supply of a mineral nutrient is limited. Root morphology is also affected (Table 1) and root hair formation is strongly enhanced under deficiency of phosphorus (Mackay & Barber, 1985) or nitrogen (Foehse & Jungk, 1983). Changes in phytohormone balance between shoots and roots are also thought to be involved in this regulation (Marschner, 1986). The well-documented stimulation of lateral root formation at zones of localised supply of mineral nutrients such as nitrogen has also been discussed in relation to hormonal activity and to shoot-derived auxin in particular (Sattelmacher & Thoms, 1989).

In cluster-rooted plants like white lupin (*Lupinus albus* L.) phosphorus deficiency enhances the formation of so-called proteoid roots (Table 2) which strongly acidify the rhizosphere (see later).

Whether this potential for self-regulation of root growth and the

Table 3. *Effect of soil bulk density on root growth, translocation and consumption of photosynthates in the roots and rhizosphere of maize (Sauerbeck & Helal, 1986).*

	Bulk density (g cm^{-3})		
	1.2	1.4	1.6
Total root length (m plant^{-1})	246	125	89
Carbon translocation to roots (percentage of net photosynthesis)	46	44	53
Carbon consumption (mg C m^{-1} root length)	4.3	6.1	7.2

shoot:root ratio can affect mineral nutrient acquisition depends upon other soil chemical and physical factors. Low pH often combined with Al toxicity, mechanical impedance and poor soil aeration are constraints to root growth which may be widely distributed under field conditions. The importance of mechanical impedance is shown in Table 3. With an increase in soil bulk density the root length per plant is drastically decreased. Such a decrease in root surface area may be expected to impair strongly the acquisition of relatively immobile mineral nutrients such as phosphorus. However, the simultaneous enhancement of root respiration and carbon input into the rhizosphere may, at least in part, compensate via higher root exudation and microbial activity in the rhizosphere for the expected impairment of the spatial nutrient availability (see later).

In dry-land species, cessation of root growth and of mineral nutrient uptake is an immediate response to waterlogging whereas shoot growth can continue for some time (Fig. 1). The decline in nutrient concentrations in the shoots associated with poor soil aeration is therefore the result of at least three factors: the inhibition of root respiration (metabolically coupled uptake), the cessation of root growth and dilution by shoot growth.

Nutrient concentrations in the rhizosphere

The rhizosphere of plants growing in the soil is characterised by concentration gradients in both the radial and longitudinal directions of an individual root. Gradients are apparent for mineral nutrients, pH, redox potential, root exudates and microbial activity and are determined by soil and plant factors which strongly affect the acquisition of mineral nutrients. They also play a key role in adaptation of plants to adverse soil chemical

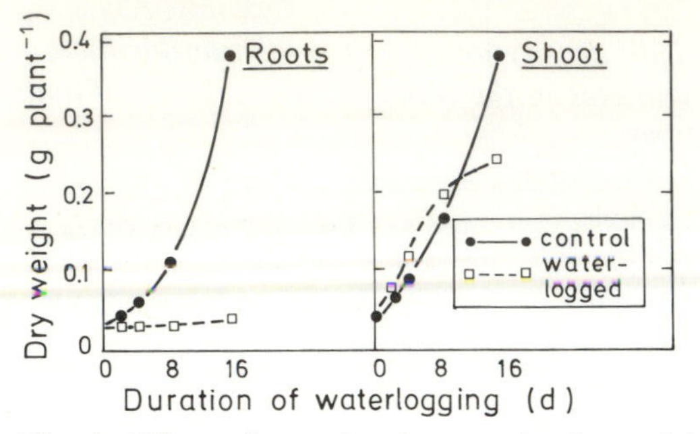

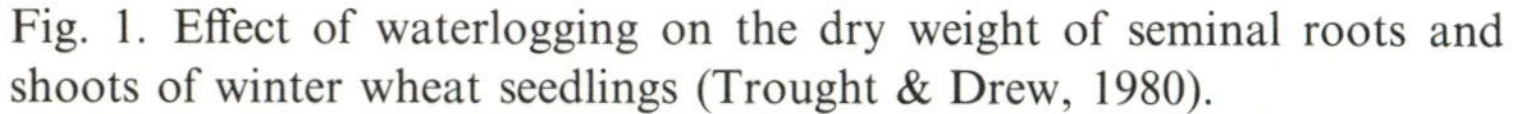
Fig. 1. Effect of waterlogging on the dry weight of seminal roots and shoots of winter wheat seedlings (Trought & Drew, 1980).

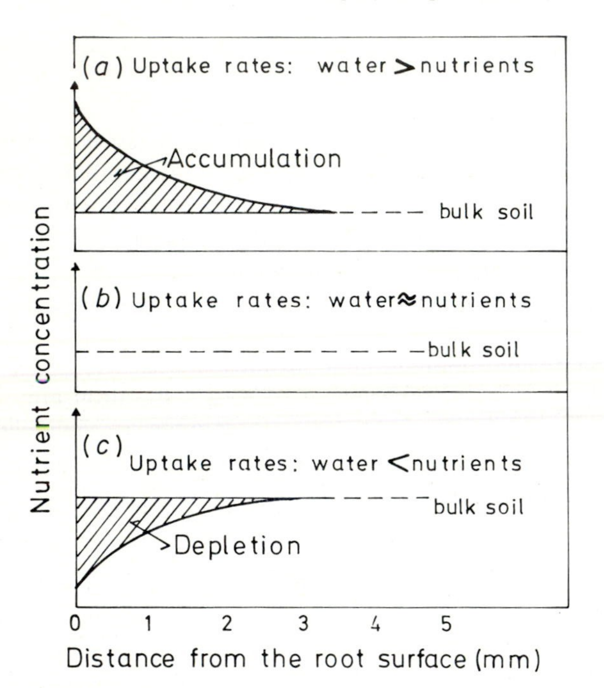

Fig. 2. Mineral nutrient concentration in the rhizosphere caused by different uptake rates of water and mineral nutrients. Schematic presentation based on a mechanistic mass flow–diffusion model.

conditions. Mineral nutrient gradients set up in the radial direction from the rhizoplane (root surface) to the bulk soil are mainly a function of their concentration in the bulk soil solution, the rate of mass flow to the root surface and rate of uptake by the roots. Taking a mechanistic approach to the development of gradients (Barber, 1984) predictions can be made of the

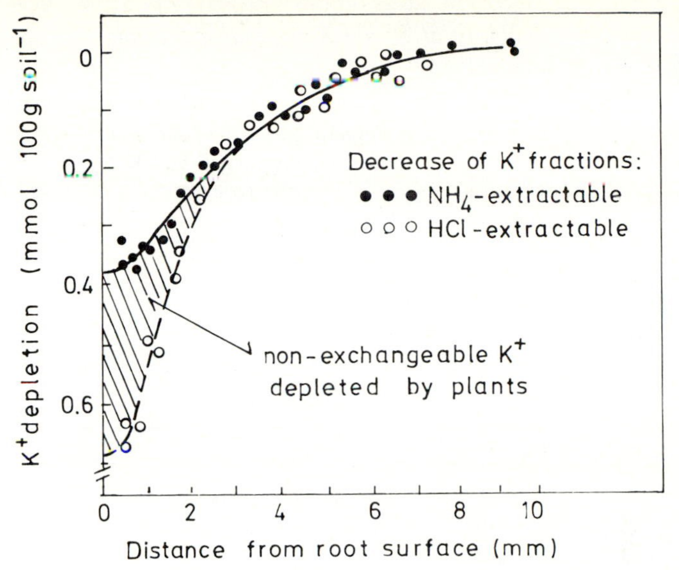

Fig. 3. Depletion of K^+ in a silt loam soil by rape seedlings 7 days after germination (Jungk & Claassen, 1986).

mineral nutrient concentration in the rhizosphere (Fig. 2). An impressive example of the accumulation and even precipitation of mineral nutrients in the rhizosphere are the so-called pedotubules formed from $CaSO_4$ and $CaCO_3$ around the roots of perennials growing in soils with high calcium concentrations in the bulk soil solution (Barber, 1984). An accumulation of soluble salts can also occur in the rhizosphere of plants grown in saline soils (Sinha & Singh, 1974) with corresponding negative effects on plant water relations. Distinct gradients of nutrients are not to be expected when rates of mineral nutrients to the roots and uptake rates of water and mineral nutrients are similar (Fig. 2*b*).

On the other hand, depletion of mineral nutrients in the rhizosphere is typical when they are present at low concentrations and thus have a low mobility in the soil relative to demand by the plants (Fig. 2*c*). This holds true in particular for phosphorus and potassium. Depletion of potassium in the rhizosphere may cause its release from the non-exchangeable K^+ fraction (Fig. 3) and even induce enhanced 'weathering' of clay minerals at the root–soil interface (Sarkar, Jenkins & Wyn Jones, 1979). This illustrates one aspect of the limitations of assessing the nutrient availability in soils by chemical extraction methods only.

For a given mineral nutrient the radial extension of the depletion zone depends on factors such as soil fertility level and soil moisture content

Table 4. *Spatial availability of P and K for maize growing in a sandy loam (Fußeder & Kraus, 1986).*

Root length density (cm cm^{-3} soil)	Percentage of the total soil volume delivering P and K to the roots	
	P	K
⩾ 2	20	50
< 2	5	12

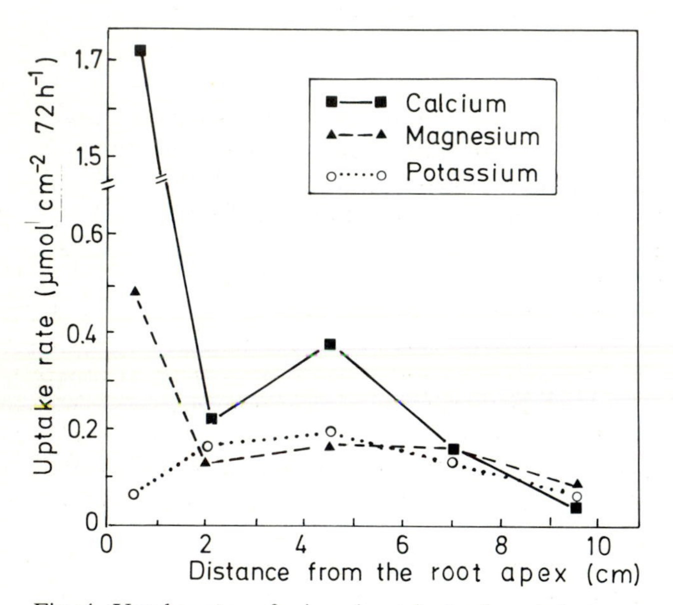

Fig. 4. Uptake rates of mineral nutrients along intact, non-mycorrhizal roots of 3-year-old Norway spruce (modified from Häußling *et al.*, 1988).

(Kuchenbuch, Claassen & Jungk, 1986; Jungk & Claassen, 1989) and plant factors, length of root hairs in particular (Hendriks, Claassen & Jungk, 1981; Jungk & Claassen, 1989). Because of the limited spatial availability of potassium and phosphorus, only a relatively small proportion of the total soil volume contributes to the delivery of these nutrients to the roots, particularly at low rooting densities (Table 4).

Gradients in mineral nutrient concentration in the rhizosphere also occur along the root axis from the apex to the base. These gradients reflect

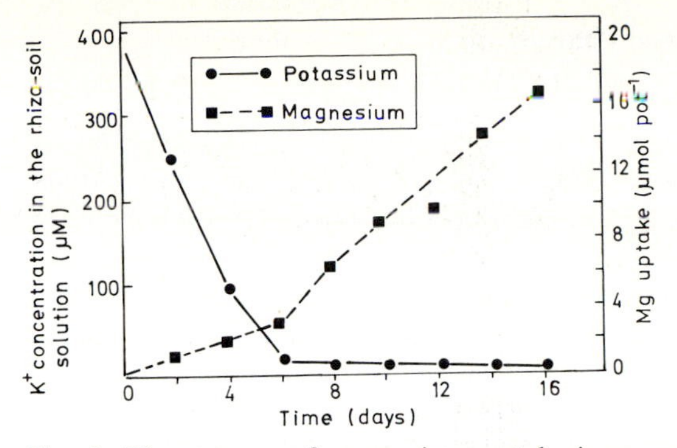

Fig. 5. Time course of magnesium uptake in ryegrass as affected by the potassium concentration in the rhizo–soil solution (Seggewiss & Jungk, 1988).

differences in characteristics of the root zones, e.g. in their respiration rates, formation of the Casparian band and the exodermis (Peterson, 1987; 1988), differentiation of the meta-xylem vessels (McCully, Canny & Van Steveninck, 1987; Sanderson, Whitbread & Clarkson, 1988) and penetration of the Casparian band by lateral roots. General conclusions about root performance based only on uptake studies in nutrient solutions where only average values for the whole root axis are obtained, can therefore be misleading unless uptake rates are measured along the root axis (Fig. 4). The apical and the basal root zones where lateral roots penetrate both, the Casparian band and the cortex, are the main uptake zones for calcium (Fig. 4) and water (Häußling *et al.*, 1988) and, to a lesser degree, for magnesium. This indicates the key role of the apoplasmic pathway for radial transport of calcium into the stele. In contrast, the main uptake zone for potassium is behind the apex.

The importance of gradients in soil mineral nutrient concentrations along the root axis for the acquisition of mineral nutrients has been demonstrated by Seggewiss & Jungk (1988) for magnesium (Fig. 5). The strong depression of the rate of magnesium uptake by potassium, which can be demonstrated readily in solution culture, occurs in soil-grown plants only as long as potassium is not depleted in the rhizosphere. After potassium depletion in the rhizosphere of the basal root zones, the rates of magnesium uptake markedly increase. Thus, the spatial separation of ions along the root axis of soil-grown plants can overcome limitations in mineral nutrition of plants caused by ion competition for uptake sites.

Root-induced changes in the rhizosphere

Roots not only act as a 'sink' for mineral nutrients supplied by the soil via mass flow and diffusion but they can also change the rhizosphere in various ways (Marschner *et al.*, 1986*c*; Marschner, Römheld & Cakmak, 1987*a*). Examples for concentration gradients have been given above. The type and extent of changes depend on soil and plant factors, such as nutrient supply, soil aeration, plant genotype and the nutritional status of the plants. In the following discussion particular emphasis is placed on conditions with low or deficient levels of mineral nutrients in the bulk soil and thus on deficiency-induced root responses which affect the rhizosphere and which are important for the enhancement of nutrient acquisition.

Rhizosphere pH

The rhizosphere pH may differ from the bulk soil pH by more than two units, depending on;

- the pH buffering capacity of the soil (Nye, 1986),
- the form of nitrogen supply, i.e. NH_4^+, NO_3^- or symbiotic N_2 fixation (Marschner & Römheld, 1983).
- the plant species (van Raij & van Diest, 1979; Marschner *et al.* 1986*c*),
- the nutritional status of the plants.

In most instances these changes in pH are brought about by differences in net excretion of H^+ due to an imbalance in the cation:anion uptake ratio (Table 5). For this, the form of the nitrogen supply has the most prominent influence (Marschner & Römheld, 1983; Nye 1986). Compared with the bulk soil pH, NH_4–N supply decreases and NO_3–N supply increases the rhizosphere pH. In neutral and alkaline soils rhizosphere acidification in plants fed with NH_4–N can enhance acquisition of phosphorus and potassium (Jungk & Claassen, 1986), boron (Reynolds, Scaife & Turner, 1987), zinc and manganese (Sarkar & Wyn Jones, 1982; Marschner, Treeby & Römheld, 1989). This enhanced acquisition of nutrients is brought about by increased solubility, exchange adsorption or by a combination of exchange and reduction.

Legumes which fix N_2 symbiotically increase the cation:anion uptake ratio and thus acidify the rhizosphere similarly to that of plants supplied with NH_4–N. Therefore, in legumes the acquisition of phosphorus from rock-phosphates (Aguilars & van Diest, 1981) and of iron and manganese from a calcareous soil (Wallace, 1982) is distinctly higher in plants fixing N_2 symbiotically compared with plants supplied with NO_3–N.

Even with the same form of nitrogen supply, considerable differences in

Table 5. *Relation between the concentrations of NH_4^+ and NO_3^- in the bulk soil solution, bulk soil pH and pH at the rhizoplane along roots of 60–80-year-old Norway spruce trees from different locations (Marschner et al., 1986a).*

Soil type (location)	Nitrogen in bulk soil solution (μM) as NH_4^+	Nitrogen in bulk soil solution (μM) as NO_3^-	Bulk soil pH (H_2O)	pH at the rhizoplane[a] root tip	pH at the rhizoplane[a] white root zone	pH at the rhizoplane[a] brown root zone
				basi-petal direction →		
Podsol (brown earth)	2.0	202	4.0	+0.25	+0.19	+0.16
Luvisol	0.8	736	4.1	+0.36	+0.31	+0.37
Podsol (brown earth)	33.5	790	3.5	+0.32	+0.18	+0.10

[a] Difference (+increase) from bulk soil pH.

rhizosphere pH exist between plant species (van Raij & van Diest, 1979; Bekele *et al.*, 1983; Marschner & Römheld, 1983; Marschner *et al.*, 1986*c*). In legumes, pronounced rhizosphere acidification is found particularly in those species which have their geographic origin in semi-arid climates (Römheld, 1986).

In acid mineral soils, rhizosphere acidification with NH_4–N supply may strongly impair the acquisition of some nutrients (e.g. molybdenum and magnesium) and furthermore may increase the risk of aluminium toxicity. However, average rhizosphere pH measured after mechanical separation of rhizosphere soil (Rollwagen & Zasoski, 1988) may be misleading because of the distinct pH gradients along the root (Fig. 6). The higher rhizosphere pH in the apical root zones of soil-grown plants even in presence of NH_4^+ (Fig. 6) may be related to the usually much greater activity of nitrate reductase in apical as compared to basal root zones (Klotz & Horst, 1988). As aluminium is toxic primarily at the root apex, via the inhibition of cell extension and division, the maintenance of a higher rhizosphere pH around the apex can be considered an important factor for root growth and the acquisition of mineral nutrients in acid mineral soils.

Root-induced changes in rhizosphere pH are also related to the nutritional status of plants. For example, a decrease in rhizosphere pH via enhanced net excretion of H^+ even with a supply of NO_3^- can be observed in the response of rape to phosphorus deficiency (Grinsted *et al.*, 1982) of

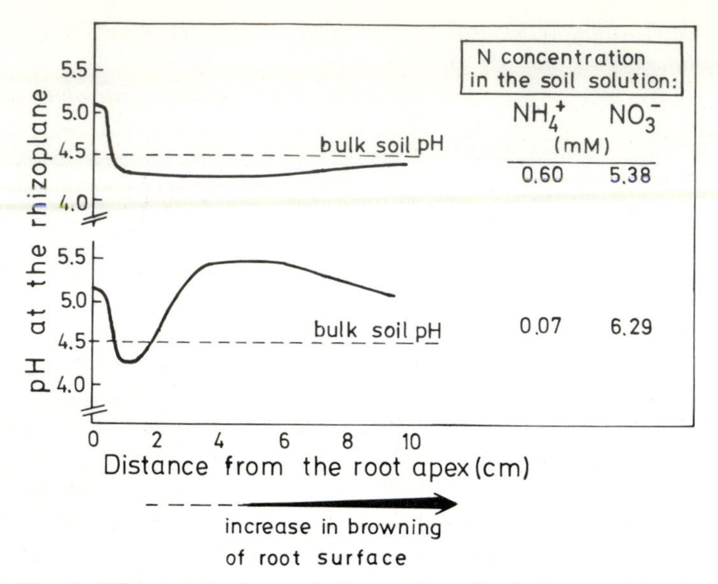

Fig. 6. Effect on the form of nitrogen in soil solution on rhizoplane pH along roots of 4-year-old Norway spruce trees grown in a Luvisol of pH 4.5_{H_2O} (Marschner, Häußling & Leisen, 1986*a*).

cotton to zinc deficiency (Marschner *et al.*, 1987*a*) and of non-graminaceous species to iron deficiency (Brown, 1978; Römheld, 1987*a, b*). In rape, the decrease in rhizosphere pH enhances the mobilisation of sparingly soluble phosphates in the rhizosphere and promotes recovery from phosphorus deficiency. Rhizosphere acidification caused by iron deficiency is a component in a coordinated sequence of events in non-graminaceous plant species enhancing the acquisition of iron from inorganic Fe^{III} compounds (Römheld, 1987*a, b*).

In white lupin the proteoid roots formed in response to phosphorus deficiency (Table 2) strongly acidify the rhizosphere by citric acid excretion (see below) and mobilise sparingly soluble phosphates, Fe^{III} compounds and Mn oxides (Gardner, Parbery & Barber, 1982; Marschner, Römheld & Kissel, 1986*b*; Marschner *et al.*, 1986*c*). In a mixed culture of white lupin and wheat growing in a phosphorus deficient soil, mobilisation of both phosphorus and manganese in the rhizosphere of proteoid roots of lupin can substantially increase the availability and uptake of manganese (Gardner & Boundy, 1983) and phosphorus (Horst & Waschkies, 1987) by wheat.

Redox potential and reducing processes in the rhizosphere

Aerated soils are not homogenous and a mosaic of anaerobic microsites occurs varying in location and time. While such microsites are most likely more abundant in the rhizosphere than in bulk soil (Fischer, Flessa & Schaller, 1989), their role in the acquisition of mineral nutrients is not clear. Poor soil aeration or waterlogging implies a higher rate of O_2 consumption in the rhizosphere by root and microbial respiration and thus enhances denitrification or manganese reduction and manganese toxicity in plants (Graven, Attoe & Smith, 1965).

In contrast, it is well known that plants adapted to submerged conditions increase the rhizosphere redox potential of the rhizosphere. The maintenance of oxidising conditions in the rhizosphere by the transport of O_2 from the shoot through the aerenchyma and release into the rhizosphere ('internal ventilation'; Armstrong, 1979) is essential to decrease phytotoxic concentrations of Fe^{2+} and Mn^{2+} in the rhizosphere by oxidation. For lowland rice there is a close positive correlation between the oxidation power of the roots and the amount of Fe^{III} precipitated at the root surface (Chen, Dixon & Turner, 1980). When mineral nutrients, particularly magnesium, potassium and phosphorus, are deficient, exudation of low molecular weight organic solutes by roots and microbial activity in the rhizosphere increase leading to enhanced O_2 consumption. This may cause iron toxicity in lowland rice (Ottow *et al.*, 1982).

In dryland species growing in well aerated calcareous soils with high pH, in contrast, both a decrease in the rhizosphere redox potential and increase in the reducing capacity of the roots are important factors for the acquisition of iron and manganese. The possible mechanisms involved in manganese reduction in the rhizosphere have been reviewed recently (Marschner, 1988; Uren & Reisenauer, 1988).

The reducing capacity of the roots increases in most dicotyledons and non-graminaceous monocotyledons in response to iron deficiency (Römheld, 1987*a*, *b*). This 'inducible reductase' (Bienfait, 1985) is located at the plasma membrane of rhizodermal cells and is strongly stimulated by low pH. As both root responses to iron deficiency, the induced reductase activity and the enhanced net H^+ excretion are located in apical root zones (Marschner *et al.*, 1986*b*; Marschner *et al.*, 1986*c*), rhizosphere acidification increases the solubility and mobility of Fe^{III} in the rhizosphere and its subsequent reduction and uptake (Table 6). This type of root response to iron deficiency (Strategy I; Marschner *et al.*, 1986*b*; Römheld, 1987*a*, *b*) may also lead to enhanced reduction of copper and manganese and increase in plant contents of these two micronutrients in iron deficient plants (Table 6). Accordingly, in calcareous soils with low availability of iron and

Table 6. *Effect or iron nutritional status of chickpea (*Cicer arietinum *L.) on reduction and uptake of iron, copper and manganese supplied separately as synthetic iron chelate (FeEDDHA),* $CuCl_2$ *and* MnO_2 *(Marschner et al., 1987a).*

Reduction and/or uptake		Fe nutritional status[a]	
		+Fe	−Fe
Iron			
Fe^{III} reduction			
(nmol g^{-1} root fresh wt h^{-1})		50	1850
Fe uptake			
(nmol g^{-1} root fresh wt h^{-1})		2.2	418
Copper			
Cu^{II} reduction			
(nmol g^{-1} root dry wt h^{-1})		30	280
Cu content			
(nmol g^{-1} root dry wt)		240	1040
Manganese			
Mn content			
(μmol g^{-1} dry wt)	roots	8.2	39.5
	shoots	1.2	13.8

[a] +Fe = sufficient; −Fe = deficient.

manganese this response mechanism to iron deficiency also favours the acquisition of manganese and may prevent manganese deficiency in plants (Moraghan, 1985), whereas in calcareous soils with only low availability of iron, manganese toxicity may occur (Moraghan, 1979; Marschner *et al.*, 1986*c*).

The efficiency of Strategy I in mobilising iron and manganese in the rhizosphere is strongly impaired by a high pH buffering capacity in the rhizosphere. In this respect high bicarbonate concentrations in the soil solution and at the root surface are of major ecological importance and are most probably the main factor responsible for 'lime-induced' chlorosis in plants (Inskeep & Bloom, 1986; Marschner *et al.*, 1986*b*).

Root exudates, organic carbon release

Up to 30% of the carbon fixed in photosynthesis is released as organic carbon into the rhizosphere (Whipps & Lynch, 1986). Various forms of stress increase this carbon release, as for example, potassium deficiency

(Kraffczyk, Trolldenier & Beringer, 1984), phosphorus deficiency (Graham, Leonard & Menge, 1981), zinc deficiency (Cakmak & Marschner, 1988), drought stress and anaerobiosis (Whipps & Lynch, 1986), mechanical impedance (Sauerbeck & Helal, 1986) and the presence of microorganisms (Schönwitz & Ziegler, 1982; Gardner, Barber & Parbery, 1983*a*).

The organic carbon released into the rhizosphere consists of three major components (Warembourg & Billes, 1979): sloughed-off cells and their lysates; high molecular weight gelatinous material (mucilage); and low molecular weight organic solutes (free exudates). Sloughed-off cells and their lysates may become important indirectly for mineral nutrient acquisition by enhancing microbial activity in the rhizosphere (see later). Mucilage, which consists mainly of polysaccharides and polyuronic acids (Morel, Mench & Guckert, 1986), is secreted by cells of the root cap and rhizodermis and may play a role in nutrient acquisition in dry soils, for example, by facilitating the transport of zinc from soil particles to the root surface (Nambiar, 1976). Whether the capacity of mucilage to form complexes with heavy metals such as copper, lead and cadmium (Morel *et al.*, 1986; Mench, Morel & Guckert, 1987) favours or inhibits transfer into the root cells is not yet clear. However, the protective function of mucilage in inhibiting the uptake of aluminium, and consequently the detrimental effect of aluminium on cell extension and division in roots, has been clearly demonstrated (Horst, Wagner & Marschner, 1982).

The low molecular weight root exudates consist mainly of sugars, organic acids, amino acids and phenolics (Mench *et al.*, 1988). Their importance for nutrient acquisition has been recently reviewed (Uren & Reisenauer 1988), and is highlighted in the following examples. The main emphasis is placed on the enhanced release of low molecular weight root exudates in response to nutrient deficiency.

In white lupin, phosphorus deficiency induces proteoid root formation (Table 2) and enhances citrate excretion in these root zones (Gardner, Barber & Parbery, 1983*b*; Marschner *et al.*, 1987*a*). In a phosphorus deficient calcareous soil, 13-week-old plants excreted 23 % of their total dry weight as citrate into the rhizosphere (Dinkelaker, Römheld & Marschner, 1989). The cation–anion balance of the plants indicate that citric acid had been excreted, presumably driven by an H^+ efflux pump. The low solubility of calcium citrate formed in the rhizosphere favours citric acid excretion. This acidification not only increased the acquisition of phosphorus from sparingly soluble calcium phosphates, but also increased the solubility of micronutrients such as iron, manganese and zinc (Table 7).

In contrast to organic acid excretion in white lupin, the rhizosphere acidification observed in response to phosphorus deficiency in rape (Grinsted *et al.*, 1982) seems to be directly related to net excretion of H^+ and

Table 7. *pH (in water), citrate and DTPA*[a] *extractable Fe, Mn and Zn in the bulk soil and in the rhizosphere soil of proteoid roots of 13-week-old white lupin (*Lupinus albus *L.) grown in a P deficient calcareous soil (20% $CaCO_3$) and supplied with N as $Ca(NO_3)_2$ (Dinkelaker et al., 1989).*

Soil	pH	Citrate (μmol g^{-1} soil)	DTPA extractable (μmol kg^{-1} soil)		
			Fe	Mn	Zn
bulk	7.5	n.d.	34	44	2.8
rhizosphere	4.8	47.7	251	222	16.8

[a] Diethylethylenetriamine pentaacetate.
n.d., not detectable.

the imbalance in cation:anion uptake ratio (Hedley, Nye & White, 1982) which is probably stoichiometrically compensated for formation and storage of organic acid anions within the plant. However, also in rape, release of organic acids contribute to the rhizosphere acidification under phosphorus deficiency (Hoffland, Findenegg & Nelemans, 1989). Furthermore, plant species like rape growing on highly calcareous soils form large amounts of $CaCO_3$ within the root cells (Jaillard, 1984; Callot & Jaillard, 1987). This may be another mechanism for rhizosphere acidification, which requires little investment of organic carbon in the rhizosphere or in the plant.

Enhanced release of phenolic compounds such as caffeic acid may occur in response to iron deficiency in many plant species with Strategy I (Olsen *et al.*, 1981; Römheld, 1987*b*). These phenolics effectively mobilise sparingly soluble inorganic Fe^{III} compounds (Marschner *et al.*, 1986*b*) as well as manganese from MnO_2, most probably by reduction in the rhizosphere.

A particular mechanism of root response to iron deficiency exists in grasses. This response mechanism (Strategy II, Marschner *et al.*, 1986*b*, *c*; Römheld, 1987*a*, *b*) is characterised by two processes, namely an increase in the release of non-proteinogenic amino acids, the so-called phytosiderophores (Takagi, Nomoto & Takemoto, 1984) and a highly efficient uptake system for Fe^{III} phytosiderophores (Fig. 7). Both processes are located in apical root zones (Marschner *et al.*, 1987*b*). The release of phytosiderophores follows a distinct diurnal rhythm and can be suppressed within 12–24 h by providing iron (Takagi *et al.*, 1984; Marschner *et al.*, 1986*b*). The

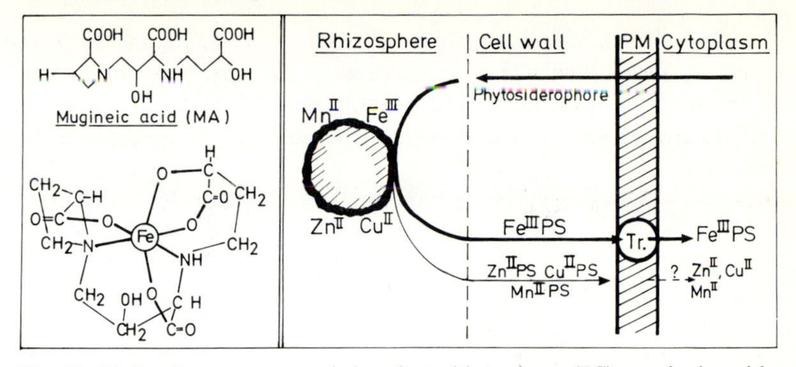

Fig. 7. Molecular structure of the phytosiderophore (PS) mugineic acid and of the corresponding Fe^{III} chelate (based on Nomoto, Sugiura & Takagi, 1987) and a proposed mechanism of PS release, mobilisation in the rhizosphere and uptake of Fe^{III} and other metal cations. Tr, translocator for Fe^{III}–PS; PM, plasma membrane.

chemical nature of the phytosiderophores may differ between plant species and even cultivars. The ecological relevance of this mechanism for the acquisition of iron is indicated by the positive correlation between phytosiderophore release of various genotypes under controlled conditions and their corresponding differences in resistance to chlorosis under field conditions (Römheld, 1987*b*). In barley grown in a calcareous soil the concentration of phytosiderophores in the rhizosphere soil is several times higher than in the bulk soil (Shi *et al.*, 1989). In the roots of grasses the Fe^{III} phytosiderophores are taken up by a specific transport system at rates which are 100–1000 times greater than those of iron from synthetic (FeEDDTA) or microbial Fe^{III} chelates such as ferrioxamine B or ferrichrome (Table 8). Transport of the Fe^{III} phytosiderophores across the plasma membrane of root cells of grasses occurs mainly as the molecular form, i.e. without splitting the chelate (Marschner *et al.*, 1989); this is similar to the mechanism of iron uptake as in the form of siderophores by microorganisms (Winkelmann, 1986).

Phytosiderophores also form chelates with other heavy metal cations such as zinc, manganese and particularly copper (Sugiura & Nomoto, 1984; Crowley, Reid & Szaniszlo, 1987) and thus mobilise not only iron from soils but also zinc, manganese and copper (Takagi, Kamei & Ming-Ho, 1988; Treeby, Marschner & Römheld, 1989). Nevertheless, in calcareous soils phytosiderophores mobilise similar or even greater amounts of Fe^{III} than microbial siderophores such as ferrioxamine B, despite the much greater stability of ferrated siderophores (Takagi *et al.*, 1988; Treeby *et al.*, 1989). Considering the high efficiency uptake system in roots of grasses for Fe^{III}

Table 8. *Rates of iron uptake by Fe sufficient (+ Fe) and Fe deficient (− Fe) barley plants (*Hordeum vulgare *L.) supplied with various* ^{59}*Fe chelates in nutrient solution at pH 6.5 (Marschner et al., 1989).*

^{59}Fe source	^{59}Fe uptake rates (μmol g^{-1} root dry wt 4 h^{-1})	
	+ Fe	− Fe
Fe phytosiderophore		
Fe deoxymugineic acid	672	3769
Synthetic Fe chelate		
Fe EDDHA	0.4	0.5
Microbial Fe chelate		
ferrioxamine B (Desferal)	2.5	1.2

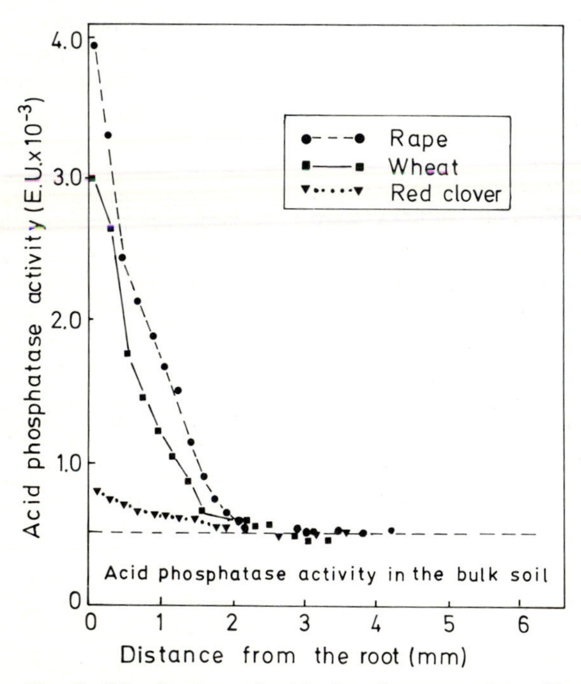

Fig. 8. Distribution of acid phosphatase activity (Enzyme Units) in the rhizosphere of different plant species grown in a silty loam. (Tarafdar & Jungk, 1987).

phytosiderophores (Table 8) it is concluded that microbial siderophores are of minor importance for iron acquisition of graminaceous species from calcareous soils. The situation might be different in non-graminaceous plant species where this high affinity uptake system is absent, but where Fe^{III} siderophores are reduced effectively by the plasma membrane-bound reductase (Römheld & Marschner, 1986).

Species such as barley or wheat also release phytosiderophores under zinc deficiency (Zhang, Römheld & Marschner, 1989). For a given genotype the same phytosiderophore (e.g. 2′-deoxymugineic acid) is released under both zinc and iron deficiency. The release in zinc deficient plants follows the same diurnal rhythm as that with iron deficiency. In calcareous soils, where the availability of both iron and zinc is often low, release of phytosiderophores may facilitate acquisition of both nutrients by increasing solubility and mobility in the rhizosphere. In addition, the transport of iron as Fe^{III} phytosiderophore across the plasma membrane of root cells is also enhanced.

Ectoenzymes

Organic phosphorus (P_{org}) compounds (e.g. sugar phosphates) are rapidly hydrolysed by plant roots. This hydrolysis is mediated by root-borne acid phosphatase (Tarafdar & Jungk, 1987), fungal acid phosphatase (Doumas *et al.*, 1986) or bacterial alkaline phosphatase (Tarafdar & Claassen, 1988). A distinct gradient in phosphatase activity therefore exists from the bulk soil to the root surface (Fig. 8). Phosphatases are adaptive enzymes and with phosphorus deficiency their activity increases at the root surface (Helal & Sauerbeck, 1988). The activity of root-borne acid phosphatase may far exceed the uptake rate of hydrolysed inorganic phosphorus (Tarafdar & Claassen, 1988). Depletion of organic phosphorus in the rhizosphere, but not of inorganic phosphorus which may accumulate in the rhizosphere of various annual species (Tarafdar & Jungk, 1987) and 80-year-old Norway spruce trees (Häußling & Marschner, 1989), illustrates the rapid hydrolysis of organic phosphorus by both root-borne and microbial phosphatases.

The high activity of phosphatases in the rhizosphere is, most likely, important for the acquisition of phosphorus particularly from soils rich in organic phosphorus, such as unfertilised agricultural soils (Sharpley, 1985) or forest soils where organic phosphorus in the topsoil may form 80–95 % of the total phosphorus (Zech *et al.*, 1987).

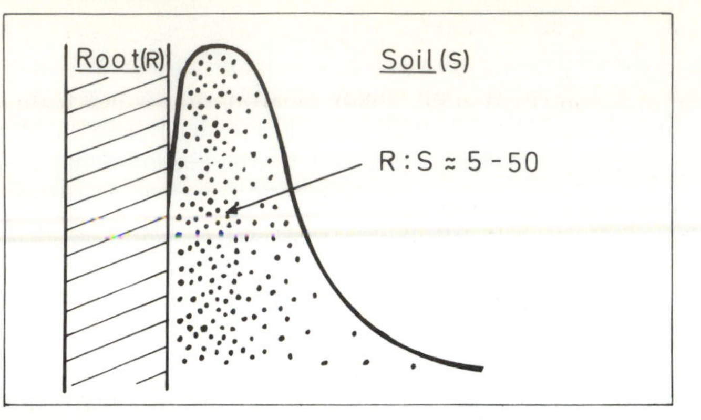

Fig. 9. Schematic presentation of the distribution of non-infecting microorganisms at the rhizoplane, in the rhizosphere and in the bulk soil. The 'R:S ratio' expresses the proportions at the rhizoplane plus rhizosphere compared to the bulk soil.

Rhizosphere microorganisms

Non-infecting rhizosphere microorganisms

As a result of the abundant supply of organic carbon from roots the population density of microorganisms, especially bacteria, is much higher in the rhizosphere than in the bulk soil (Fig. 9). The proportion depends on plant species, age, nutritional status and position along the root axis. Nevertheless, only a small percentage of the root surface (rhizoplane) is occupied by microorganisms (Schönwitz & Ziegler, 1982; Rovira, Bowen & Foster, 1983). In general, the root cap is free from microorganisms, which proliferate most behind the root hair zone where lateral roots emerge (Vuurde & Schippers, 1980). Microbial activity in the rhizosphere may affect the acquisition of mineral nutrients by roots either directly via effects on mobilisation and/or immobilisation or indirectly via effects on root morphology and/or physiology.

Acquisition of mineral nutrients may be decreased if, for example, in response to phosphorus or iron deficiency roots exude organic acids or phytosiderophores which are utilised by rhizosphere microorganisms. However, in the rhizosphere of soil-grown plants the spatial separation of the main sites of exudation and of microbial activity (see above) may allow root exudates to diffuse a considerable distance before being degraded by microorganisms (Gardner *et al.*, 1983*c*; Helal & Sauerbeck, 1986).

On the other hand, rhizosphere microorganisms may increase mineral nutrient acquisition if, for example, sugars in root exudates or sloughed-off cells are utilised by microorganisms which produce metal chelators such as organic acids or siderophores. For example, 10–50 times higher siderophore concentrations have been found in the rhizosphere soil of *Pinus ponderosa*, than in the bulk soil (Reid *et al.*, 1984). Siderophores which form chelates of high stability with Fe^{III} and other heavy metals, as well as aluminium (Takagi *et al.*, 1988), have been considered to be of crucial importance for the iron nutrition of annual species (Crowley *et al.*, 1987). However, this assumption has to be questioned for graminaceous species (Table 8).

Furthermore, non-infecting rhizosphere microorganisms are of particular importance for the manganese nutrition of plants (Marschner, 1988). Between pH 6.0 and 7.5, microorganisms which oxidise manganese are far more effective than any non-biological oxidation of Mn^{2+}. A large proportion of manganese oxidising microorganisms in the rhizosphere may therefore either decrease the risk of manganese toxicity in plants growing in poorly aerated or waterlogged soils, or increase the risk of manganese deficiency in well-aerated calcareous soils. The proportion of manganese oxidising microorganisms in the rhizosphere may differ between plant species and even cultivars within a species. In oat (*Avena sativa* L.) genotypical differences in susceptibility to manganese deficiency are related to the proportion of manganese oxidising bacteria in the rhizosphere (Timonin, 1946). Their importance for manganese acquisition by other cereals has been confirmed by Bromfield (1978) and Huber & Wilhelm (1988). As root infections with pathogens such as *Gaeumannomyces graminis* ('take-all') are negatively correlated with the manganese concentration in the roots (Graham & Rovira, 1984), the proportion of manganese oxidising to manganese reducing microorganisms in the rhizosphere may have to be considered in breeding programmes aimed at manganese efficiency in crop species (Graham, 1988).

In the rhizosphere, N_2 fixing (diazotrophic) bacteria are also present. These bacteria may form, in some instances, rather specific rhizosphere associations, for example *Azospirillum* in *Zea mays* L. In C_4 species (e.g. maize or sorghum) grown in nitrogen deficient soils in the tropics and subtropics the fixation of N_2 by rhizosphere bacteria can contribute substantially to the nitrogen nutrition of the host plant (Boddey & Döbereiner, 1988). However, in temperate climates this contribution is rather small because of low soil temperatures (Jain, Beyer & Rennie, 1987) and higher levels of bound soil nitrogen (Jagnow, 1987; Martin *et al.*, 1989). Nevertheless, inoculation with diazotrophic bacteria often stimulates plant growth, even with much bound nitrogen in the soil (Rynders & Vlassak, 1982). This stimulation is most probably the result of hormonal effects, as many diazotrophic bacteria (*Azospirillum* sp. in particular) produce auxin (Inbal & Feldman, 1982). Inoculation with *Azospirillum* therefore effects root growth in general and root morphology in particular (Fig. 10).

Table 9. *Interaction between* Rhizobium *and VA mycorrhiza (*Glomus fasciculatus*) on shoot and root growth and nitrogenase activity (ARA) in alfalfa plants (Piccini, Ocampo & Bedmar, 1988).*

Treatment	Dry wt (mg plant^{-1}) shoot	Dry wt (mg plant^{-1}) roots	Root: shoot	ARA (μmol plant^{-1} h^{-1})
Control	63	105	1.67	0
+VAM	170	296	1.74	0
+*Rhiz.*	113	194	1.72	0.51
+VAM+*Rhiz.*	185	199	1.07	1.17

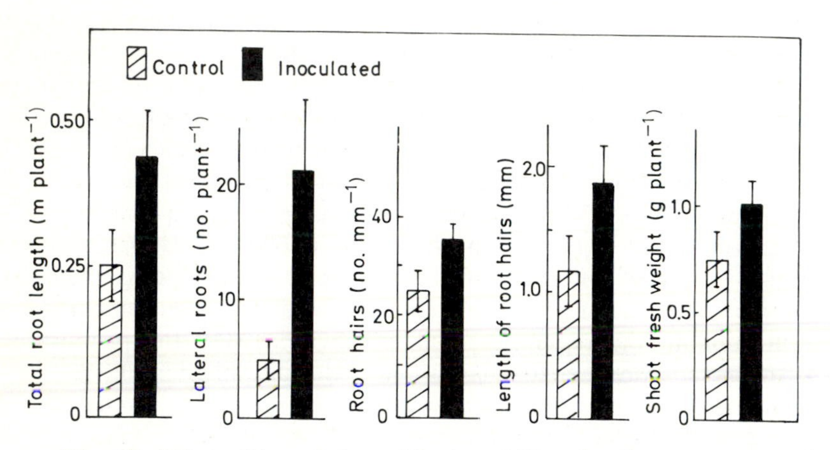

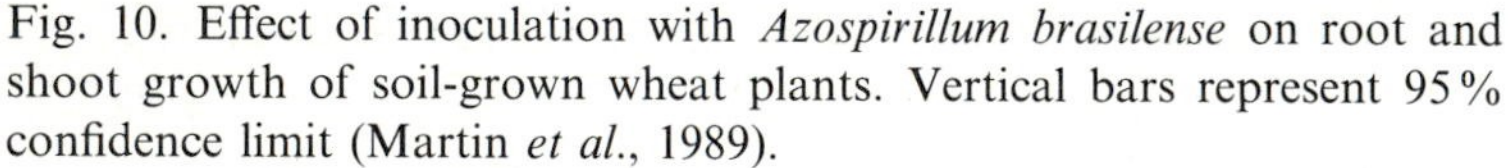
Fig. 10. Effect of inoculation with *Azospirillum brasilense* on root and shoot growth of soil-grown wheat plants. Vertical bars represent 95% confidence limit (Martin *et al.*, 1989).

These effects on root morphology may favour the acquisition of sparingly soluble mineral nutrients such as phosphorus. Furthermore, in legumes, a stimulating effect of *Azospirillum* on the root infection with *Rhizobium* (Martin *et al.*, 1989) may partly be related to changes in root morphology (e.g. root hair development).

Infecting rhizosphere microorganisms

The most widespread associations between microorganisms and higher plants are mycorrhizas, and the roots of most soil-grown plants are usually mycorrhizal. The fungus is usually strongly or wholly dependent on the plant which may or may not benefit. In some instances mycorrhizas are essential for growth. There are two major groups of mycorrhizae, endo- and ectomycorrhizas.

The most common group of *endomycorrhizas* is the vesicular arbuscular mycorrhiza (VAM), which infects the root cortex and extends its external mycelium into the soil. The resulting increase in root surface area favours phosphorus acquisition, particularly in soils with low phosphorus levels and for plant species with a relatively small root surface area, for example cassava (*Manihot esculentum* L.) and some legumes (Robson, O'Hara & Abbott, 1981; Howeler, Sieverding & Saif, 1987). Formation of polyphosphates in the external hyphae (Lapeyrie, Chilvers & Douglass, 1984) presumably plays a key role providing efficient uptake and a large transport capacity of the hyphae for phosphorus. Acquisition of zinc (Lambert, Baker & Cole, 1979) and copper (Gildon & Tinker, 1983; Manjunath & Habte, 1988) are also occasionally enhanced; this may be related to their binding to polyphosphates in the hyphae.

In legumes, nodulation and N_2 fixation are severely impaired under phosphorus deficiency. Therefore, dual infection with *Rhizobium* and VAM is an attractive approach for adapting legumes to phosphorus deficient soils. However, as both types of infection are important sinks for photosynthates, the dual infection may decrease plant dry weight and N_2 fixation compared to *Rhizobium* plus fertiliser phosphorus (Pacovsky, Bethlenfalvay & Paul, 1986). In legumes with dual infection the demand by microorganisms for photosynthates can strongly depress root growth and decrease the root:shoot ratio (Table 9). Under field conditions this decreased ratio may substantially decrease the drought resistance of the plants.

In mixed stands of legumes and grasses VAM hyphae may connect plant species and transfer symbiotically fixed nitrogen from the legume directly to the grass (Haystead, Malajczuk & Grove, 1988): such nutrient transfer is well documented in trees infected with ectomycorrhizae.

Ectomycorrhizas dominate in trees particularly on soils of low fertility in temperate climates. Fungal hyphae penetrate the intercellular space of the cortex and form a network of mycelium (Hartig net) and an interwoven mantle around the root (sheath-forming mycorrhizas); frequently, the infected roots are fully mantled, short and highly branched (Kottge & Oberwinkler, 1986). This effect on branching is presumably caused by auxin production of the fungus (Gay, 1986). With such a stunted root system the external mycelium becomes particularly important for mineral nutrient acquisition. The external mycelium differs considerably, depending on the type of ectomycorrhizal fungi; in some instances mycelium strands are formed (rhizomorphes) which probably transport water and mineral nutrients more efficiently to the host root.

For the acquisition of phosphorus, fungal acid phosphatases (Mousain & Salsac, 1986) and phytases (Heinrich, Mulligan & Patrick, 1988) are important. Some ectomycorrhizal fungi excrete oxalic acid (Malajczuk & Cromack, 1982) and may, for example, effectively mobilise sparingly soluble calcium phosphates. Production of siderophores may enhance uptake of iron and copper (Reid, 1984), and the release of cellulases and proteases increases the utilisation of organically bound nitrogen (protein-N) in ectomycorrhizal plants (Read, 1987). However, the host plant has to allocate a substantial proportion of its photosynthates to support the ectomycorrhizal fungus (Söderström & Read, 1987). Although ectomycorrhizas are certainly an important component in forest ecosystems, there is no substantial experimental evidence to support speculations (Schlechte, 1986) of a causal relationship between a decrease in ectomycorrhizas and forest decline in Central Europe.

Summary

The availability of mineral nutrients for plants is determined by the chemical, physical and biological characteristics of the bulk soil, as well as root-induced changes in the rhizosphere. Mineral nutrient supply to the root surface is mediated mainly by mass flow (e.g. calcium) or diffusion (e.g. phosphorus). Accordingly, root surface area is important for the acquisition of phosphorus. Acquisition of most nutrients is particularly dependent upon root-induced changes in rhizosphere pH, redox potential and the release of low molecular weight organic solutes. Depending on their chemical nature, these solutes can mobilise mineral nutrients directly, in the case of organic acids and phytosiderophores, or indirectly, by enhancing the activity of non-infecting microorganisms in the rhizosphere. Rhizosphere microorganisms may, however, also decrease micronutrient availability (e.g. manganese) or affect nutrient acquisition via changes in root morphology caused by hormones. Mineral nutrient uptake can be considerably increased by the infection of plant roots with VA mycorrhiza (e.g. phosphorus uptake) or, in perennials, infection with ectomycorrhiza. Root-induced changes in the rhizosphere vary between genotypes, and are affected by the nutritional status of the plant (e.g. phosphorus or iron deficiency). The extent of root-induced changes in the rhizosphere is important for adaptation of higher plants to soils with extreme chemical and physical properties.

Acknowledgement

The author is grateful to Ernest A. Kirkby for valuable comments and correction of the English text.

References

Aguilars, A. & van Diest, A. (1981). Rock-phosphate mobilization induced by the alkaline uptake pattern of legumes utilizing symbiotically fixed nitrogen. *Plant and Soil*, **61**, 27–42.

Anghinoni, I. & Barber, S.A. (1980). Phosphorus influx and growth characteristics of corn roots as influenced by phosphorus supply. *Agronomy Journal*, **72**, 685–8.

Armstrong, W. (1979). Aeration in higher plants. *Advances in Botanical Research*, **7**, 225–332.

Barber, S.A. (1984). *Soil Nutrient Bioavailability. A Mechanistic Approach.* New York: John Wiley and Sons.

Bekele, T., Cino, B.J., Ehlert, P.A.I., Van der Maas, A.A. & van Diest, A. (1983). An evaluation of plant-borne factors promoting the solubilization of alkaline rock phosphates. *Plant and Soil*, **75**, 361–78.

Bienfait, H.F. (1985). Regulated redox processes at the plasmalemma of plant root cells and their function in iron uptake. *Journal of Bioenergetics and Biomembranes*, **17**, 73–83.

Boddey, R.M. & Döbereiner, J. (1988). Nitrogen fixation associated with grasses and cereals: Recent results and perspectives for future research. *Plant and Soil*, **108**, 53–65.

Bromfield, S.M. (1978). The effect of manganese-oxidizing bacteria and pH on the availability of manganous ions and manganese oxides to oats in nutrient solutions. *Plant and Soil*, **49**, 23–39.

Brown, J.C. (1978). Mechanism of iron uptake by plants. *Plant, Cell and Environment*, **1**, 249–57.

Cakmak, I. & Marschner, H. (1988). Increase in membrane permeability and exudation in roots of zinc deficient plants. *Journal of Plant Physiology*, **132**, 356–61.

Callot, G. & Jaillard, B. (1987). Apports de la loupe binoculaire à l'étude des interfaces sol/racine et sol/champignon. In *Micro-morphologie des Sols*, eds. N. Fedoroff *et al.*, pp. 73–9.

Chapin, F.S. III (1988). Ecological aspects of plant mineral nutrition. In *Advances in Plant Nutrition*, Vol. 3. eds. B. Tinker & A. Läuchli, pp. 161–91.

Chen, C.C., Dixon, J.B. & Turner, F.T. (1980). Iron coatings on rice roots: Mineralogy and quantity influencing factors. *Soil Science Society America Journal*, **44**, 635–9.

Crowley, D.E., Reid, C.P.P. & Szaniszlo, P.J. (1987). Microbial siderophores as iron sources for plants. In *Iron Transport in Microbes, Plants and Animals* eds. G. Winkelmann, D. Van der Helm & J.B. Neilands, pp. 375–86. Weinheim, FRG: VCH Verlagsgesellschaft.

Dinkelaker, B., Römheld, V. & Marschner, H. (1989). Citric acid secretion and precipitation of calcium citrate in the rhizosphere of white lupin (*Lupinus albus* L.). *Plant, Cell and Environment*, **12**, 285–92.

Doumas, P., Berjoud, C., Calléja, M., Coupé, M., Espian, C. & d'Auzac, J. (1986). Phosphatases extracellulaires et nutrition phosphatée chez les champignons ectomycorhiziens et les plantes hotes. *Physiology Végétale*, **24**, 173–84.

Fischer, W., Flessa, H. & Schaller, G. (1989). pH values and redox potentials in microsites of the rhizosphere. *Zeitschrift für Pflanzenernährung und Bodenkunde*, **152**, 191–5.

Foehse, D. & Jungk, A. (1983). Influence of phosphate and nitrate supply on root hair formation of rape, spinach and tomato plants. *Plant and Soil*, **74**, 359–68.

Fußeder, A. & Kraus, M. (1986). Individuelle Wurzelkonkurrenz und Ausnutzung der immobilen Makronährstoffe im Wurzelraum von Mais. *Flora*, **178**, 11–18.

Gardner, W.K., Barber, D.A. & Parbery, D.G. (1983*a*). Non-infecting rhizosphere micro-organisms and the mineral nutrition of temperate cereals. *Journal of Plant Nutrition*, **6**, 185–99.

Gardner, W.K., Barber, D.A. & Parbery, D.G. (1983*b*). The acquisition of phosphorus by *Lupinus albus*. III. The probable mechanism by which phosphorus movement in the soil/root interface is enhanced. *Plant and Soil*, **70**, 107–24.

Gardner, W.K. & Boundy, K.A. (1983). The acquisition of phosphorus by *Lupinus albus* L. IV. The effect of interplanting wheat and white lupin on the growth and mineral composition of the two species. *Plant and Soil*, **70**, 391–402.

Gardner, W.K., Parbery, D.G. & Barber, D.A. (1982). The acquisition of phosphorus by *Lupinus albus* L. II. The effect of varying phosphorus supply and soil type on some characteristics of the soil/root interface. *Plant and Soil*, **68**, 33–41.

Gardner, W.K., Parbery, D.G., Barber, D.A. & Swinden, L. (1983*c*). The acquisition of phosphorus by *Lupinus albus* L. V. The diffusion of exudates away from roots: a computer simulation. *Plant and Soil*, **72**, 13–29.

Gay, G. (1986). Effect of glucose on indole-3-acetic acid production by ectomycorrhizal fungus *Hebeloma hiemale* in pure culture. *Physiology Végétale*, **24**, 185–92.

Gildon, A. & Tinker, P.B. (1983). Interactions of vesicular-arbuscular mycorrhizal infections and heavy metals in plants. II. The effects of infection on uptake of copper. *New Phytologist*, **95**, 263–8.

Graham, J.H., Leonard, R.T. & Menge, J.A. (1981). Membrane-mediated decrease in root exudation responsible for phosphorus inhibition of vesicular-arbuscular mycorrhizal formation. *Plant Physiology*, **68**, 548–52.

Graham, R.D. (1988). Genotypic differences in tolerance to manganese deficiency. In *Manganese in Soils and Plants*, eds. R.D. Graham, R.J. Hannam & N.C. Uren, pp. 261–76. Dordrecht: Kluwer Academic Publishers.

Graham, R.D. & Rovira, A.D. (1984). A role for manganese in the resistance of wheat plants to take-all. *Plant and Soil*, **78**, 441–4.

Graven, E.H., Attoe, O.J. & Smith, D. (1965). Effect of liming and flooding on manganese toxicity in alfalfa. *Soil Science Society of America Proceedings*, **29**, 702–6.

Grinsted, M.M., Hedley, M.J., White, R.E. & Nye, P.H. (1982). Plant-induced changes in the rhizosphere of rape (*Brassica napus* var. Emerald) seedlings. I. pH change in the increase in P concentration in the soil solution. *New Phytologist*, **91**, 19–29.

Häußling, M., Jorns, C.A., Lehmbecker, G., Hecht-Buchholz, Ch. & Marschner, H. (1988). Ion and water uptake in relation to root development in Norway spruce (*Picea abies* (L.) Karst.). *Journal of Plant Physiology*, **133**, 486–91.

Häußling, M. & Marschner, H. (1989). Organic and inorganic soil phosphates and acid phosphatase activity in the rhizosphere of 80 year old Norway spruce (*Picea abies* (L.) Karst.) trees. *Biology and Fertility of Soils*, **8**, 128–33.

Haystead, A., Malajczuk, N. & Grove, T.S. (1988). Underground transfer of nitrogen between pasture plants infected with vesicular–arbuscular mycorrhizal fungi. *New Phytologist*, **108**, 417–23.

Hedley, M.J., Nye, P.H. & White, R.E. (1982). Plant-induced changes in the rhizosphere of rape (*Brassica napus* var. Emerald) seedlings. II. Origin of the pH change. *New Phytologist*, **91**, 31–44.

Heinrich, P.A., Mulligan, D.R. & Patrick, J.W. (1988). The effect of ectomycorrhizas on the phosphorus and dry weight acquisition of *Eucalyptus* seedlings. *Plant and Soil*, **109**, 147–9.

Helal, H.M. & Sauerbeck, D. (1986). Effect of plant roots on carbon metabolism of soil microbial biomass. *Zeitschrift für Pflanzenernährung und Bodenkunde*, **149**, 181–8.

Helal, H.M. & Sauerbeck, D. (1988). Phosphataseaktivität von Pflanzenwurzeln in Abhängigkeit von der P-Versorgung. *VDLUFA Schriftenreihe*, **23**, 195–210.

Hendriks, L., Claassen, N. & Jungk, A. (1981). Phosphatverarmung des wurzelnahen Bodens und Phosphataufnahme von Mais und Raps. *Zeitschrift für Pflanzenernährung und Bodenkunde*, **144**, 486–99.

Hoffland, E., Findenegg, G.R. & Nelemans, J.A. (1989). Solubilization of rock phosphate by rape. II. Local root exudation of organic acids as a response to P-starvation. *Plant and Soil*, **113**, 161–5.

Horst, W.J., Wagner, A. & Marschner, H. (1982). Mucilage protects root meristems from aluminium injury. *Zeitschrift für Pflanzenphysiologie*, **105**, 435–44.

Horst, W.J. & Waschkies, Ch. (1987). Phosphatversorgung von Sommerweizen (*Triticum aestivum* L.) in Mischkultur mit weißer Lupine (*Lupinus albus* L.). *Zeitschrift für Pflanzenernährung und Bodenkunde*, **150**, 1–8.

Howeler, R.H., Sieverding, E. & Saif, S. (1987). Practical aspects of mycorrhizal technology in some crops and pastures. *Plant and Soil*, **100**, 249–83.

Huber, D.M. & Wilhelm, N.S. (1988). The role of manganese in resistance to plant diseases. In *Manganese in Soils and Plants*, eds. R.D. Graham, R.J. Hannam & N.C. Uren, pp. 155–73. Dordrecht: Kluwer Academic Publishers.

Inbal, E. & Feldman, M. (1982). The response of a hormonal mutant of common wheat to bacteria of the genus *Azospirillum*. *Israel Journal of Botany*, **31**, 257–63.

Inskeep, W.P. & Bloom, P.R. (1986). Effects of soil moisture on soil pCO_2, soil solution bicarbonate, and iron chlorosis in soybeans. *Soil Science Society American Journal*, **50**, 946–52.

Jagnow, G. (1987). Inoculation of cereal crops and forage grasses with nitrogen-fixing rhizosphere bacteria: Possible causes of success and failure with regard to yield response – a review. *Zeitschrift für Pflanzenernährung und Bodenkunde*, **150**, 361–8.

Jaillard, B. (1984). Mise en évidence de la néogenase de sables calcaires sous l'influence des racines: incidence sur la gramulométrie du sol. *Agronomie*, **4**, 91–100.

Jain, D.K., Beyer, D. & Rennie, R.J. (1987). Dinitrogen fixation (C_2H_2 reduction) by bacterial strains at various temperatures. *Plant and Soil*, **103**, 233–7.

Jungk, A. & Claassen, N. (1986). Availability of phosphate and potassium as the result of interactions between root and soil in the rhizosphere. *Zeitschrift für Pflanzenernährung und Bodenkunde*, **149**, 411–27.

Jungk, A. & Claassen, N. (1989). Availability in soil and acquisition by plants as the basis for phosphorus and potassium supply to plants. *Zeitschrift für Pflanzenernährung und Bodenkunde*, **152**, 151–7.

Klotz, F. & Horst, W.J. (1988). Genotypic differences in aluminium tolerance of soybean (*Glycine max* L.) as affected by ammonium and nitrate nitrogen nutrition. *Journal of Plant Physiology*, **132**, 702–7.

Kottge, I. & Oberwinkler, F. (1986). Mycorrhiza of forest trees – structure and function. *Trees*, **1**, 1–24.

Kraffczyk, I., Trolldenier, G. & Beringer, H. (1984). Soluble root exudates of maize: Influence of potassium supply and rhizosphere microorganisms. *Soil Biology and Biochemistry*, **16**, 315–22.

Kuchenbuch, R., Claassen, N. & Jungk, A. (1986). Potassium availability in relation to soil moisture. II. Calculations by means of a mathematical simulation model. *Plant and Soil*, **95**, 233–43.

Lambert, D.H., Baker, D.E. & Cole Jr., H. (1979). The role of mycorrhizae in the interactions of phosphorus with zinc, copper, and other elements. *Soil Science Society American Journal*, **43**, 976–80.

Lapeyrie, F.F., Chilvers, G.A. & Douglass, P.A. (1984). Formation of metachromatic granules following phosphate uptake by mycelial hyphae of an ectomycorrhizal fungus. *New Phytologist*, **98**, 345–60.

Mackay, A.D. & Barber, S.A. (1985). Effect of soil moisture and phosphate level on root hair growth on corn roots. *Plant and Soil*, **86**, 321–31.

Malajczuk, N. & Cromack Jr., K. (1982). Accumulation of calcium oxalate in the mantle of ectomycorrhizal roots of *Pinus radiata* and *Eucalyptus marginata*. *New Phytologist*, **92**, 527–32.

Manjunath, A. & Habte, M. (1988). Development of vesicular–arbuscular mycorrhizal infection and the uptake of immobile nutrients in *Leucaena leucocephala*. *Plant and Soil*, **106**, 97–103.

Marschner, H. (1986). *Mineral Nutrition of Higher Plants*. London: Academic Press.

Marschner, H. (1988). Mechanism of manganese acquisition by roots from soils. In *Manganese in Soils and Plants*, eds. R.D. Graham, R.J. Hannam & N.C. Uren, pp. 191–204. Dordrecht: Kluwer Academic Publishers.

Marschner, H., Häußling, M. & Leisen, E. (1986*a*). Rhizosphere pH of Norway spruce trees grown under both controlled and field conditions. In *Proceedings of CEC Workshop Jülich 1985 on Effects of Air Pollution on Terrestrial and Aquatic Ecosystems*, pp. 113–18.

Marschner, H. & Römheld, V. (1983). In-vivo measurement of root-induced pH changes at the soil–root interface: Effect of plant species and nitrogen source. *Zeitschrift für Pflanzenphysiologie*, **111**, 241–51.

Marschner, H., Römheld, V. & Cakmak, I. (1987*a*). Root-induced changes of nutrient availability in the rhizosphere. *Journal of Plant Nutrition*, **10**, 1175–84.

Marschner, H., Römheld, V., Horst, W.J. & Martin, P. (1986*c*). Root-induced changes in the rhizosphere: Importance for the mineral nutrition of plants. *Zeitschrift für Pflanzenernährung und Bodenkunde*, **149**, 441–56.

Marschner, H., Römheld, V. & Kissel, M. (1986*b*). Different strategies in higher plants in mobilization and uptake of iron. *Journal of Plant Nutrition*, **9**, 695–713.

Marschner, H., Römheld, V. & Kissel, M. (1987*b*). Localization of phytosiderophore release and of iron uptake along intact barley roots. *Physiologia Plantarum*, **71**, 157–62.

Marschner, H., Treeby, M. & Römheld, V. (1989). Role of root-induced changes in the rhizosphere for iron acquisition in higher plants. *Zeitschrift für Pflanzenernährung und Bodenkunde*, **152**, 237–40.

Martin, P., Glatzle, A., Kolb, W., Omay, H. & Schmidt, W. (1989). N_2-fixing bacteria in the rhizosphere: Quantification and hormonal effects on root development. *Zeitschrift für Pflanzenernährung und Bodenkunde*, **152**, 237–45.

McCully, M.E., Canny, M.J. & Van Steveninck, R.F.M. (1987). Accumulation of potassium by differentiating metaxylem elements of maize roots. *Physiologia Plantarum*, **69**, 73–80.

Mench, M., Morel, J.L. & Guckert, A. (1987). Metal binding properties of high molecular weight soluble exudates from maize (*Zea mays* L.) roots. *Biology and Fertility of Soils*, **3**, 165–9.

Mench, M., Morel, J.L., Guckert, A. & Guillet, B. (1988). Metal binding with root exudates of low molecular weight. *Journal of Soil Science*, **39**, 521–7.

Moraghan, J.T. (1979). Manganese toxicity in flax growing on certain calcareous soils low in available iron. *Soil Science Society of America Journal*, **43**, 1177–80.

Moraghan, J.T. (1985). Manganese deficiency in soybeans as affected by FeEDDHA and low soil temperature. *Soil Science Society of America Journal*, **49**, 1584–6.

Morel, J.L., Mench, M. & Guckert, A. (1986). Measurement of Pb^{2+}, Cu^{2+} and Cd^{2+} binding with mucilage exudates from maize (*Zea mays* L.) roots. *Biology and Fertility of Soils*, **2**, 29–34.

Mousain, D. & Salsac, L. (1986). Utilisation du phytate et activités phosphatases acides chez *Pisolithus tinctorus*, basidiomycete mycorrhizien. *Physiologie Végétale*, **24**, 193–200.

Nambiar, E.K.S. (1976). The uptake of zinc-65 by oats in relation to soil water content and root growth. *Australian Journal of Soil Research*, **14**, 67–84.

Nomoto, K., Sugiura, Y. & Takagi, S. (1987). Mugineic acids, studies on phytosiderophores. In *Iron Transport in Microbes, Plants and Animals*, eds. G. Winkelmann, D. van der Helm & J.B. Neilands, pp. 401–25. Weinheim, FRG: VCH Verlagsgesellschaft.

Nye, P.H. (1986). Acid–base changes in the rhizosphere. In *Advances in Plant Nutrition 2*, eds. B. Tinker & A. Läuchli, pp. 129–53. New York: Praeger Scientific.

Olsen, R.A., Bennett, J.H., Blume, D. & Brown, J.C. (1981). Chemical aspects of the Fe stress response mechanism in tomatoes. *Journal of Plant Nutrition*, **3**, 905–21.

Ottow, J.C.G., Benckiser, G., Santiago, S. & Watanabe, I. (1982). Iron toxicity of wetland rice (*Oryza sativa* L.) as a multiple nutritional stress. In *Proceedings of the 9th Plant Nutrition Colloquium, Warwick*, ed. A. Scaife, pp. 454–60. Commonwealth Agricultural Bureaux.

Pacovsky, R.S., Bethlenfalvay, G.J. & Paul, E.A. (1986). Comparisons between P-fertilized and mycorrhizal plants. *Crop Science*, **26**, 151–6.

Peterson, C.A. (1987). The exodermal Casparian band of onion roots blocks the apoplastic movement of sulphate ions. *Journal of Experimental Botany*, **38**, 2068–81.

Peterson, C.A. (1988). Exodermal Casparian bands: their significance for ion uptake by roots. *Physiologia Plantarum*, **72**, 204–8.

Piccini, D., Ocampo, J.A. & Bedmar, E.J. (1988). Possible influence of

Rhizobium on VA mycorrhiza and metabolic activity in double symbiosis of alfalfa plants (*Medicago sativa* L.) grown in a pot experiment. *Biology and Fertility of Soils*, **6**, 65–7.

Raij, B. van & Diest, A. van (1979). Utilization of phosphate from different sources by six plant species. *Plant and Soil*, **51**, 577–89.

Read, D.J. (1987). In support of Frank's organic nitrogen theory. *Angewandte Botanik*, **61**, 25–37.

Reid, C.P.P. (1984). Mycorrhizae: a root–soil interface in plant nutrition. In *Microbial–Plant Interactions*, eds. R.L. Todd & J.E. Giddens, pp. 29–50. ASA Special Publication No. 47.

Reid, R.K., Reid, C.P.P., Powell, P.E. & Szaniszlo, P.J. (1984). Comparison of siderophore concentrations in aqueous extracts of rhizosphere and adjacent bulk soils. *Pedobiologia*, **26**, 263–6.

Reynolds, S.B., Scaife, A. & Turner, M.K. (1987). Effect of nitrogen form on boron uptake by cauliflower. *Communications in Soil Science and Plant Analysis*, **18**, 1143–54.

Robson, A.D., O'Hara, A.O. & Abbott, L.K. (1981). Involvement of phosphorus in nitrogen fixation by subterranean clover (*Trifolium subterraneum* L.). *Australian Journal of Plant Physiology*, **8**, 427–36.

Rollwagen, B.A. & Zasoski, R.J. (1988). Nitrogen source effects on rhizosphere pH and nutrient accumulation by Pacific Northwest conifers. *Plant and Soil*, **105**, 79–86.

Römheld, V. (1986). pH-Veränderungen in der Rhizosphäre verschiedener Kulturpflanzenarten in Abhängigkeit vom Nährstoffgehalt. *Kali-Briefe*, **18**, 13–30.

Römheld, V. (1987*a*). Different strategies for iron acquisition in higher plants. *Physiologia Plantarum*, **70**, 231–4.

Römheld, V. (1987*b*). Existence of two different strategies for the acquisition of iron in higher plants. In *Iron Transport in Microbes, Plants and Animals*, eds. G. Winkelmann, D. van der Helm & J.B. Neilands, pp. 353–74. Weinheim, FRG: VCH Verlagsgesellschaft.

Römheld, V. & Marschner, H. (1986). Evidence for a specific uptake system for iron phytosiderophores in roots of grasses. *Plant Physiology*, **80**, 175–80.

Rovira, A.D., Bowen, G.D. & Foster, R.C. (1983). The significance of rhizosphere microflora and mycorrhizas in plant nutrition. In *Encyclopedia of Plant Physiology, New Series, Vol. 15A*, eds. A. Läuchli & R.L. Bieleski, pp. 61–89. Berlin: Springer-Verlag.

Rynders, L. & Vlassak, K. (1982). Use of *Azospirillium brasilense* as biofertilizer in intensive wheat cropping. *Plant and Soil*, **66**, 217–23.

Sanderson, J., Whitbread, F.C. & Clarkson, D.T. (1988). Persistent xylem cross-walls reduce the axial hydraulic conductivity in the apical 20 cm of barley seminal root axes: implications for the driving force for water movement. *Plant, Cell and Environment*, **11**, 247–56.

Sarkar, A.N., Jenkins, D.A. & Wyn Jones, R.G. (1979). Modification to mechanical and mineralogical composition of soil within the rhizosphere. In *The Soil–Root Interface*, eds. J.L. Harley & R. Scott Russell, pp. 125–36. London: Academic Press.

Sarkar, A.N. & Wyn Jones, R.G. (1982). Effect of rhizosphere pH on availability and uptake of Fe, Mn and Zn. *Plant and Soil*, **66**, 361–72.

Sattelmacher, B. & Thoms, K. (1989). Root growth and ^{14}C-translocation into the roots of maize (*Zea mays* L.) as influenced by local nitrate supply. *Zeitschrift für Pflanzenernährung und Bodenkunde*, **152**, 7–10.

Sauerbeck, D.R. & Helal, H.M. (1986). Plant root development and photosynthate consumption depending on soil compaction. *Transactions of the 13th Congress International Society of Soil Science*, Hamburg, **3**, 948–9.

Schlechte, G. (1986). Zur Mykorrhizapilzflora in geschädigten Forstbeständen. *Zeitschrift für Mykologie*, **42**, 225–32.

Schönwitz, R. & Ziegler, H. (1982). Exudation of water-soluble vitamins and of some carbohydrates by intact roots of maize seedlings (*Zea mays* L.) into a mineral nutrient solution. *Zeitschrift für Pflanzenphysiologie*, **107**, 7–14.

Schubert, S. & Mengel, K. (1989). Important factors in nutrient availability: root morphology and physiology. *Zeitschrift für Pflanzenernährung und Bodenkunde*, **152**, 169–74.

Seggewiss, B. & Jungk, A. (1988). Einfluß der Kaliumdynamik im wurzelnahen Boden auf die Magnesiumaufnahme von Pflanzen. *Zeitschrift für Pflanzenernährung und Bodenkunde*, **151**, 91–6.

Sharpley, A.N. (1985). Phosphorus cycling in unfertilized and fertilized agricultural soils. *Soil Science Society of America Journal*, **49**, 905–11.

Shi, W., Chino, M., Youssef, R.A., Mori, S. & Takagi, S. (1989). The occurrence of mugineic acid in the rhizosphere soil of barley plant. *Soil Science and Plant Nutrition*, **34**, 585–92.

Sinha, B.K. & Singh, N.T. (1974). Effect of transpiration rate on salt accumulation around corn roots in a saline soil. *Agronomy Journal*, **66**, 557–60.

Söderström, B. & Read, D.J. (1987). Respiratory activity of intact and excised ectomycorrhizal mycelial systems growing in unsterilized soil. *Soil Biology and Biochemistry*, **19**, 231–6.

Sugiura, Y. & Nomoto, K. (1984). Phytosiderophores: Structures and properties of mugineic acids and their metal complexes. *Structure and Bonding*, **58**, 107–35.

Takagi, S., Kamei, S. & Ming-Ho, Y. (1988). Efficiency of iron extraction from soil by mugineic acid family phytosiderophores. *Journal of Plant Nutrition*, **11**, 643–51.

Takagi, S., Nomoto, K. & Takemoto, T. (1984). Physiological aspect of mugineic acid, a possible phytosiderophore of graminaceous plants. *Journal of Plant Nutrition*, **7**, 469–77.

Tarafdar, J.C. & Claassen, N. (1988). Organic phosphorus compounds as a phosphorus source for higher plants through the activity of phosphatases produced by plant roots and microorganisms. *Biology and Fertility of Soils*, **5**, 308–12.

Tarafdar, J.C. & Jungk, A. (1987). Phosphatase activity in the rhizosphere and its relation to the depletion of soil organic phosphorus. *Biology and Fertility of Soils*, **3**, 199–204.

Timonin, M.I. (1946). Microflora of the rhizosphere in relation to the manganese deficiency disease of oats. *Soil Science Society of America Proceedings*, **11**, 284–292.

Treeby, M., Marschner, H. & Römheld, V. (1989). Mobilization of iron and other micronutrients from a calcareous soil by plant-borne, microbial, and synthetic metal chelators. *Plant and Soil*, **114**, 217–26.

Trought, M.C.T. & Drew, M.C. (1980). The development of water logging damage in young wheat plants in anaerobic solution cultures. *Journal of Experimental Botany* **31**, 1573–85.

Uren, N.C. & Reisenauer, H.M. (1988). The role of root exudates in nutrient acquisition. In *Advances in Plant Nutrition 3*, eds. B. Tinker & A. Läuchli, pp. 79–114. New York: Praeger Publishers.

Vuurde, J.W.L., van & Schippers, B. (1980). Bacterial colonization of seminal wheat roots. *Soil Biology and Biochemistry*, **12**, 559–65.

Wallace, A. (1982). Effect of nitrogen fertilizer and nodulation on lime-induced chlorosis in soybean. *Journal of Plant Nutrition*, **5**, 363–8.

Warembourg, F.R. & Billes, G. (1979). Estimating carbon transfers in the plant rhizosphere. In *The Soil–Root Interface*, eds. J.L. Harley & R. Scott Russell, pp. 183–96. London: Academic Press.

Whipps, J.M. & Lynch, J.M. (1986). The influence of the rhizosphere on crop productivity. *Advances in Microbial Ecology*, **6**, 187–244.

Winkelmann, G. (1986). Iron complex products (siderophores). In *Biotechnology*, Vol. 4, eds. H.J. Rehm and G. Reed, pp. 216–243. Weinheim, FRG: VCH Verlagsgesellschaft.

Zech, W., Alt, H.G., Haumaier, L. & Blasek, R. (1987). Characterization of phosphorus fractions in mountain soils of the Bavarian Alps by ^{31}P NMR spectroscopy. *Zeitschrift für Pflanzenernährung und Bodenkunde*, **150**, 119–23.

Zhang, R., Römheld, V. & Marschner, H. (1989). Effect of zinc deficiency in wheat on the release of zinc and iron mobilizing root exudates. *Zeitschrift für Pflanzenernährung und Bodenkunde*, **152**, 205–10.

I.H. RORISON

Ecophysiological aspects of nutrition

Introduction

The timing of this conference marked the 21st Anniversary of the BES 9th Symposium 'Ecological Aspects of the Mineral Nutrition of Plants', 1–5 April 1968 (Rorison, 1969). This could be said to be the true forerunner of the present volume in which the organisers have also gone to great pains to attract contributions from physiologists and soil scientists as well as ecologists. Much work has been published since 1968 but in some respects little has changed. We still have a preponderance of contributions from crop physiologists, most of whom are concerned with annuals and have yet to embrace comparisons with natural systems. Nevertheless, with their relative simplicity, crop studies are making a significant contribution to our understanding of plant responses to environmental factors (e.g. Porter, Klepper & Belford, 1986; Wild, Jones & Macduff 1987; also Marshall & Porter and Groot & Spiertz, 1991, both this volume).

A preoccupation with relationships between form and function is also paramount among many experimental ecologists, but it is their involvement with wild species, on a comparative basis (Bradshaw, 1987), which is their most valuable contribution.

It is therefore the aim of this paper to emphasise the heterogeneity of the problem facing the plant ecologist compared with the crop physiologist and to illustrate the value of the comparative approach in furthering our overall understanding of the influence of nutrients on plant growth.

It is also important to stress that the increasing economic pressures which give rise not only to major problems of pollution but also to those of land-use, focus attention on the *tolerance* of plants as much as on their *productivity*. We are, all of us, concerned more than ever with survival both of plants and of ourselves.

In considering tolerance and survival we are dealing with dynamic processes and need to consider their limits, not under steady states, but under *fluctuating* conditions. We must also be aware of responses to change which may be, not only irregular, but unexpected. The study of 'wild' species provides both an extensive genetic reservoir and an unrivalled breath of response and thus allows us to bring a wider ecological perspective to present-day issues which transcends the boundaries of agriculture and physiology.

In order to ensure survival of the landscape as we would like it, we need to consider these native species and their range of response to nutrients, climate and management. We should include habitats which justify only low financial inputs, such as set-aside land and others that are, by agricultural standards, deficient and unbalanced in their mineral nutrient content and likely to remain so.

Thus, we still need to know more about why many uncultivated species grow in one place and not in another and why they may co-exist in what, on first inspection, are unlikely circumstances. That they do so is increasingly being related to the *heterogeneity* of the environments in which they survive and the fluctuations of nutrient availability in both time and space which are part of this heterogeneity.

Given this background, the role of the eco-physiologist so far has been in the measurement of responses to single factors or a few interacting factors, often studied in the laboratory rather than in the field. Several advantages are immediately apparent in as far as the control, precision and reproducibility of laboratory experiments is concerned. Under such methodology, flexibility of response may show both at the individual plant and at the inter-specific level (Rorison, 1987). Yet evidence must be weighed against patterns of behaviour in the field – at whatever scale. So, the following questions arise. Can simple laboratory tests be used to predict field behaviour? Can a database revealing 'pecking orders' in response to laboratory tests be confirmed from known behaviour of plants in the field? Some problems arising from these questions and some results of recent experiments designed to answer them will now be discussed.

Simple 'yield' responses

First, I would like to describe some nutritional experiments which form part of a species 'MOT-test', which is currently employed at the Unit of Comparative Plant Ecology (UCPE). The Integrated Screening Programme (ISP) is a series of simple, usually short-term, tests aimed at producing quantitative estimates of plants' response to environmental factors: the tests are standardised to provide a basis for inter- and intra-specific comparison and hopefully, provide results compatible with the response of the same plants in the field.

Like several other major environmental influences, mineral nutrition presents a severe challenge. The ISP brief was to cover deficiency, adequacy,

Table 1. *Solutions used to measure responses of seedlings to nutrient depletion.*[a]

Solution[b]	Individual elements concentration			
	N	P	K	Ca
Control	4 mM	1 mM	2 mM	2 mM
Control × 0.01	40 μM	10 μM	20 μM	20 μM
Low N	40 μM	1 mM	2 mM	2 mM
Low P	4 mM	10 μM	2 mM	2 mM
Low K	4 mM	1 mM	20 μM	2 mM
Low Ca	4 mM	1 mM	2 mM	20 μM
High Ca	4 mM	1 mM	2 mM	16 mM

[a] Details of composition are given in Hewitt (1966), Table 30(c), 'Rorison Solution'.
[b] Full strength (unless otherwise stated).

toxicity or imbalance of nutrients, in order to predict seedling response to soils of different fertility. Results are available for *c.* 16 species examined to date. In order to set them in the broader context they will be interpreted with reference to data, from our own field surveys, from our own field experimental analyses, and from the literature. Nutrient concentrations used were based on our past experience of the responses of seedlings to nutrient depletion (Table 1). Measurements were of dry weight yield after 21 days and nutrient content of root and shoot. They are expressed as a percentage of the control and dry weight data for four contrasted species are given in Fig. 1.

The two fast-growing species, *Epilobium hirsutum* and *Chenopodium album*, both respond to separate deficiencies of the individual elements N, P and K. The effect of overall dilution appears to be cumulative. The most obvious difference between the two species is in their response to low calcium supply. *E. hirsutum*, a calcicolous plant (Grime, Hodgson & Hunt, 1988), is particularly susceptible, while *C. album*, one of a group tolerant of saline conditions, is not significantly affected. By contrast, *Deschampsia flexuosa* and *Eriophorum vaginatum*, two slow-growing species, are much less sensitive to nutrient depletions, and their requirements to maintain optimum growth are small. Of the elements tested, nitrogen exerts a major influence, whilst the response to Ca is particularly variable.

In the soil calcium is present most commonly as $CaCO_3$ and is a critical component of the electro-chemical balance in soils and plants (Rorison & Robinson, 1984). Its relative presence or absence affects soil reactions and

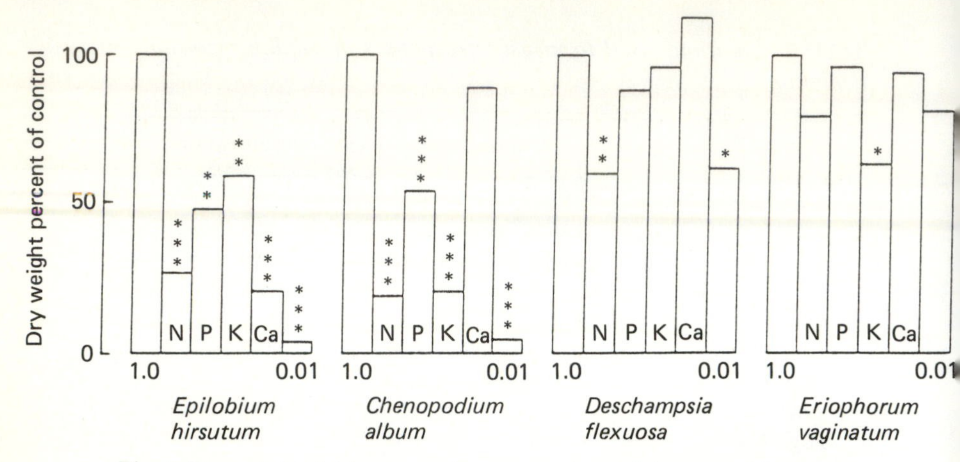

Fig. 1. Dry weight yield expressed as a percentage of the control (treatment 1.0) of *Epilobium hirsutum*, *Chenopodium album*, *Deschampsia flexuosa* and *Eriophorum vaginatum* after 21 days growth in solutions containing 100-fold dilutions of either N, P, K or Ca or a 100-fold dilution of all components (treatment 0.01). Differences between individual treatments and controls: * ($p < 0.05$); ** ($p < 0.01$); *** ($p < 0.001$).

therefore the availability of other inorganic elements. Knowledge of its availability (together with the soil's cation exchange capacity) enables us to understand that particular soil's potential for leaching and its response to liming. Normally a plant requires only trace amounts of Ca but the distribution of many native species is very much related to the calcareousness of the soil, i.e. indicating differential tolerance, if not requirement. Therefore the growth response of seedlings to Ca might indicate requirement for establishment and, because of implications from other field factors, their degree of tolerance to acidic/calcareous conditions.

Results from the ISP tests for 16 species indicate specific differences (Fig. 2). Ecologists will be interested in the order of species, whereas crop physiologists might be interested in both the extent and in the *range* of responses to reduced Ca. In Fig. 2 there is an overall tendency for low demand for Ca to be linked with low Ca sensitivity and vice versa, for the monocotyledons to be less sensitive to Ca deficiency than dicotyledons and for this division to over-shadow the calcicole–calcifuge spectrum. Shipley & Keddy (1988) found a similar general relationship between relative growth rate (RGR) and overall nutrient sensitivity after using Hoagland v. 0.1 Hoagland solutions.

What do such results mean in terms of explaining spatial and temporal distribution of plants? UCPE's Mineral Nutrient Survey of 1969 provides a good starting point to answer such questions since it included nutrient

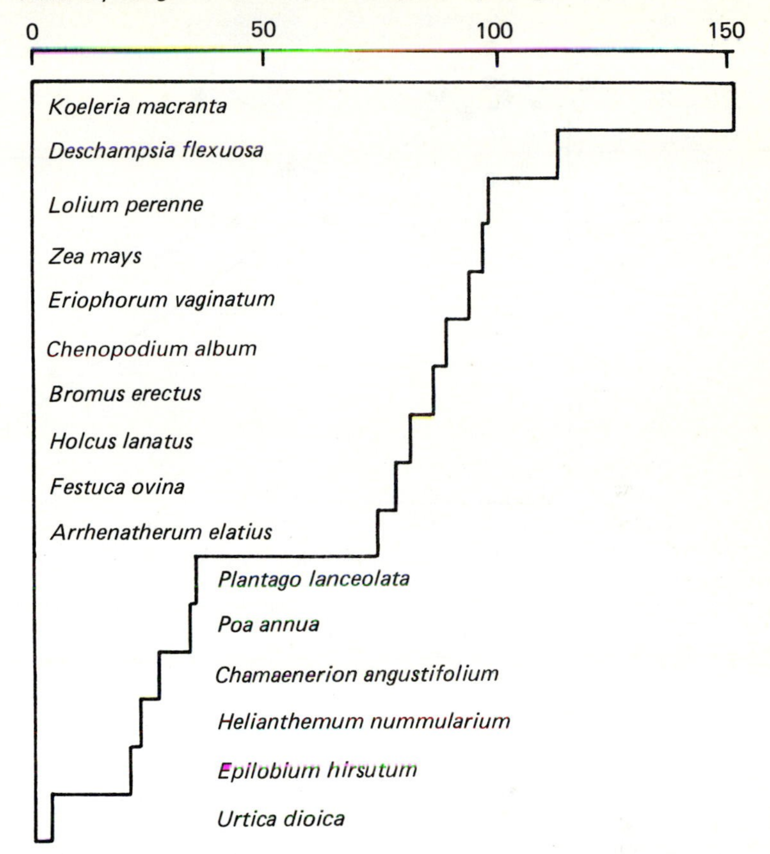

Fig. 2. Dry weight yield of 16 species grown in low Ca solution for 21 days, expressed as a percentage of the control and arranged in order of increasing sensitivity down the page.

analysis of both plants and soils from a wide range of habitats. For example, concentrations of Ca in mature green leaves were correlated with exchangeable Ca in the soil and, despite the differences in concentrations at different sites, a significant difference in behaviour between monocotyledons and dicotyledons and calcifuges and calcicoles was apparent (Fig. 4 in Rorison & Robinson, 1984).

How do these data from field-grown plants correspond with the ISP results on seedling survival? Fig. 3 shows that the higher the Ca concentration in the leaves of plants growing in the field the more sensitive to low Ca supply is the seedling in the ISP test. The primary division along the regression line remains that between monocots and dicots and within

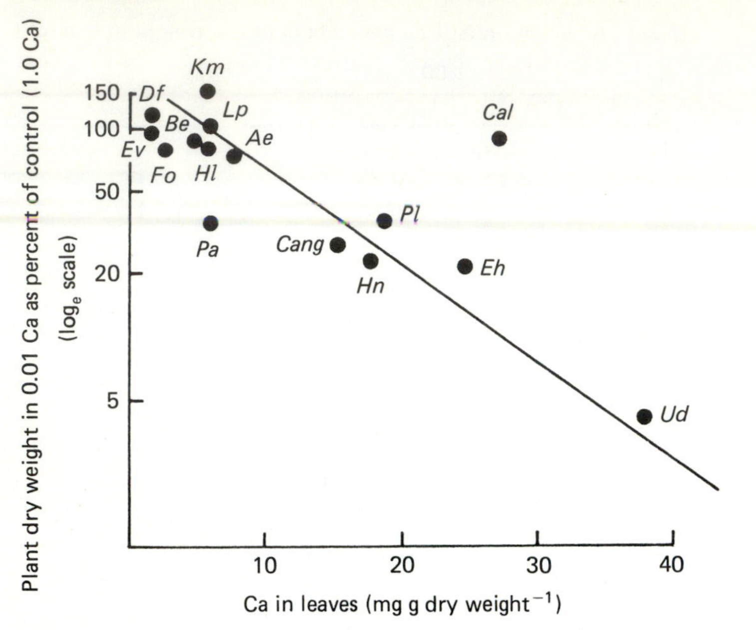

Fig. 3. Relationship between seedling sensitivity to calcium depletion and calcium content of leaves of plants growing in the field. Regression equation; $y = 47.75 - 9.12 \log_e x$ ($p < 0.01$), $r^2 = 0.78^{**}$, ($n-2 = 13$). Species key: *Df, Deschampsia flexuosa; Ev, Eriophorum vaginatum; Fo, Festuca ovina; Be, Bromus erectus; Km, Koeleria macrantha; Lp, Lolium perenne; Hl, Holcus lanatus; Cal, Chenopodium album; Ae, Arrhenatherum elatius; Pl, Plantago lanceolata; Pa, Poa annua; Cang, Chamaenerion angustifolium; Hn, Helianthemum nummularium; Eh, Epilobium hirsutum; Ud, Urtica dioica.*

these two groups, the tendency of calcifuges to be less Ca-demanding than the calcicoles. *Koeleria macrantha* is something of an exception among the grasses – although thought of as a calcicole, its demands are low. *Chenopodium album*, on the other hand, has a very low Ca demand despite its preponderance in fertile sites with relatively high Ca.

While in uncultivated soils phosphate may be most limiting, nitrogen is the inorganic nutrient required in the greatest amounts for the growth of plants. A response to N supply should indicate relative requirements; i.e. the degree of nitrophily expected in the field and, in general, the degree of fertility required for optimum growth. Results from the ISP reveal somewhat different relationships for N than for Ca. Sensitivity can still be significant at the $P < 0.001$ level but without the calcicole–calcifuge, monocot–dicot divisions (cf. Fig. 1). Rather, there is a stronger relationship

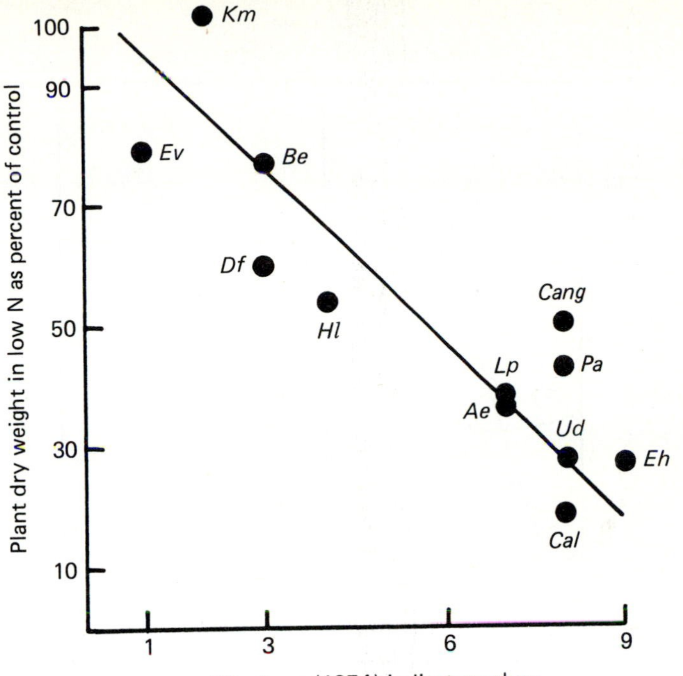

Fig. 4. The relationship between seedling sensitivity to nitrogen depletion and Ellenberg's (1974) index of nitrophily. Regression equation; $Y = 10.85 - 0.10x$ ($p < 0.001$), $r^2 = 0.88$***, ($n-2 = 10$). Species key: *Km, Koeleria macrantha; Ev, Eriophorum vaginatum; Be, Bromus erectus; Df, Deschampsia flexuosa; Hl, Holcus lanatus; Cang, Chamaenerion angustifolium; Pa, Poa annua; Lp, Lolium perenne; Ae, Arrhenatherum elatius; Ud, Urtica dioica; Eh, Epilobium hirsutum; Cal, Chenopodium album.*

between indices of nitrophily and the level of nitrate reductase (Lee & Stewart, 1978), and between RGR and plant strategy (Grime *et al.*, 1988).

One valuable correlation with field behaviour comes from Ellenberg (1974) who proposed indicator values for 2000 species of vascular plants, based on 40 years work of many researchers. Ellenberg provides a numerical index of response to 'nitrogen' (NH_4^+ and/or NO_3^- supply) on a scale 1–9, with a 1 denoting occurrence only in soils very poor in mineral nitrogen up to 9, which denotes occurrence in soils very rich in nitrogen 'indicating pollution, manure deposits or similar' (Ellenberg, 1974).

Several of the ISP species are described by Ellenberg as indifferent to nitrogen with a wide amplitude of response or, differential behaviour in different regions, so they do not appear in Fig. 4. For the species other than

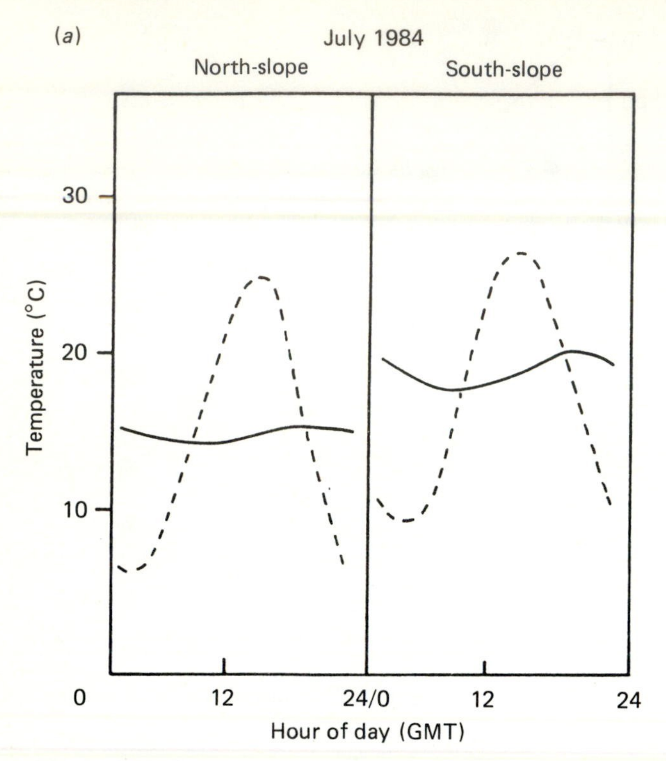

Fig. 5(*a*). Daily temperature cycles. Smoothed curves of the means of half-hourly values for July 1984. (*a*) At 200 mm above (---) and 200 mm below (——) the soil surface on the N-facing and S-facing slopes of a Derbyshire dale; (*b*) as (*a*) at 20 mm.

those designated as indifferent, sensitivity in terms of dry weight production in the ISP test is significantly correlated with the index of distribution based on plant affinity for nitrogen (see Fig. 4). Field data on nitrate reductase activity (NRA) are more fragmentary but measurements from Lee & Stewart (1978) tally with the UCPE RGR data to produce a general relationship between RGR and NRA that is highly significant. Thus there are encouraging signs of the relevance of the ISP tests to findings in the field.

As with several other environmental pressures, experience has shown that the response of plants to certain combinations of soil chemical conditions can be inferred from their response to one or more critical components of that combination. Therefore, responses to soil reactions can be inferred from responses to nitrogen source and calcium. But in hypothesising such close links between laboratory results and field behaviour one is running the

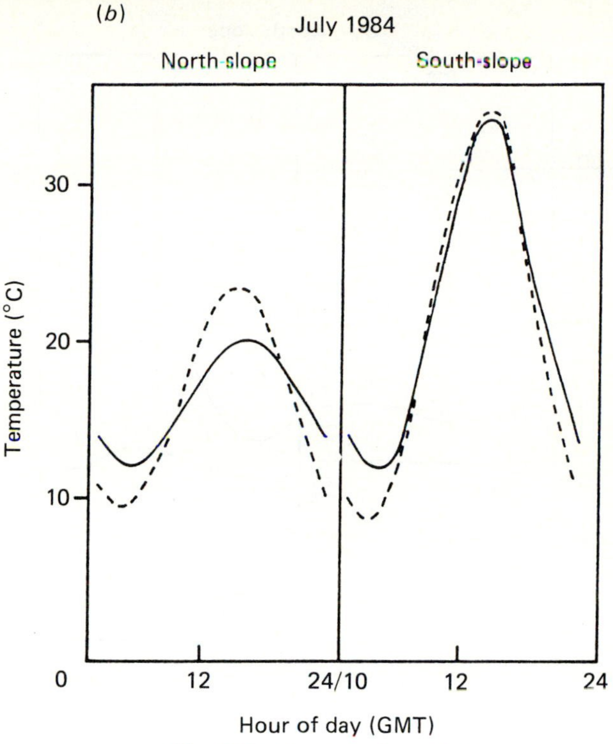

Fig. 5(*b*). For legend see facing page.

gauntlet, and has to be ever aware of, the gap between responses to single factors and the complexity of a managed or semi-natural community. In other words, yield in response to a standard test lacks the sensitivity a test involving temporal and spatial variation provides. An example from investigations of temperature profiles illustrates one important aspect of the problem.

Temperature profiles

We know from the pioneering work reported by Geiger (1965), Larcher (1980) and Monteith & Unsworth (1989) that the temperature of a plant is not the same either all over or all day or every day. Also, inter-specific differences occur in the way that plants respond to the temperature profiles of the soil and air in which they grow, via their phenology and morphology.

In order to quantify such profiles, half-hourly measurement of soil and air temperatures on plateau, north- and south-facing slopes of a Derbyshire Dale have been made by the UCPE for almost twenty years. Situated in the

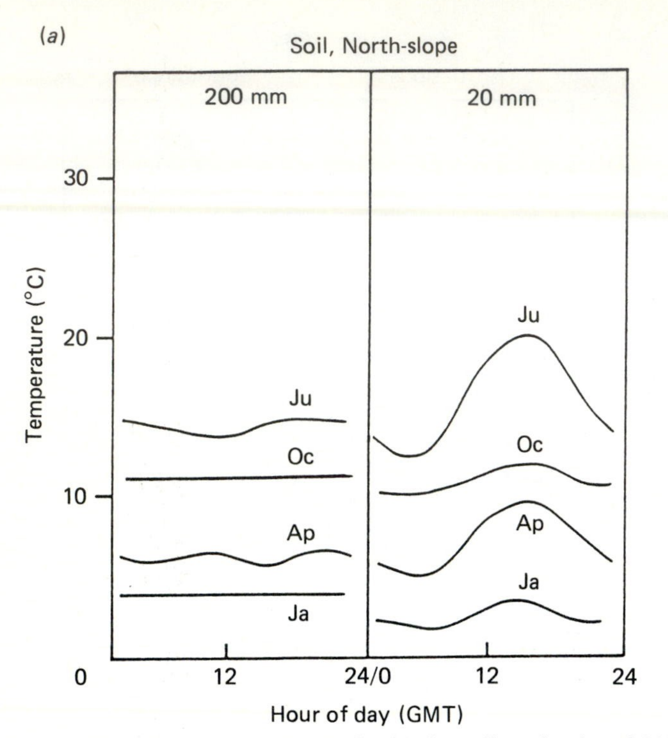

Fig. 6(*a*) Daily temperature cycles in the soil at depths of 200 mm and 20 mm. Records for four individual months in 1984. (*a*) On the N-facing slope of a Derbyshire dale; (*b*) on the S-facing slope. Abbreviations: Ju, July; Oc, October; Ap, April; Ja, January.

Peak District National Park, the long axis of this Carboniferous limestone dale runs roughly east–west. It is reasonably uniform topographically and has areas of recently grazed, semi-natural grassland on both north and south slopes (used in this paper to mean north- and south-facing slopes). In contrast to standard Stevenson screen readings (1.2 m above ground level), they provide a unique set of data from the immediate vicinity of the plant (Rorison, Sutton & Hunt, 1986 and unpublished). They may be used to identify long-term patterns, such as the daily means for July 1984, recorded in four positions, 200 and 20 mm above the soil surface and 20 and 200 mm below. Fig. 5*a* illustrates temperatures at the two extreme positions, 200 mm above and below the soil surface. The upper leaves of plants experienced a 15 °C diurnal fluctuation, while below ground the lowest roots were more buffered about the mean temperature, which differed 2–3 °C between the north and south slopes. Within 20 mm of the surface diurnal variation is very similar in both soil and in air (Fig. 5*b*). This is particularly

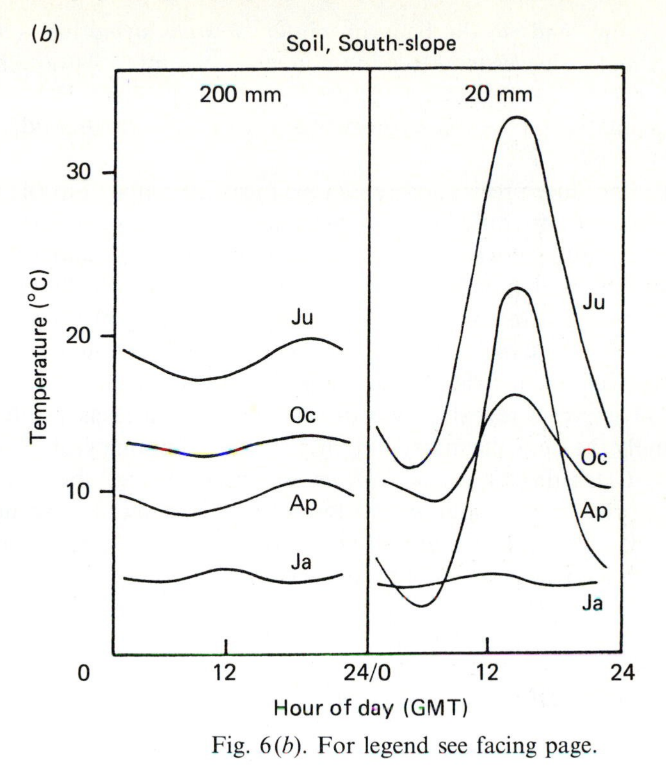

Fig. 6(*b*). For legend see facing page.

marked on the south slope (*c*. 25 °C) where increased direct radiation compared with the north slope resulted in higher maximum temperatures.

Throughout the year temperature also varies with soil depth. Fig. 6*a* and *b* show means for January, April, July and October. At 200 mm there is little diurnal fluctuation and a peak is reached in July. At 20 mm diurnal fluctuations vary from *c*. 20 °C in July to *c*. 1 °C in January.

Temporal and spatial variation

Using these data as guidelines for setting up temperature regimes, we have attempted to simulate field behaviour under controlled conditions.

Phenological evidence of interspecific differences in the onset of shoot growth of native species in spring and throughout the 'growing season' has been available for many years. More recently attempts have been made to quantify phenological patterns of root growth as well as shoot growth (Fitter, 1986; Rorison, Gupta & Spencer, 1983). Of particular interest is the difference in patterns shown by species which co-exist, particularly if they have different rates of growth. Two such species occur together in the Derbyshire Dales. *Arrhenatherum elatius* is potentially fast-growing, deep-rooted and tends to die back in winter; *Festuca ovina* has a slower growth rate and its leaves tend to remain green over winter. Although their native soil is as rich as fertile agricultural soil in total N and P, very little of this is available to the plants growing in it even with the aid of VA mycorrhiza (Birch, 1989).

In order to produce precise and replicated measurements on whole plants and to distinguish any differences in patterns of growth between the two species, an experiment was run for two successive years under natural climatic conditions. Undue heterogeneity and predation in the field was avoided by sowing plants in replicated individual tubes 200 mm high and containing 1 litre of fresh soil. This represented the maximum soil depth and volume likely to be available to each plant in the field.

Sequential harvests revealed two distinct growth patterns resulting in approximately the same mean dry weight per plant after one year (Rorison, Peterkin & Clarkson, 1983). *A. elatius*, sown in spring, showed rapid summer growth, die-back and weight loss of root and shoot in winter but rapid recovery in spring. *F. ovina* showed slower more continuous growth throughout and no net loss of weight over winter. Peterkin (1981) has called this, aptly, the 'tortoise and hare syndrome'. Controlled environment experiments also revealed that die-back in *A. elatius* was not related to short days or low temperatures.

A further factorial experiment with initial additions of N and/or P indicated a significant response to P (Fig. 5 in Rorison, 1987) but not to N. Growth and nutrient uptake (including N) was significantly increased in plants receiving additional P. Even *A. elatius* continued to gain weight over winter in response to an enhancement of P supply (Rorison *et al.*, 1983). Therefore, the limitation on growth was nutritional rather than climatic as in the case of white clover in North Wales (Bradshaw, 1987).

This pattern for *A. elatius* and *F. ovina* was further examined by Rose (1987). In the first year following autumn germination, the roots of both species continued to grow slowly over winter, as did the shoots.

In the second winter, *A. elatius* showed a net loss of shoot but not of root, and no increase in root weight in the following spring until shoot weight had reached its maximum of the previous year. In contrast, *F. ovina* did not show a net loss of shoot in the second winter and both root and shoot growth resumed early in spring. By May of the second year *F. ovina* had achieved a biomass equal to that of *A. elatius* (Fig. 7), with its root growth throughout the profile contributing to the total. This was in contrast to the first stage of establishment when *F. ovina* had the shallower root system and therefore most of its roots were exposed to the low temperatures near the soil surface in the first winter.

One reason for *F. ovina*'s rapid growth response in spring could be the

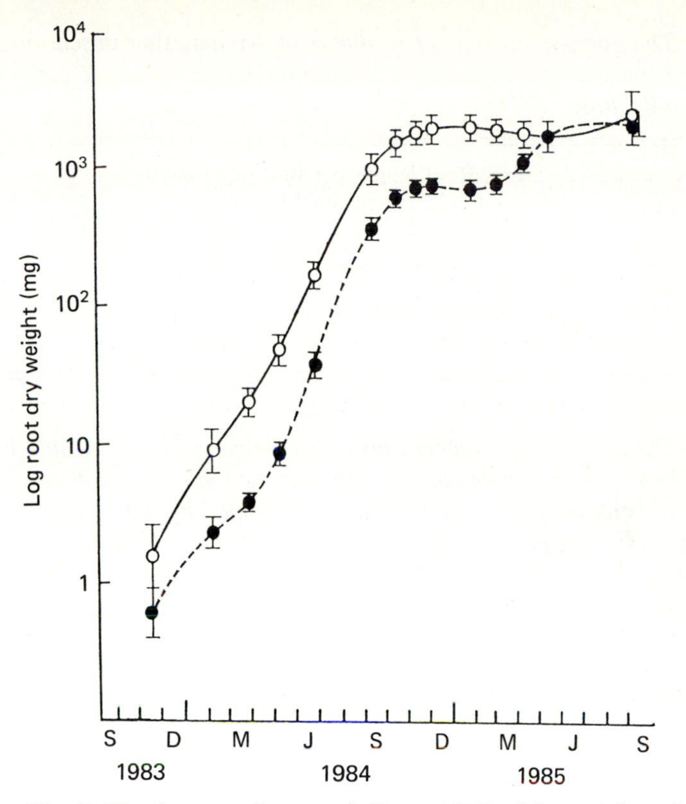

Fig. 7. Fitted curves (Parsons & Hunt, 1981) of dry weights of roots for (○) *Arrhenatherum elatius* and (●) *Festuca ovina* grown from September 1983 to September 1985 under field conditions in 1 litre tubes. Vertical bars represent 95% confidence limits (Rose, 1987).

early warming of its root mass which lay close to the surface. Although this was a smaller system in total weight than that of *A. elatius* it comprised finer roots (Table 2). Thus the difference in area of absorptive surfaces between the two species was smaller than might be implied by their dry weights. P uptake in the upper and lower sections of the tubes varied with root mass and temperature but was too variable to provide detailed patterns. This became possible in a laboratory simulation.

Using mean temperatures recorded in the field, Rose (1987) grew *A. elatius* and *F. ovina* in a nutrient film culture which provided for separate temperature regimes for shoots and for upper and lower root systems. On the premise that temperatures near the soil surface fall in autumn more rapidly than those further down the profile, the upper half of each root

Table 2. *The fineness of roots of seedlings of* Arrhenatherum elatius *and* Festuca ovina *growing in the upper and lower halves of 1 litre capacity tubes of soil (Rose, 1987)*

	Root length per unit dry weight (cm mg^{-1} (s.e.))	
	Upper	Lower
A. elatius	13.4 (1.1)	10.2 (0.8)
F. ovina	27.1 (1.4)	44.3 (4.8)

Table 3. *The phosphate contents (μg) of seedlings of* A. elatius *and* F. ovina *grown with their roots in four different temperature profiles using nutrient film culture for 12 days. Figures not joined by horizontal underlining are significantly different ($p < 0.05$) (Rose, 1987).*

Temperature of profile (°C)		Upper / Lower	3 / 3	3 / 10	10 / 3	10 / 10
Phosphate (μg)	Initial		After 12 days			
A. elatius	124		154	283	390	465
F. ovina	77		74	119	144	165

system was maintained at 3 °C and the lower at 10 °C. For 'spring' conditions the temperature regimes were reversed. Air temperatures were maintained at 15 °C day and 5 °C night throughout.

In all treatments dry matter accumulation by *A. elatius* was greater than that by *F. ovina* and this reflected their relative responses to temperature. The growth of both species was strongly inhibited by a root temperature of 3 °C. The upper root temperature was more influential for total plant growth and phosphate uptake (Table 3). At overall root temperatures of 3 °C phosphate uptake virtually ceased in both species (Table 3). Root extension was greater in *F. ovina* than *A. elatius* due to the production of a finer root system (*cf.* Table 2). Both species showed compensatory growth in length and root number in the warm, upper root when only the lower root was cold. In this treatment *F. ovina* responded more markedly than *A. elatius*. No compensatory growth occurred with temperatures reversed. Shoot growth was maintained in all cases except for *A. elatius* with cold upper roots, in which case there was no detectable growth of shoots.

The response to warming of the upper root system enabled growth to begin early in spring and the more marked response by *F. ovina* suggests several reasons why its growing season is longer than that of *A. elatius* in the field. This is because its growth is relatively insensitive to low temperature and permits continuous slow growth in winter, with an increased expansion of the root system in the top of the profile in spring. Also, there is an increase in the specific absorption rate in warm roots.

These characteristics, combined with a predominantly shallow root system allow an early increase in spring growth. There is less evidence for characteristics to prolong growth in autumn. The faster-growing *A. elatius* showed less compensatory growth in early spring, mainly because it was less responsive to initial rises in surface temperatures. In this way the slow-growing species is potentially able to take up nutrients when competition from faster-growing species is less intense. Thus, at least in moderately stressed environments they are able to co-exist.

Nutrients, wider implications

A study of wild plants indicates that there is no universal minimum requirement of nutrients for growth (Bradshaw *et al.*, 1958; 1964; Rorison, 1968). Confirmation for this comes from time studies involving both depletion and enrichment.

Depletion

Recent work of Kachi & Rorison (1990) illustrates the reduction in relative growth rate (RGR) of the potentially fast-growing species *Holcus lanatus* below that of the slow-growing *F. ovina*, if the external supply of nutrients was sufficiently reduced. In this instance, after seven days in a nutrient solution of adequate concentration, seedlings were placed in a solution of twenty-fold dilution (*c.* 20 mmol m^{-3}) for a further 21 days. Fig. 8 illustrates the shift in RGR with time for the two species at the two temperature regimes. After *c.* 14 days depletion of the RGR of *H. lanatus* was reduced to that of *F. ovina* and fell significantly below it within a further seven days. In part, this reflected the time taken to achieve an optimal root:shoot ratio (Kachi & Rorison, 1989) in response to a depleted supply.

Enrichment

In a longer and more detailed study of pulse exploitation, Campbell & Grime (1989) compared nitrogen capture and dry matter production in *A. elatius* ssp. *bulbosus* and *F. ovina*. This solution culture experiment extended over 42 days and involved ten treatments, in which seven episodes of

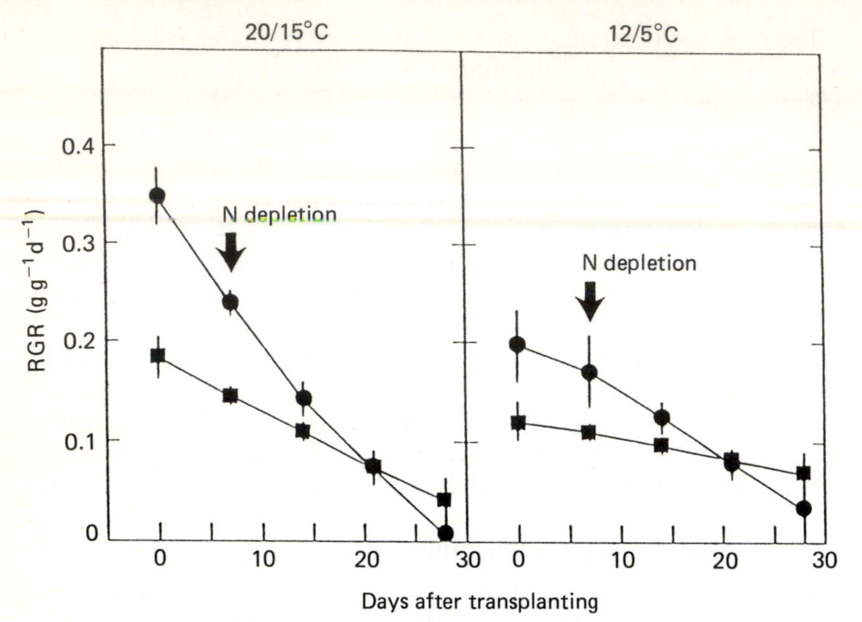

Fig. 8. Relative growth rates (RGR) of *Holcus lanatus* (●) and *Festuca ovina* (■) subjected to nutrient depletion at warm 20/15 °C and cool 12/5 °C day/temperatures. Vertical bars indicate 95 % confidence limits of each mean (Kachi & Rorison, 1990).

mineral nutrient enrichment, applied to the whole root system during successive six-day cycles throughout the experiment, varied in duration from 60 seconds to six days.

Their results (Fig. 4 in Grime, 1991, this volume) expressed as relative growth rates and specific N-absorption rates, revealed, in accordance with its confinement in the field to infertile soils, that the performance of *F. ovina* was superior to that of the potentially fast-growing *A. elatius* in treatments which provided brief pulses (0.1–10 h duration). Above this range, the effect of increasing the length of the enrichment episode was to switch the advantage sharply in favour of *A. elatius*.

Visual inspection of the plants of *A. elatius* exposed to brief nutrient pulses revealed symptoms of nitrogen deficiency in the shoots and prematurely senescent foliage. Shoot tissue turnover rates under nutrition stress were clearly higher in *A. elatius* than in *F. ovina* and it seems reasonable to assume that there was also faster deterioration of *A. elatius* roots resulting in lower specific N-absorption rates. This interpretation is consistent with the theoretical prediction that a capacity to maintain viable tissues during periods of chronically low nutrient availability is an important component of the survival mechanism of plant species attuned to nutrient-poor soils. As we have seen earlier, this capacity is also valuable in the over-wintering of slow-growing species.

Conclusions

Ecophysiological experiments comparing the response of different species can help us to refine our ideas about the role of nutrients in the survival of plants. Nothing is static. There may be pulses of supply in spring and autumn and there is also likely to be heterogeneity in the placement and movement of individual nutrients in soil profiles, particularly uncultivated ones.

Plants respond in different ways. At a first approximation these responses may be grouped according to whether the plants are fast-growing as individuals and/or potentially dominant in a fertile community, or slow-growing and potentially dominant in an infertile habitat.

Some slow-growing species have been shown to be not only better husbanders of resources, but also more efficient in their utilization and re-distribution of resources, than faster-growing species (Robinson, 1991, this volume). Efficiency is manifest morphologically in the growth of fine roots and root hairs 'where it matters', physiologically in the greater flexibility of specific absorption rates and phenologically in extending the season of growth.

In contrast, fast-growing species are seen as gross feeders lacking the flexibility to cope when supplies are depleted below a certain level. Such are the guidelines suggested by results from autecological studies. Their implementation and implications for community ecology are beyond the scope of the present chapter and are dealt with by Grime (1991, this volume).

Acknowledgements

I am indebted to J.C. Rose for permission to use unpublished material.

References

Birch, C.P.D. (1989). *Relationships between Nutrient Concentration Mycorrhizal Infection and Plant Growth in Semi-natural Grassland.* Ph.D. thesis, University of Sheffield.

Bradshaw, A.D. (1987). Comparison – its scope and limits. *New Phytologist*, **106**, 3–21.

Bradshaw, A.D., Lodge, R.W., Jowett, D. & Chadwick, M.J. (1958). Experimental investigations into the mineral nutrition of several grass species. 1. Calcium level. *Journal of Ecology*, **46**, 749–57.

Bradshaw, A.D., Chadwick, M.J., Jowett, D. & Snaydon, R.W. (1964). Experimental investigations into the mineral nutrition of several grass species. IV. Nitrogen level. *Journal of Ecology*, **52**, 665–76.

Campbell, B.D. & Grime, J.P. (1989). A comparative study of plant responsiveness to the duration of episodes of mineral nutrient enrichment. *New Phytologist*, **112**, 261–7.

Ellenberg, H. (1974). *Scripta Geobotanica. Indicator values of vascular plants in Central Europe*. Gottingen: Verlag Erick Goltze KG.

Fitter, A.H. (1986). Spatial and temporal patterns of root activity in a species-rich alluvial grassland. *Oecologia (Berlin)*, **69**, 594 – 9.

Geiger, R. (1965). *The Climate near the Ground.* Cambridge, Mass.: Harvard University Press.

Grime, J.P. (1991). Nutrition, environment and plant ecology: an overview. In *Plant Growth: Interactions with Nutrition and Environment*, eds. J.R. Porter & D.W. Lawlor, Society for Experimental Biology, Seminar Series 43, pp. 249–67. Cambridge University Press.

Grime, J.P., Hodgson, J.G. & Hunt, R. (1988). *Comparative Plant Ecology: A Functional Approach to Common British Species*. London: Unwin Hyman.

Groot, J.R.R. & Spiertz, J.H.J. (1991). The role of nitrogen in yield formation and achievement of quality standards in cereals. In *Plant Growth: Interactions with Nutrition and Environment*, eds. J.R. Porter & D.W. Lawlor, Society for Experimental Biology, Seminar Series 43, pp. 227–47. Cambridge University Press.

Hewitt, E.J. (1966). *Sand and Water Culture Methods Used in the Study of Plant Nutrition*, 2nd edn. Technical Communication No. 22, Commonwealth Agricultural Bureaux, Farnham Royal.

Kachi, N. & Rorison, I.H. (1989). Optimal partitioning between root and shoot in plants with contrasted growth rates in response to nitrogen availability and temperature. *Functional Ecology*, **3**, 549–59.

Kachi, H. & Rorison, I.H. (1990). Effects of nutrient depletion on growth and on the capacity of roots to absorb nitrogen in *Holcus lanatus* and *Festuca ovina* at warm and cool temperatures. *New Phytologist* **115**, 531–7.

Larcher, W. (1980). *Physiological Plant Ecology*, 2nd edn. Berlin: Springer-Verlag.

Lee, J.A. & Stewart, G.R. (1978). Ecological aspects of nitrogen assimilation. *Advances in Botanical Research*, **6**, 1–42.

Marshall, B. & Porter, J.R. (1991). Concepts of nutritional and environmental interactions determining plant productivity. In *Plant Growth: Interactions with Nutrition and Environment*, eds. J.R. Porter & D.W. Lawlor, Society for Experimental Biology, Seminar Series 43, pp. 99–124. Cambridge University Press.

Monteith, J.L. & Unsworth, M. (1989). *Principles of Environmental Physics*, 2nd edn. London: Edward Arnold.

Parsons, I.T. & Hunt, R. (1981). Plant growth analysis: a program for the fitting of lengthy series of data by the method of B-splines. *Annals of Botany*, **48**, 341–52.

Peterkin, J.H. (1981). *Plant Growth and Nitrogen Nutrition in Relation to Temperature*. Ph.D. Thesis, University of Sheffield.

Porter, J.R., Klepper, B. & Belford, R.K. (1986). A model (WHTROOT) which synchronizes root growth and development with shoot development for winter wheat. *Plant and Soil*, **92**, 133–145.

Robinson, D. (1991). Strategies for optimising growth in response to nutrient supply. In *Plant Growth: Interaction with Nutrition and Environment*, eds. J.R. Porter & D.W. Lawlor, Society for Experimental Biology, Seminar Series 43, pp. 177–205. Cambridge University Press.

Rorison, I.H. (1968). The response to phosphorus of some ecologically distinct plant species. *New Phytologist*, **67**, 913–23.

Rorison, I.H. (ed.) (1969). *Ecological Aspects of the Mineral Nutrition of Plants*, BES Symposium No. 9. Oxford: Blackwell Scientific Publications.

Rorison, I.H. (1987). Mineral nutrition in time and space. *New Phytologist*, **106** (Suppl.), 79–92.

Rorison, I.H., Gupta, P.L. & Spencer, R.E. (1983). Nutrient balance and soil exhaustion in relation to plant growth and uptake by contrasted species. In *Annual Report of the Unit of Comparative Plant Ecology (NERC) 1983*, pp. 26–7. University of Sheffield.

Rorison, I.H., Peterkin, J.H. & Clarkson, D.T. (1983). Nitrogen source, temperature and the growth of herbaceous plants. In *Nitrogen as an Ecological Factor*, ed. J.A. Lee, S. McNeill & I.H. Rorison, pp. 189–209. Oxford: Blackwell Scientific Publications.

Rorison, I.H. & Robinson, D. (1984). Calcium as an environmental variable. *Plant, Cell and Environment*, **7**, 381–90.

Rorison, I.H., Sutton, F. & Hunt, R. (1986). Local climate, topography and plant growth in Lathkill Dale NNR. I. A twelve-year summary of solar radiation and temperature. *Plant, Cell and Environment*, **9**, 49–56.
Rose, J.C. (1987). *The Comparative Physiology of Grasses in Response to Climatic Gradients*. Ph.D. Thesis, University of Sheffield.
Shipley, B. & Keddy, P.A. (1988). The relationship between relative growth rate and sensitivity to nutrient stress in twenty-eight species of emergent macrophytes. *Journal of Ecology*, **76**, 1101–10.
Wild, A., Jones, L.H.P. & Macduff, J.H. (1987). Uptake of mineral nutrients and crop growth: the use of flowing nutrient solutions. *Advances in Agronomy*, **41**, 171–219.

DAVID ROBINSON

Strategies for optimising growth in response to nutrient supply

Introduction: genes, phenotypes and strategies

As anyone who has tried to grow 'uniform' plant material will know, plants are extremely variable in many of their physiological and morphological features: they exhibit *phenotypic plasticity* in response to their environment (Bradshaw & Hardwick, 1989; Schlichting, 1989). This makes it difficult to relate phenotypic expression in plants to their potential reproductive success, an essential step if the reasons for the evolution of certain phenotypic characters are to be found. The following hypothetical example illustrates the problem.

Suppose that in each of the two environments, E_1 and E_2, plants with long unbranched roots each leave, on average, 10 and 2 descendants, respectively; the corresponding numbers of descendants of plants with short branched roots are 6 and 8, say. In E_1, long unbranched roots are apparently more suitable for plant growth and reproduction than short branched ones; in E_2, the converse is true. To translate this vague idea of 'suitability' into genetic terms, we need to know how the plants' phenotypes are expressed. If plants with long unbranched roots could produce no other kind, genes coding for this character would gradually increase in frequency in the population growing in E_1, but would decline in E_2. Similarly, if plants with short branched roots produced only this type, genes coding for this character would increase in frequency in E_2 and fall in E_1 (Kimura & Ohta, 1971; Harper, 1977). If, however, the same plants produced long unbranched roots in E_1 but short branched roots in E_2, it would not be possible to say how gene frequencies would change in relation to phenotypic expression as genes 'for' both phenotypes would be carried by the same individuals.

A similar problem has been addressed in studies of certain behaviour patterns in animals and of their potential contribution to individuals' reproductive success. It is often more informative to think not of genes coding for specific characters of behaviour, but coding for *sets* of characters.

Such sets are known as 'strategies' (Levins, 1968; Maynard Smith, 1982; Dawkins, 1982). One strategy can encompass a single character or a multiplicity of characters (Levins, 1968). 'Behaviour' in animals can be considered analogous to 'phenotypic plasticity' in plants (Tomlinson, 1982; Bradshaw & Hardwick, 1989). This analogy is particularly apt for the obvious parallel between the foraging strategies of animals as they search for food (Stephens & Krebs, 1987) and the growth of leaves and roots as they acquire resources from the environment (Grime, Crick & Rincon, 1986).

The most successful strategies are those which become and remain most frequent in a population. Success depends strongly on the physico-chemical peculiarities of the local habitat and on the biological peculiarities of the organisms in it (Givnish, 1986*a*). It is important to note that there is no reason for assuming that a phenotypic response to a certain environmental factor will always make a positive contribution to reproductive success (Partridge & Harvey, 1988). Experiments in which no apparent relation was found between phenotypic expression in plants and environmental variation have been reported by Boutin & Morisset (1988) and Taylor & Aarssen (1988), for *Chrysanthemum leucanthemum* and *Agropyron repens*, respectively.

In principle, there is no reason why plants' phenotypic responses to their environment should not be treated as strategies and their influence on potential reproductive success analysed on a basis similar to that used in studies of animal behaviour. In the hypothetical example described above, at least four possible strategies can be identified. The first two are *pure* strategies: 'long unbranched roots' and 'short branched roots'; here 'strategy' is the same as 'phenotype' (Maynard Smith, 1982) as no plasticity occurs in response to the environment. In contrast, two *mixed* or *conditional* strategies exist in which phenotypic expression depends on the environment: 'if E_1, then long unbranched roots; if E_2 then short branched roots' and the reverse of this: 'if E_1, then short unbranched roots; if E_2, then long unbranched roots'. Imaginary strategies which are not actually expressed may be specified – the combinations of long branched roots and short unbranched roots, for example – to test their possible effects on reproductive success in comparison with 'real' strategies, using the techniques of game theory (Maynard Smith, 1982). It is possible to define or identify a strategy without assuming that it has any particular effect (positive, negative or neutral) on reproduction. Phenotypic plasticity means that plants can have many mixed or conditional strategies with which to respond facultatively to non-uniform environments. As Bradshaw & Hardwick (1989) noted, even though such responses are facultative, the response systems themselves are constitutive.

This type of approach has, however, been applied only rarely to plants. Plant biologists have generally taken another route to the study of phenotypic strategies. Significant correlations, based on data from many experiments and field observations, have been found between certain groups of phenotypic characters and environments. These groups have been identified as strategies (Grime, 1979, and this volume; Calow & Townsend, 1981*a*; Grime, Hodgson & Hunt, 1988; Southwood, 1988). This invaluable approach allows strategies to be identified statistically, in effect, without the need to analyse precisely *how* a certain phenotypic response functions and contributes to potential reproductive success (Stephens & Krebs, 1987). Instead, such contributions are inferred from the presence or absence of a certain phenotype in different taxa occurring in a similar environment, compared with its presence or absence in similar taxa in different environments; this might be termed the 'phylogenetic' approach to the study of biological strategies (Hailman, 1988). However, precise analyses of phenotypic form and function are needed. Otherwise, to return to our example, it is not possible to explain *why* plants with long unbranched roots are more common in E_1 than in E_2, nor why the combinations of long branched roots and short unbranched roots are apparently not possible components of any strategy in either environment. Such explanations – obtained from a 'functional' approach rather than a phylogenetic one – are important agriculturally. Genes are amenable to artificial manipulation at the molecular level to produce agronomically useful phenotypes, but how can 'useful' phenotypes be defined *a priori* without a 'functional' understanding of their operation? This question becomes especially relevant if the environments in which future genotypes will have to grow will be radically different from those under which their ancestors – both wild and cultivated – were selected (Crawford *et al*., 1989). 'Functional' explanations are also important to biology as a whole because they offer answers to what Williams (1966) called 'The central biological problem ... not survival as such, but design for survival.'

One of the main challenges in biology is to find generally applicable principles to explain biological phenomena and to reconcile these with the diversity of such phenomena. The generality of theories, based on genetic models, seeking to explain the *evolution* of phenotypic strategies (Via & Lande, 1985; Wagner, 1988) contrasts with the specificity of the many reports describing the *operation* of particular strategies. Comparatively little progress has been made in the middle of this continuum which remains largely *terra incognita* (McGraw & Wulff, 1983; Roughgarden, May & Levin, 1989). The main theme of this chapter is to consider in general how the operation of phenotypic responses of plants to their environment can influence their potential reproductive success.

Growth rates and potential reproductive success

The functional links between physiological processes that are the components of a plant's strategies and the plant's ultimate reproductive success are long, tenuous and uncertain, especially in natural communities. Reproductive success in wild plants depends as much on external factors – the occurrence of gaps in the vegetation or of resource inputs, the destruction of a dominant competitor or pathogen, for example – as it does on their phenotypic responses to their environment (Harper, 1977). It is sometimes more useful and expedient to consider instead the influence of physiological processes on *potential* reproductive success, provided that the latter can be represented by a suitable phenotypic indicator or analogue, and that the important distinction between *potential* and *actual* success is recognised.

The growth rates of individual organisms are often used as phenotypic indicators of potential reproductive success (Calow & Townsend, 1981*a*; McGraw & Wulff, 1983). Throughout this chapter, 'growth' is used to mean an increase in dry weight of whole plants and which results, ultimately, from meristematic activity. This distinguishes 'growth' from 'reproduction' – the production of a new individual from a single cell originating sexually or asexually (see Harper, 1977; Dawkins, 1982). 'Growth rate' is used synonymously with *relative* growth rate, RGR (Hunt, 1982*a*). RGR is preferred over absolute growth rate as it accounts for differences in individuals' sizes (Hunt, 1982*a*) and it 'provides a convenient integration of the combined performances of the various parts of the plant' without, however, revealing anything about the 'causal processes which contribute to the plants' gross performance' (Hunt, 1982*a*). Inter-specific comparisons between RGR are central to phylogenetic definitions of phenotypic strategies (see above). The limited *explanatory* power of measurements of the net or realised performances of whole plants, of which RGR is the most useful indicator, reflects the somewhat 'dead-end' nature of a purely phylogenetic approach to strategy theory. This is one reason why pluralistic approaches that incorporate mechanistic or functional models or physiological processes are needed.

Givnish (1986*a*) justified the use of individuals' growth rates as indicators of potential reproductive success: 'Natural selection should favour plants whose form and physiology tend to maximise their net rate of carbon gain in a particular competitive context, because such plants generally have the most resources with which to reproduce and compete for additional space'; essentially the same statement can be found in Calow & Townsend (1981*b*) and Tilman (1988). In other words, a strategy allowing plants to grow faster than others has, on average, a better chance of success than one which constrains plants to grow relatively slowly. How can this be reconciled with the fact that plants with widely different potential rates of growth have evolved and can co-exist?

Givnish's statement contains an important qualifier: 'in a particular competitive context'. To understand Givnish's argument it is necessary to consider first a simple 'competitive context': a local population of even-aged plants growing as a monoculture. In this population, variation in potential growth rate will exist, as it does for many phenotypic characters. If the environment is stable and predictable, plants which can grow faster than others will gain access more quickly to whatever resources are available, thereby depriving competitors of the opportunity of obtaining those resources. More rapid vegetative growth can also restrict the growth of competitors by means of shading or inter-root competition. The capacity of plants to grow faster than their neighbours might be expected to evolve as a consequence of a type of 'arms race' (Dawkins, 1982; Westoby, 1984). The tendency for plants to evolve even faster potential growth rates will be constrained within the bounds of physiological reality by conflicting selection pressures: the need to allocate some resources to storage, maintenance or defence, rather than to immediate growth, for example. In such a population, therefore, one might expect the frequency distribution of potential growth rates to be biased towards the upper extreme.

The only detailed intra-population survey of potential growth rates is that by Burdon & Harper (1980). They measured RGRs of individuals of *Trifolium repens* (white clover) from a population in a permanent grassland. The frequency distribution of RGR (Fig. 1) shows that the modal RGR of the sample was significantly closer to the upper extreme of the range than the lower. This seems to support the idea that RGR tends to be maximised in a population. However, as no other data sets are available to confirm this, and as the dynamics of growth, mortality, size- and age-distribution are exceedingly complex, even in monocultures of even-aged plants (Silvertown, 1982; Benjamin & Hardwick, 1986), some caution is perhaps required in the interpretation of Burdon & Harper's data.

There is another reason why relatively fast growth rates should be selectively favoured within a local population. In many species, a certain 'threshold' size (or age) must be reached by the parent before resources can be allocated to reproduction (Fig. 2) (Harper, 1977; McGraw & Wulff, 1983; Lacey, 1986; Weiner, 1988). There are some exceptions to this, notably in herbaceous weeds. *Chenopodium rubrum*, for example, can flower immediately after seed germination, in a short-day photoperiod (Wareing & Phillips, 1970). This threshold, and the time taken to reach it, varies widely between species and between habitats, depending on the predominant environmental factors influencing plant growth; many examples

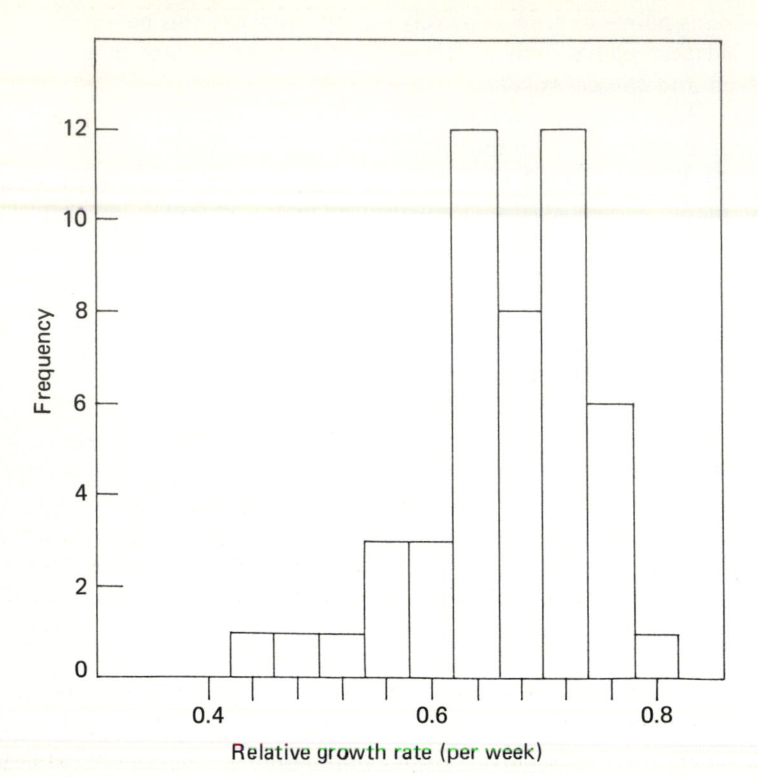

Fig. 1. The frequency distribution of relative growth rate in a population sample of *Trifolium repens* (after Burdon & Harper, 1980).

are given by Harper (1977), Grime (1979), Lacey (1986) and Grime *et al.* (1986). Within a local population, however, more rapid vegetative growth to maximise opportunities for resource acquisition would tend to allow any threshold for effective reproduction to be attained sooner than in comparable individuals with a slower potential rate of growth. In addition, the total amount of resources allocated to reproduction would be greater in faster growing plants, increasing by simple allometry with plant size (Fig. 2).

It is surprisingly difficult, however, to find firm evidence in the literature of a direct relation between a high vegetative growth rate and a correspondingly higher seed production, for example, but a recent experiment (D. Robinson, D.J. Linehan, S. Caul & D. McRae, unpublished) does provide circumstantial evidence for such a link. Spring wheat (*Triticum aestivum* cv. Wembley) grown in pots without nitrogen fertiliser had an initial RGR of 0.032 per day; the corresponding RGR of plants

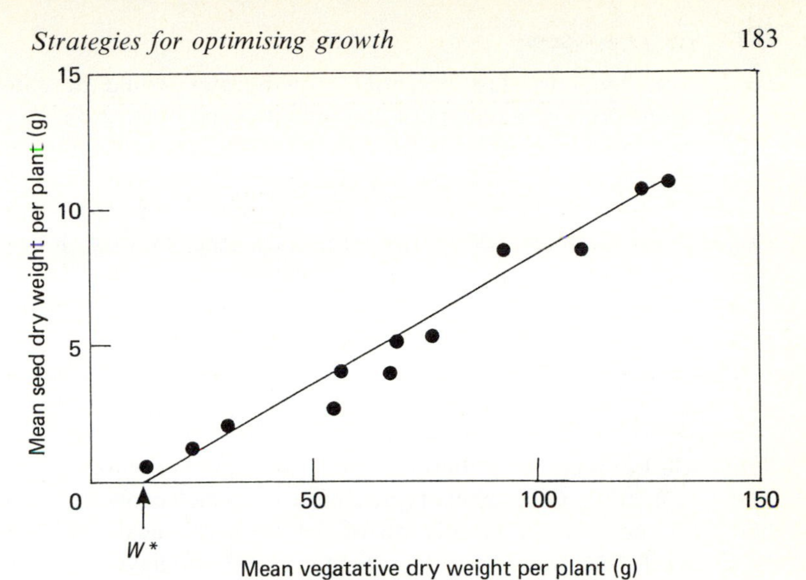

Fig. 2. Seed production in relation to plant weight in the monocarpic herb *Verbascum thapsus* (Common mullein), showing the threshold weight, W^*, for reproduction; $r^2 = 0.96$ (after Weiner, 1988; original data of J.A. Reinartz).

with nitrogen fertiliser was 0.072 per day, a 2.3-fold difference in response to N. RGRs in both treatments declined linearly with time during the experiment. The difference between treatments of the rate of change in RGR was 2.4-fold. The final yields of ears were 0.485 g and 1.12 g per pot without and with nitrogen fertiliser, respectively, a difference of 2.3-fold. The close correspondence between the differences in vegetative growth rates and in final reproductive output (as approximated by ear yield) induced by the availability of nitrogen is striking, but it does not prove a direct *functional* link between the two, and more explicit data are needed. Note that a high reproductive output *per se* does not guarantee reproductive success: some of the rarest species are also some of the most fecund (Harper, 1977).

A fast growth rate would *not* be an advantage in certain competitive contexts. Suppose that, at a given time, the largest plants in an even-aged monoculture (i.e. those with potentially rapid growth rates) were destroyed before reproductive growth could begin (Schlichting, 1989). If this selective mortality was relatively infrequent, plants with slower-than-average growth rates would be able to reproduce perhaps with greater success than in the presence of their faster growing neighbours. This would allow a range of potential growth rates to persist in the population, as in Burdon & Harper's sample of *T. repens* (Fig. 1). If, however, such mortality was a regular occurrence, genes for fast potential growth rates would be eliminated progressively from the gene pool, and plants capable of growing rapidly would disappear from the population. Over time, the maximum potential growth rate of individuals in the population would become that which had previously been at the lower end of the range (cf. Fig. 1, for example). Conversely, Westoby (1988) discusses how the average RGR of a population can be *increased* by selective mortality.

Even this, in many ways, is too simple a picture. To take herbivory as an exemplary agent of selective mortality: the grazing of foliage by birds or mammals can produce rapid changes in the frequencies in a population of plants of different growth habit, favouring prostrate forms over erect ones (Harper, 1977; Grime, 1979; McNaughton, 1983; Tilman, 1988; Westoby, 1988), without necessarily altering the distribution of potential growth rates in the population. Other phenotypic characters which evolve in response to herbivory, notably the production of defensive chemicals and structures (Janzen, 1981; McNaughton, 1983; Schultz, 1988), do have implications for individuals' potential growth rates (Coley, 1987; Fagerstrom, Larsson & Tenow, 1987). Large herbivores can influence nutrient cycling and availability in any ecosystem through excretal returns (Westoby, 1988). The precise influences that herbivores have on plant populations depends on the type of herbivory, whether the animals feed on leaves, stems, roots, seeds, fruits or flowers and the extent of the damage inflicted at any one time (McNaughton, 1983). It is difficult to predict how the combination of these factors will influence the success of alternative strategies.

When more realistic communities are considered, those comprising many species and age distributions of plants, it becomes more difficult to see how selection might favour relatively fast rates of growth *within populations*. It is important to remember that potential growth rate is just one phenotypic component of any strategy influencing potential reproductive success. In 'a particular competitive context', members of the community that 'opt out' of an inter-specific arms-race towards faster potential rates of growth (Tilman, 1988) might be as reproductively successful as those which have evolved to compete in such a race. Instead, modifications to their phenology might confer selective advantages, as might a shift towards longer life-spans. These phenotypes might allow the maximum potential growth rate in populations to be *reduced* without compromising their members' potential reproductive success. Indeed, some reduction in maximum potential growth rate would be inevitable if certain strategies evolved in which greater amounts of resources were allocated to defence, storage or maintenance, as opposed to their being used immediately in growth (Coley, 1987; Fagerstrom *et al.*, 1987). In any event, the net result is the possibility of minimising direct competition for resources within a community, at least between plants which have evolved widely different strategies. Even so, *within* each component population of such a community, the aforementioned selective advantages associated with growing *relatively* faster would still apply, even if the maximum potential growth rates *between* populations in the community were very different: some competition for resources (including pollinators, establishment sites, symbionts, etc.) between neighbouring plants with similar strategies could not be avoided, unless individuals were effectively isolated from each other.

If this interpretation is correct, it answers (for plants, at least) the question posed by Calow & Townsend (1981*b*): 'Is growth rate maximised or optimised?' In general, growth rates tend to be maximised during ontogeny; potential growth rates tend to be optimised during phylogeny. Physiological features have, in general, evolved such that a plant's growth rate is maximised as far as the prevailing environmental conditions allow. The *potential* rate at which a plant could grow will undoubtedly change during its ontogeny as its proportion of non-photosynthetic tissue increases with age; generally, the potential rate should decline over time. Perhaps it is more accurate to say that natural selection tends to optimise the *ontogeny* of the potential rate of growth. It is surprising that no one has tried to explain mechanistically the complex progressions of whole-plant growth rates such as those measured by Hunt (1982*b*), for example (Fig. 3).

'Optimisation' in this context is a relative term. No implication is intended of any 'long-distance target of perfection to serve as a criterion of selection... the criterion of selection is always short-term, either simple survival, or more generally, reproductive success' (Dawkins, 1986). If the attainment of a certain growth rate significantly improves (directly or indirectly) actual reproductive success, individuals not capable of attaining that rate will probably become less frequent in the population, other things being equal. Similarly, any plants that are physiologically capable of growing far faster than this rate may be selectively disadvantaged because of lower investment of resources into anti-herbivore defences, for example. These plants would, too, decline in frequency in a population subjected to grazing. By default, therefore, the illusion is given of an optimisation process, one which in any case operates under the constraints of limited genetic variation in a population at any given time, genetic linkage, time-lags, the availability of totipotent cells and essential resources, allometry, phylogeny and chance (see Levins, 1968; Niklas, 1976; Calow & Townsend, 1981*a*; Givnish, 1986*a*; Dawkins, 1986; Trewavas, 1986*a*; Stephens & Krebs, 1987). Theoretically, in an ideal steady-state population and stable environment, every individual should eventually come to have the same potential rate of growth. In reality, this could never happen. Some variation in maximum potential growth rate is maintained in a population, as seen in

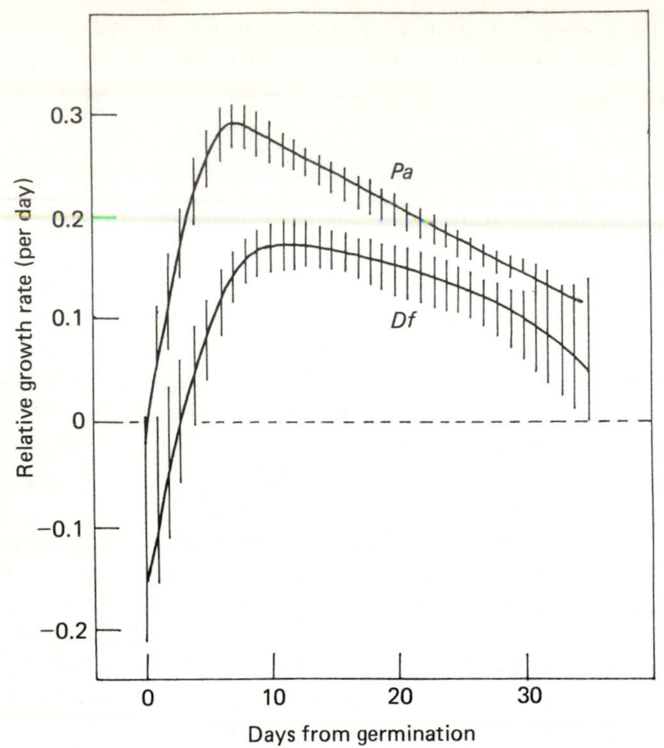

Fig. 3. Progressions of relative growth rate in seedlings of the grasses *Deschampsia flexuosa* (*Df*) and *Poa annua* (*Pa*). Plants were grown in a controlled environment. Curves are derived from splines fitted to primary data of daily harvests; bars represent 95% confidence limits (after Hunt, 1982*b*).

Fig. 1. If reproductive success was determined entirely by factors beyond genetic influence – as in an artificial programme of random selection, as opposed to one using some phenotypic criterion – it is highly improbable that any phenotypic features would show evidence of having been 'optimised' by the selection process (Dawkins, 1986).

To summarise, Givnish's principle that individuals should maximise their rates of growth is generally applicable, even within communities containing co-existing species of plants capable of growing at very different maximum rates. The capacity to grow as fast as the local physical, chemical and biotic environments allow is likely to form part of any individual's phenotypic strategy, irrespective of species or location. So, it is important to relate observed phenotypic responses to individuals' growth rates if the likely functional significance of these responses is to be found. The remainder of this chapter will consider the extent to which phenotypically plastic responses to shortages of essential nutrients can help to maximise relative growth rates and, in so doing, contribute to potential reproductive success.

A 'balanced' approach to phenotypic plasticity

Little is known about the physiological mechanisms inducing phenotypic plasticity; less is known about the mechanisms that co-ordinate phenotypic change in a single plant. One reason for this is that not enough is known about the molecular basis of the response system (Bradshaw & Hardwick, 1989) although a number of significant advances have been made in this area (Cullis, 1984; McClintock, 1984; Wessler, 1988; Benfey, Ren & Chua, 1989). Another reason is that no theoretical framework is available that can be used to interpret experimental measurements of the phenotypic responses of whole plants in ways that, ultimately, make evolutionary sense (Hardwick, 1983; 1986; McGraw & Wulff, 1983).

The ideal of such a theory would relate reproductive success to certain (preferably simple) 'rules' of plant *construction* and of physiological *operation* that underlie plants' phenotypic strategies. A large body of literature covers one or the other of these aspects. Tomlinson (1982), Givnish (1986*b*), Hardwick (1986), Green (1987), Niklas (1988) and Crawford & Young (1990) have considered some of the rules governing the architecture of shoot systems, while Fitter (1987) is one of the few to have done the same for root systems. Perhaps the most successful analyses of rules relating to the operation of physiological processes are those by Cowan (1982) on the role of stomata in balancing carbon gain by, and water loss from, single leaves. Niklas (1988) is probably alone in incorporating elements of both constructional and operational rules in his models of the evolution of shoot architecture in response to the dual constraints of maximising photosynthetic carbon gain by a canopy and minimising the carbon costs of maintaining the structural stability of that canopy. It will probably be some time before these, or similar, models include the activities of roots and the implications for potential reproductive success.

The inter-dependent activities of roots, shoots and their components, have their own extensive theoretical literature. Despite this work, few generally applicable rules have emerged to account for the phenotypic responses of plants to changes in their above- and below-ground environments. As argued in the previous section, a general 'rule' that might be a useful starting point here is the maximisation of whole-plant growth rate. Related to this is the idea of 'balanced' growth between the root and shoot systems.

'Balanced' growth is the condition in which changes in *either* the above- *or* the below-ground environments lead to changes in the plant's growth rate and, usually, in other phenotypic characters also. In other words, the plant's growth rate is limited jointly by the activities of both the shoot and root systems. This is not a new idea: explicit reference to it can be found in the work of H.L. White (cited in Trewavas, 1986*a*), R. Brouwer (cited in Russell, 1977), Davidson (1969), and Tilman (1982). Iwasa & Roughgarden (1984) and Robinson (1986*a*) have modelled the dynamics of balanced growth. Trewavas (1986*a*) suggested how balanced growth might be regulated metabolically within a plant. He also pointed out that complex metabolic systems in a steady state are likely to be controlled by a network of related processes rather than by a single 'rate limiting' step (Trewavas, 1986*b*). Therefore, any mathematical description of such systems should be based on the idea of joint control of rate processes, as is the case in balanced plant growth.

This concept contrasts with that of 'unbalanced' growth, in which a plant's growth rate is determined solely by events above ground or below at any given time, depending on whichever imposes the greater restriction on growth. If the rate of plant growth is determined by the rate of nutrient uptake, say, a change in the atmospheric concentration of CO_2 should not affect growth rate. Similarly, if the plant was growing in deep shade with ample nutrients such that irradiance was the factor limiting growth, the application of fertiliser would not affect growth rate. This concept can be traced to Blackman's (1905) influential paper on optima and limiting factors, itself based on Liebig's 'Law of the Minimum' (Wild, 1988). It has been applied to whole-plant growth by Robinson (1986*a*; 1989) and Tilman (1988). Inevitably, this approach gives a less realistic picture of a steady-state process, but it probably applies to transient-state growth occurring after a sudden, extreme change in the environment (Iwasa & Roughgarden, 1984; Robinson, 1986*a*).

The reason why maximising whole-plant growth rate is related to balanced growth is that the latter is a prerequisite of the former (Robinson, 1986*a*). Any sudden perturbation of the environment (such as defoliation or a change in nutrient availability) leads, initially, to unbalanced growth and a decrease in the plant's RGR. RGR can subsequently be increased by appropriate phenotypic changes (Robinson & Rorison, 1983), the increase being greatest in plants expressing phenotypes allowing balanced growth to be re-established (Robinson, 1986*a*).

Modelling balanced growth

The model described by Robinson (1986*a*) uses as environmental inputs the instantaneous photon flux (PF, mol m^{-2} s^{-1}) incident on leaves, and the nitrate concentration (C_l, mol m^{-3}) in the bulk soil solution. It has been extended to allow plasticity to occur as a function of changes in these

Table 1. *Phenotypic features of a 'standard' plant used in calculations, as described in the text.*

Phenotypic feature	Numerical value (and where appropriate, maximum or minimum limits of variation)	Reference
Maximum rate of net photosynthesis	30 μmol m^{-2} (leaf) s^{-1}	1
Quantum yield of photosynthesis	0.05 mol CO_2 mol^{-1} (photons) (0.02 mol CO_2 mol^{-1} minimum)	1
Leaf thickness	200 μm (100 μm minimum)	2
Fractional leaf area	0.5 (1.0 maximum)	—
Fractional root length	0.5 (1.0 maximum)	—
Nitrogen concentration in plant tissue	3 mmol g^{-1} (dry wt) (1.0 mmol g^{-1} minimum)	3
Root radius	200 μm (50 μm minimum)	4
Proportional increase in root radius due to root hairs	3.5 (10 maximum)	4
Root:shoot ratio	0.5 (1.5 maximum)	4
Nitrate concentration in soil solution at the root surface	0.02 mol m^{-3}	5

References: 1, Gutschick & Weigel (1988); 2, derived from the data of Hunt, Weber & Gates (1985); 3, Jarvis (1987); 4, Robinson & Rorison (1988); 5, Burns (1980).

environmental factors. Full details of this model will be published elsewhere. The input data required to specify a plant's initial phenotype are listed in Table 1. It is assumed that soil's water content remains constant, that the nitrate supply to the roots is dominated by diffusion, and therefore, that the mass flow of soil solution makes a negligible contribution to nitrate supply. The model is not dynamic, it does not simulate 'growth' over time, being more of an investigative tool to explore the simple mathematical relations between phenotypic variables. The plant's dry weight is fixed at 1 g, which is partitioned in various ways. The model calculates the plant's instantaneous RGR and, if phenotypic plasticity occurs, values for its various

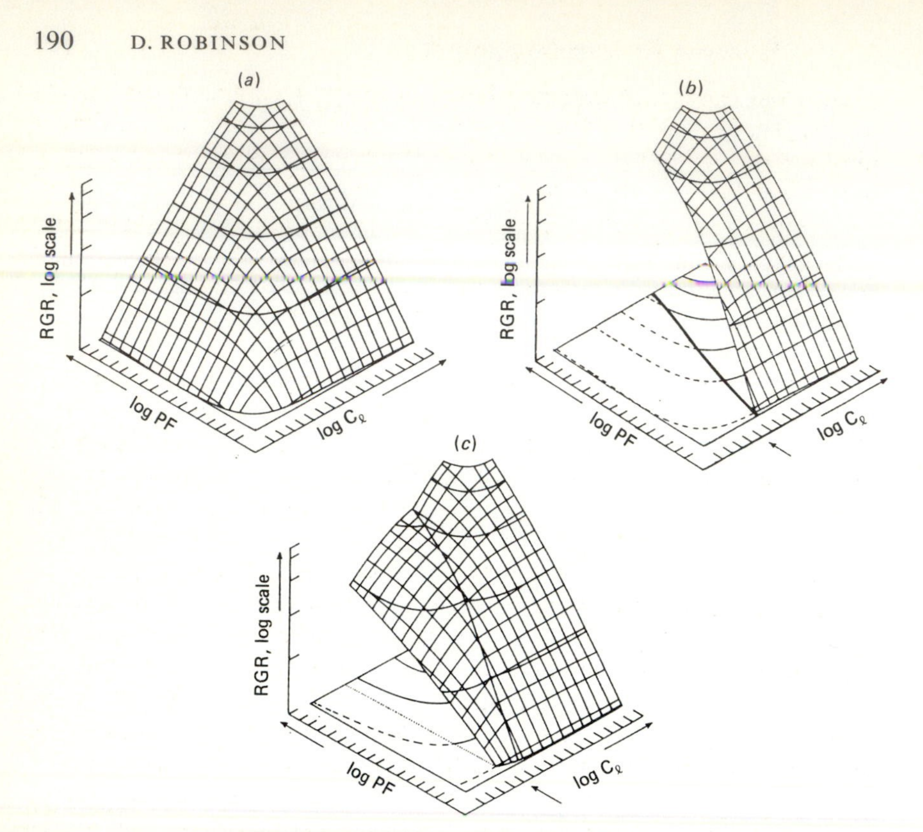

Fig. 4. Isometric projections of a generalised response surface or 'phenotypic landscape' produced by the model described in the text. The horizontal axes are log C_l (nitrate concentration) and log PF (photon flux), and the vertical axis is RGR on a log scale; their values increase in the directions shown by the arrows. The surface is plotted using 17 values of C_l and of PF; six 'contours' of constant RGR are also shown on the surface. (*a*) The complete response surface assuming no constraints on the attainment of balanced growth. A single phenotype is capable of attaining balanced growth at all points on the surface, with RGR varying as shown. The phenotype varies only in its rates of nitrate uptake and photosynthesis, which are related directly to C_l and PF, respectively. (*b*) If the diffusivity of nitrate in soil constrains the uptake of nitrogen, balanced growth of a constant phenotype is impossible at combinations of C_l and PF lying below the bold curve mapped on to the (PF, C_l) plane. This curve corresponds to the edge of the response surface remaining from that shown in (*a*). As PF approaches zero, the bold curve approaches the concentration of nitrate at the root surface (arrowed). Contours of RGR are also mapped on to the (PF, C_l) plane; those lying below the bold curve and which do not correspond to points on the response surface are shown as broken lines (see also Fig. 6). (*c*) If phenotypic plasticity occurs when nitrate diffusivity constrains the attainment of balanced growth, as

phenotypic features. Nitrogen inflow rate plays a dominant part in the model. The potential inflow rate (I_0) is calculated from the maximum diffusive flux of nitrate ions to root (Robinson, 1986*b*). If the rate of nitrogen inflow (I^*) required to maintain balanced growth in a specified environment is greater than I_0, the model plant's phenotype is changed until I^* is equal to, or less than, I_0. I^* is a function of a plant's phenotype (Robinson, 1986*a*). Each phenotypic feature of the plant is changed by the same fraction, $(I^* - I_0)/I^*$. When $I^* < I_0$, no phenotypic plasticity is required to attain balanced growth.

The phenotypic features of plants are not infinitely plastic, but it is difficult to define objectively lower or upper limits to their sizes and/or rates (Niklas, 1976). It is easier to define such limits empirically, but these may or may not represent the true physical limits to which a feature could vary and remain physiologically viable. This is a weakness of this and other such models, and reflects our poor understanding of the biophysical aspects of developmental physiology in plants (cf. Green, 1987; Niklas, 1988).

The RGRs calculated by this model as a function of C_1 and PF are shown schematically in Fig. 4*a*. This generalised 'phenotypic landscape' (cf. 'epigenetic landscape' (Wareing & Phillips, 1970), 'selective landscape' (Sibly & Calow, 1983), or 'adaptive landscape' (Niklas, 1988; Wagner, 1988)) is the type of response surface possible only if there are no constraints on the attainment of balanced growth. Such constraints are those imposed on nitrogen uptake rate by nitrate diffusivity in soil and by the existence of a 'compensation point' for nitrate uptake – a concentration of nitrate at the root surface below which no net gains of nitrogen can be made by the plant. These modify the ideality of Fig. 4*a* in ways which are now explored.

Static or plastic phenotypes?

What is the minimum C_1 for nitrate at which balanced growth can be attained (ignoring for the moment the rate of that growth) without any change in phenotype? It is necessary to define an initial 'standard' phenotype against which others can be compared; the one chosen is defined

---

described in the text, the response surface is extended beyond that shown in (*b*). Thus, balanced growth becomes possible at availabilities of nitrate which would prevent the attainment of balanced growth by a constant phenotype. Over the extended portion of the response surface, the phenotype varies continuously, in contrast to the constancy of the phenotype on the original surface. The lower limit of the new response surface is the concentration of nitrate at the root surface (arrowed). The edge of the surface is mapped on to the (PF, C_1) plane (dotted line), as are contours of RGR which are shown as broken lines where they project beyond the edge of the response surface (see also Fig. 7).

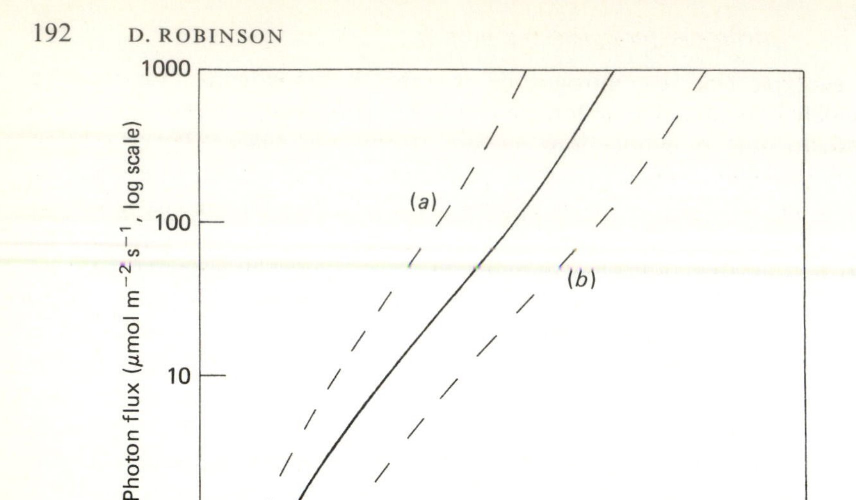

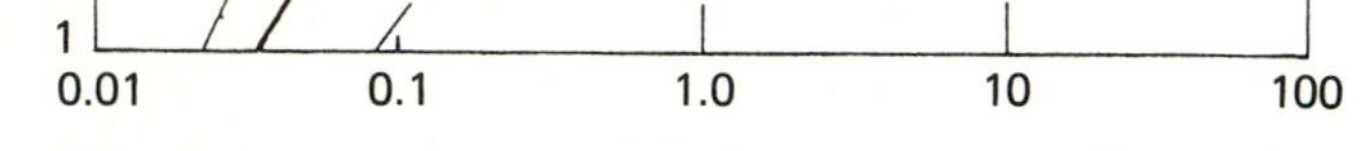

Fig. 5. The minimum concentrations (log scale) of nitrate in the bulk soil solution and photon fluxes (log scale) at which the 'standard' phenotype (Table 1) can attain balanced growth (bold curve). The broken curves are the corresponding concentrations for a phenotype whose mean root radius is half (*a*) or double (*b*) that of the 'standard'.

in Table 1. It is important to emphasise that there is no particular significance in the choice of phenotype for the present purpose.

The minimum values of C_l at which the standard phenotype can attain balanced growth are shown in Fig. 5. These limiting values are determined by the constraint $I^* < I_0$ and depend strongly on PF. A plant growing at a PF of 300 μmol m^{-2} s^{-1} could maintain balanced growth down to a C_l of 3.2 mol m^{-3}; at only 30 μmol m^{-2} s^{-1}, it could do so down to 0.5 mol m^{-3}.

Changing the phenotype from that shown in Table 1 affects the limiting C_l value at a given PF. By halving or doubling each of the values in Table 1 independently, the minimum C_l value at which balanced growth is possible is decreased or increased relative to that for the standard phenotype. Changes in mean root radius have the largest effect (shown in Fig. 5). At a PF of 300 μmol m^{-2} s^{-1}, the minimum C_l value is reduced to 0.83 mol m^{-3} when the root radius is 100 μm, but increased to 13 mol m^{-3} when it is 400 μm. Changing the other phenotypic features sometimes has the opposite effect: doubling the fractional root length, the leaf thickness, the increase caused by root hairs to the effective root radius, or the root:shoot ratio, causes the minimum C_l values to decrease compared with the corresponding ones for the standard phenotype.

A plant that maintains itself in balanced growth at low nitrogen availabilities should, in general, have the following phenotype: a high proportion of active fine roots, a high root:shoot ratio, many root hairs, and a slow maximum rate of net photosynthesis. These characteristics are common among plants from nutrient deficient soils (Grime, 1979; Sibly & Grime, 1986). The reasons suggested for plants evolving such attributes are the need to conserve available nutrients within the biomass, to produce unpalatable leaves, and to make efficient use of photosynthates. These features also, it seems, reflect the requirements for plants to maintain balanced (and hence maximal) growth rates under conditions of poor nutrient supply.

How fast can a plant grow without being plastic?

Fig. 6*a* shows for the standard phenotype, the variations in the minimum values of PF and C_1 needed to sustain certain RGRs. The resulting curves are reminiscent of Tilman's (1982, 1988) supply/consumption isoclines for two essential resources. Not surprisingly, the highest RGRs can be attained if C_1 and PF are high. However, the curve derived above defining the minimum C_1 at which balanced growth could be attained constrains RGR: it cannot exceed this limit unless its phenotype changes. The only alternative to phenotypic plasticity for a plant subjected to progressively decreased C_1 is to follow the curve defining the limiting C_1 for balanced growth, with the precipitous drop in RGR that this would entail (Fig. 6*b*).

The constraint imposed on the attainment of balanced growth by the condition $I^* < I_0$ has a considerable effect on the 'phenotypic landscape' shown in Fig. 4*a* which now appears as shown in Fig. 4*b*.

Is phenotypic plasticity beneficial?

Suppose the standard phenotype is allowed to change as C_1 decreases in order that the balanced growth is sustained. The consequences for RGR are shown in Fig. 7. When the rate of nitrate diffusion becomes limiting to uptake by the standard phenotype, complete compensation for this limitation is not possible, that is, RGR declines (Fig. 7*b*). This is because the types of phenotypic response that would maintain balanced growth with diffusion-limited nutrition are precisely those which involve producing roots at the expense of leaves, the result being balanced growth at the expense of growth rate. H.L. White's seminal work in 1937 on the interaction between irradiance and nitrogen availability on the growth of *Lemna minor* led to a similar conclusion (see Trewavas, 1986*a*). So, what *is* the benefit of phenotypic plasticity?

The benefit is seen by comparing the RGR attainable by a plant which

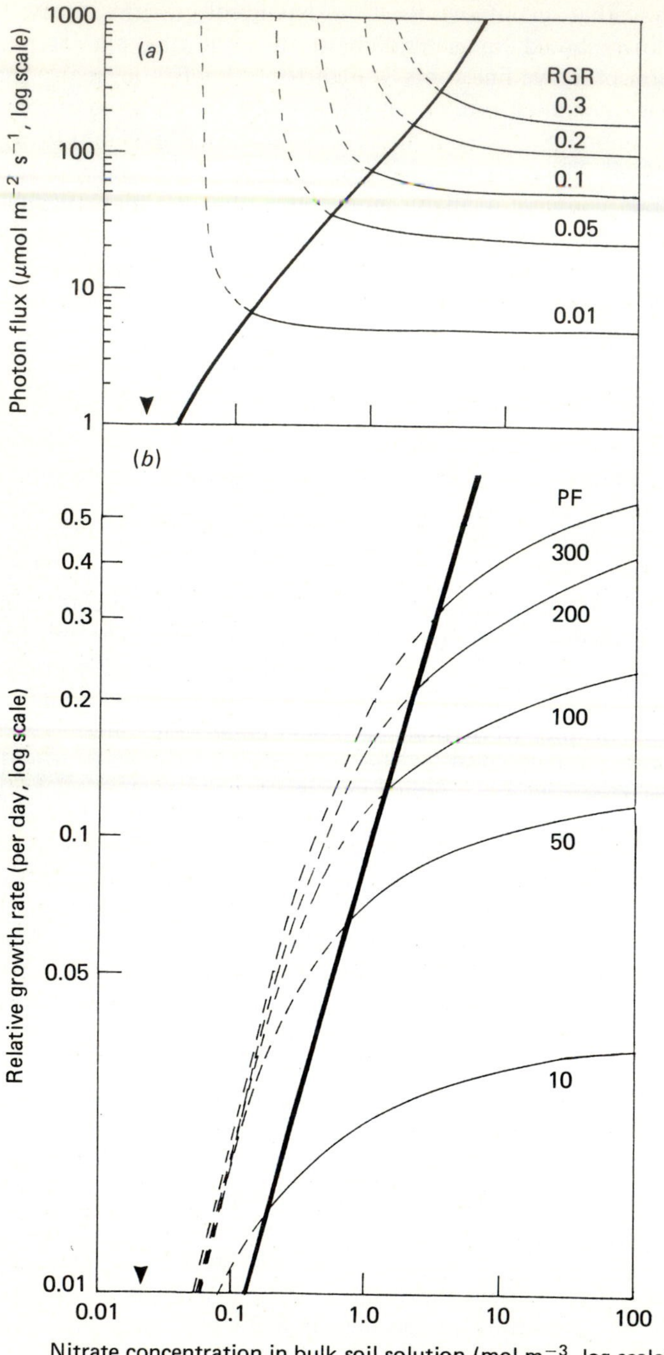
1000
(a)
RGR
0.3
0.2
0.1
0.05
0.01
100
10
1
Photon flux (µmol m⁻² s⁻¹, log scale)
(b)
PF
300
200
100
50
10
0.5
0.4
0.3
0.2
0.1
0.05
0.01
Relative growth rate (per day, log scale)
0.01
0.1
1.0
10
100

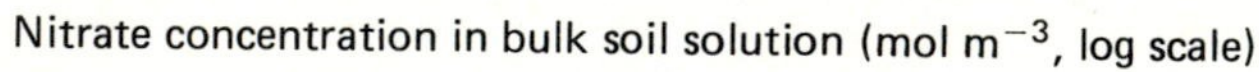
Nitrate concentration in bulk soil solution (mol m⁻³, log scale)

can alter its phenotype, with that of one without this ability – the bold line in Fig. 7*b*. At a given C_1, the RGR of the plant with the variable phenotype is always greater than that of the plant that cannot change its phenotype, a conclusion reached also by Tilman (1988). At low values of C_1, the minimum concentration to which nitrate can be depleted at the root surface imposes a lower bound on the attainment of RGR, even if phenotypic plasticity occurs. The effect of this on the generalised phenotypic landscape referred to previously is shown in Fig. 4*c*.

It is difficult to judge how realistic the response curves in Fig. 7 are, as no experiments have been reported in which RGR has been measured over a wide range of C_1. Such curves have been found, however, for the response of RGR to PF, those presented for *Cirsium palustre* and *Geum urbanum* by Lambers & Dijkstra (1987) for example.

Does a low growth rate mean low plasticity?

Plants with inherently slow potential rates of growth are more common in nutrient deficient habitats than on potentially productive sites. Such plants are, it is often said, less morphologically and physiologically plastic than plants with potentially higher rates of growth (Grime, 1979; Chapin, 1987). Robinson & Rorison (1988) pointed out, however, that this opinion depends on which phenotypic features are considered. Ecologists usually think of phenotypic plasticity in terms of very obvious features – root: shoot ratio, leaf shape, growth habit, etc. These features are relatively unresponsive to nutrient availability in slow-growing species, as Robinson & Rorison (1983, 1988) showed for *Deschampsia flexuosa* (wavy hair-grass). If more subtle responses are measured, such as the production of root hairs, slow-growing plants can show greater phenotypic plasticity than faster growing species (Robinson & Rorison, 1988). Why have such plants evolved an inherently slow growth rate?

Part of the reason for their slow potential rates of growth is their relatively greater investment of resources into anti-herbivore defences (see

---

Fig. 6. Relative growth rate of the standard phenotype (Table 1) in relation to nitrate concentration (C_1) and photon flux (PF), without plasticity (see also Fig. 4*b*). (*a*) 'Contours' of RGR (per day). The bold curve is that shown in Fig. 5. Balanced growth is not possible at combinations of PF and C_1 lying beyond this curve. Projections of the RGR contours into this region are shown as broken lines. (*b*) RGR at various constant values of PF ($\mu mol\ m^{-2}\ s^{-1}$), corresponding to the variations in RGR shown in (*a*). The concentration of nitrate at the root surface is arrowed.

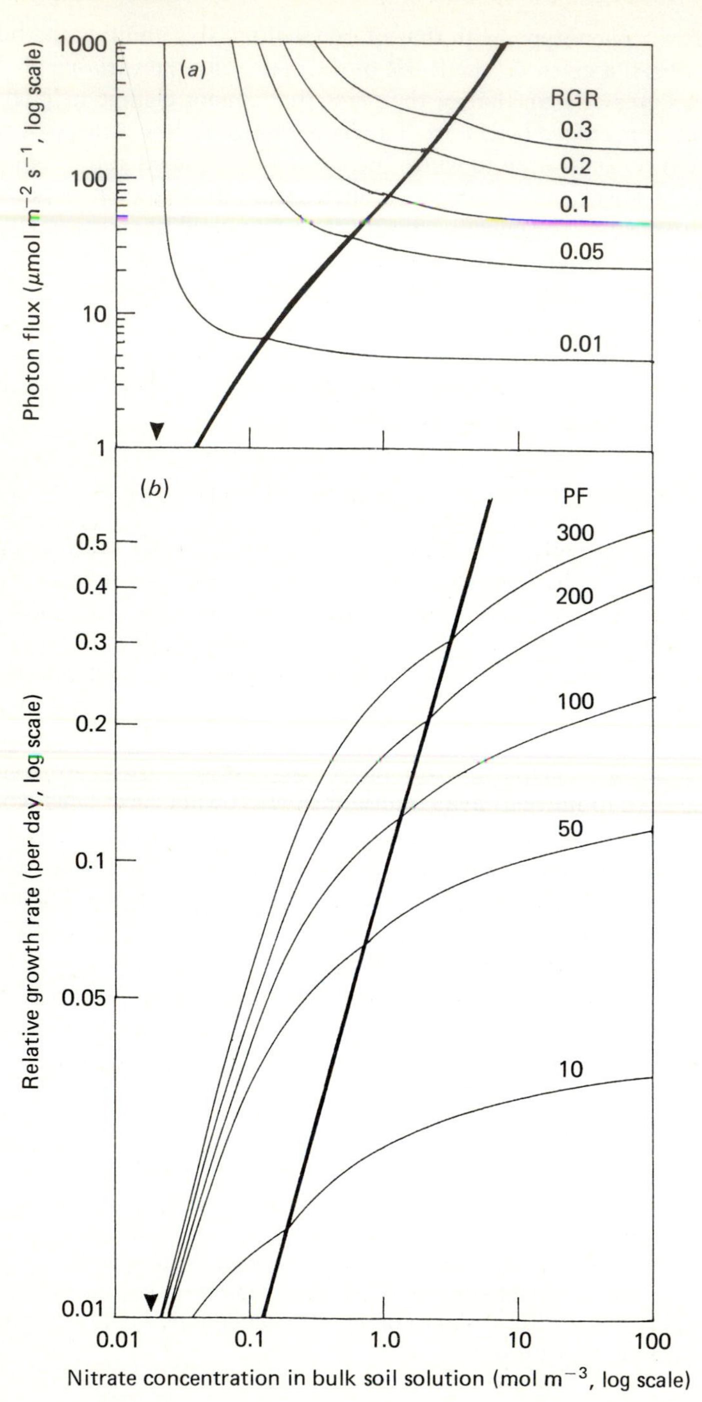
(a)
RGR
0.3
0.2
0.1
0.05
0.01
Photon flux (μmol m⁻² s⁻¹, log scale)
1000
100
10
1
(b)
PF
300
200
100
50
10
Relative growth rate (per day, log scale)
0.5
0.4
0.3
0.2
0.1
0.05
0.01
0.01
0.1
1.0
10
100
Nitrate concentration in bulk soil solution (mol m⁻³, log scale)

above) and their slow turnover rates of leaves (Grime, 1979; Sibly & Grime, 1986) and, presumably, of roots. These characteristics limit the rate at which growth can occur even when nutrients are not in short supply. This, however, does not explain their apparently low plasticity in response to environmental change.

Instead of using the standard phenotype as the input to the model, an 'extreme' one was defined by setting quantum yield, fractional root length and leaf area, leaf thickness, root radius, increase in root radius due to root hairs, root:shoot ratio and the concentration of nitrogen in the plant to their respective minima or maxima shown in Table 1. The maximum rate of net photosynthesis was set to half that of the standard phenotype, and the other features were unchanged. The RGRs of both standard and extreme phenotypes are shown in Fig. 8 for a single PF.

At a high C_1, the extreme phenotype has a slower RGR than the standard. The extreme phenotype is intrinsically unresponsive to increases in nitrate availability. Its phenotype can remain unchanged over a wide range of C_1 and still maintain the plant in balanced growth. Furthermore, this phenotype can allow faster RGRs to be attained than in the standard phenotype when the growth of both is limited by nitrate diffusion. Rorison (1968) and Campbell & Grime (1989) found such a pattern of response in a number of herbaceous species in response to nutrient availability, as did Mahmoud & Grime (1974) in response to irradiance. Some of Shipley & Keddy's (1988) data for the growth of semi-aquatic macrophytes show the 'cross-over' of RGR at low availabilities of nutrients, while others do not. None of the herbaceous species examined by Tilman & Cowan (1989) responded as shown in Fig. 8. Clearly, such a response, although having great appeal for ecologists seeking to explain how some slow-growing species can colonise infertile soils to the exclusion of other, faster-growing plants (see Shipley & Keddy, 1988), is far from being universal. The results shown in Fig. 8 suggest that it is a theoretical possibility, but one which carries the cost of an intrinsically slow RGR under nutrient-rich conditions.

The association between a capacity to grow relatively well on nutrient-poor soils and a low potential growth rate may cause problems if the breeding of crop genotypes suitable for low input agriculture is to be a realistic possibility. Breeding for bigger root systems, as Chapin (1987)

Fig. 7. As for Fig. 6, but here phenotypic plasticity occurs if nitrate diffusivity limits the uptake rate of nitrogen. RGR is attainable without plasticity at values of C_1 (nitrate concentration) to the right of the bold curves; to the left of the bold curves it is attainable only with plasticity (see also Fig. 4*c*) (*a*) Contours of RGR. (*b*) Variations in RGR at different values of PF (photon flux).

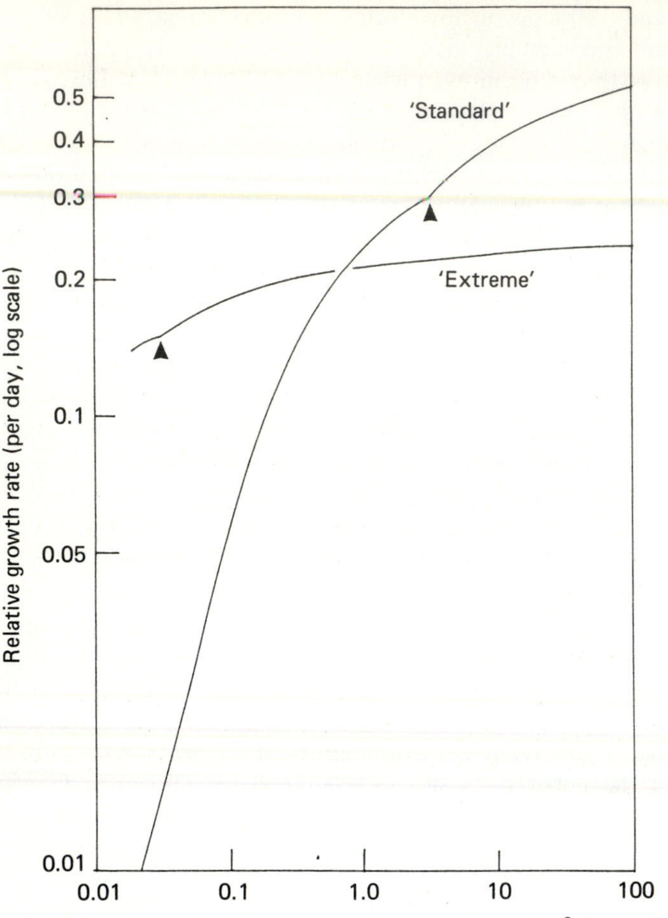

Fig. 8. Variations in RGR at a single photon flux (300 μmol m^{-2} s^{-1}) if two phenotypes, the 'standard' (Table 1) and an 'extreme' phenotype (see text). The curve for the standard phenotype is that shown in Fig. 7*b*. The points at which nitrate diffusivity limits nitrogen uptake and below which balanced growth is possible only with phenotypic plasticity are arrowed.

suggested, might not overcome the problem: selecting genotypes that allocate greater resources to roots would depress further their potential growth rates and productivities. One possible solution might be to uncouple the development and form of certain key features (such as root diameter) from the growth of the whole plant. This will not be easy; Chapin (1987) is optimistic that it is possible.

Concluding remarks

One of the main limitations of this type of approach is that the effects of fluctuations in environmental conditions have yet to be incorporated. It has become widely appreciated that many wild plants are more sensitive to temporal and spatial variations in nutrient availability than to chronically low, uniform supplies (Grime *et al.*, 1986; Rorison, 1987; 1991, this volume; Campbell & Grime, 1989; Grime, 1991, this volume). Diurnal variations in PF, and perhaps C_1, mean that plants must track daily across the surface of phenotypic landscapes similar to those shown in Fig. 4, but which themselves vary with environmental conditions. Another limitation is a failure to include, so far, the effects of 'compromise' responses (Schulze & Chapin, 1987; Robinson, 1989) to several factors that might impose conflicting demands on a plant's phenotype: changes in availability of mobile and immobile nutrient ions, for example.

Despite these – and other – limitations, simple models of the type outlined above can be used to answer questions about the strategies of phenotypic plasticity and growth rates that are not answerable by direct experimentation.

Acknowledgements

It is a pleasure to thank Professors I.H. Rorison and J.R. Hillman, Drs Naoki Kachi, Denis Linehan, John Porter and David Lawlor, and Ruth Hasty, for comments and/or discussions about some of the ideas presented in this chapter, and Joyce Davidson for typing the manuscript.

References

Benfey, P.N., Ren, L. & Chua, N.-M. (1989). The CaMV 35S enhancer contains at least two domains which can confer different developmental and tissue-specific expression patterns. *The EMBO Journal*, **8**, 2195–202.

Benjamin, L.R. & Hardwick R.C. (1986). Sources of variation and measures of variability in even-aged stands of plants. *Annals of Botany*, **58**, 757–78.

Blackman, F.F. (1905). Optima and limiting factors. *Annals of Botany*, **19**, 281–95.

Boutin, C. & Morisset, P. (1988). Etude de la plasticité phenotypique chez le *Chrysanthemum leucanthemum*. I. Croissance, allocation de la biomasse et reproduction. *Canadian Journal of Botany*, **66**, 2285–98.

Bradshaw, A.D. & Hardwick, K. (1989). Evolution and stress – genotypic and phenotypic components. In *Evolution, Ecology and Environmental Stress*, eds. P. Calow & R.J. Berry, pp. 137–55. London: Academic Press.

Burdon, J.J. & Harper, J.L. (1980). Relative growth rates of individual members of a plant population. *Journal of Ecology*, **68**, 953–7.

Burns, I.G. (1980). Influence of the spatial distribution of nitrate on the uptake of N by plants: a review and a model for rooting depth. *Journal of Soil Science*, **31**, 155–73.

Calow, P. & Townsend, C.R. (1981*a*). Energetics, ecology and evolution. In *Physiological Ecology: An Evolutionary Approach to Resource Use*, eds. C.R. Townsend & P. Calow, pp. 3–19. Oxford: Blackwell Scientific Publications.

Calow, P. & Townsend, C.R. (1981*b*). Resource utilization and growth. In *Physiological Ecology: An Evolutionary Approach to Resource Use*, eds. C.R. Townsend & P. Calow, pp. 220–44. Oxford: Blackwell Scientific Publications.

Campbell, B.D. & Grime, J.P. (1989). A comparative study of plant responsiveness to the duration of episodes of mineral nutrient enrichment. *New Phytologist*, **112**, 261–7.

Chapin, F.S. (1987). Adaptations and physiological responses of wild plants to nutrient stress. In *Genetic Aspects of Plant Mineral Nutrition*, eds. H.W. Gabelman & B.C. Loughman, pp. 15–25. Dordrecht: Martinus Nijhoff.

Coley, P.D. (1987). Interspecific variation in anti-herbivore properties: the role of habitat quality and rate of disturbance. In *Frontiers of Comparative Plant Ecology*, eds. I.H. Rorison, J.P. Grime, R. Hunt, G.A.F. Hendry & D.H. Lewis, pp. 251–63. London: Academic Press.

Cowan, I.R. (1982). Regulation of water use in relation to carbon gain in higher plants. In *Encyclopedia of Plant Physiology, Vol. 12B, Physiological Plant Ecology II*, eds. O.L. Lange, P.S. Nobel, C.B. Osmond & H. Zeigler, pp. 589–613. Berlin: Springer-Verlag.

Crawford, J.W., Duncan, J.M., Ellis, R.P., Griffiths, B.S., Hillman, J.R., MacKerron, D.K.L., Marshall, B., Ritz, K., Robinson, D., Wheatley, R.E., Woodford, J.A.T. & Young, I.M. (1989). *Global Warming. The Implications for Agriculture and Priorities for Research*. Dundee: Scottish Crop Research Institute.

Crawford, J.W. & Young, I.M. (1990). A multiple scaled fractal tree. *Journal of Theoretical Biology*, **145**, 199–206.

Cullis, C.A. (1984). Environmental induction of heritable changes in flax: defined environments inducing changes in rDNA and peroxidase isozyme band patterns. *Heredity*, **47**, 87–94.

Davidson, R.L. (1969). Effect of root/leaf temperature differentials on root/shoot ratios in some pasture grasses and clover. *Annals of Botany*, **33**, 561–9.

Dawkins, R. (1982). *The Extended Phenotype*. San Francisco: W.H. Freeman.

Dawkins, R. (1986). *The Blind Watchmaker*. London: Longman.

Fagerstrom, T., Larsson, S. & Tenow, O. (1987). On optimal defence in plants. *Functional Ecology*, **1**, 73–81.

Fitter, A.H. (1987). An architectural approach to the comparative ecology of plant root systems. In *Frontiers of Comparative Plant Ecology*, eds. I.H. Rorison, J.P. Grime, R. Hunt, G.A.F. Hendry & D.H. Lewis, pp. 61–77. London: Academic Press.

Givnish, T.J. (1986*a*). Introduction. In *On the Economy of Plant Form and Function*, ed. T.J. Givnish, pp. 1–9. Cambridge University Press.

Givnish, T.J. (1986*b*). Biomechanical constraints on crown geometry in forest herbs. In *On the Economy of Plant Form and Function*, ed. T.J. Givnish, pp. 525–83. Cambridge University Press.

Green, P.B. (1987). Inheritance of pattern: analysis from phenotype to gene. *American Zoologist*, **27**, 657–73.

Grime, J.P. (1979). *Plant Strategies and Vegetation Processes*. Chichester: John Wiley.

Grime, J.P. (1991). Nutrition, environment and plant ecology: an overview. In *Plant Growth: Interactions with Nutrition and Environment*, eds. J.R. Porter & D.W. Lawlor, Society for Experimental Biology, Seminar Series 43, pp. 249–67. Cambridge University Press.

Grime, J.P., Crick, J.C. & Rincon, J.E. (1986). The ecological significance of plasticity. In *Plasticity in Plants*, eds. D.H. Jennings & A.J. Trewavas, pp. 5–29. Cambridge: The Company of Biologists.

Grime, J.P., Hodgson, J.G. & Hunt, R. (1988). *Comparative Plant*

Ecology: A Functional Approach to Common British Species. London: Unwin Hyman.

Gutschick, V.P. & Weigel, F.W. (1988). Optimizing the canopy photosynthetic rate by patterns of investment in specific leaf mass. *American Naturalist*, **132**, 67–86.

Hailman, J.P. (1988). Operationalism, optimality and optimism: suitabilities versus adaptations of organisms. In *Evolutionary Processes and Metaphors*, eds, M.-W. Ho & S.W. Fox, pp. 85–116. New York: Wiley.

Hardwick, R.C. (1983). Why do seed legumes lose their leaves? In *Interactions Between Nitrogen and Growth Regulators in the Control of Plant Development, Monograph 9*, ed. M.B. Jackson, pp. 61–74. Wantage: British Plant Growth Regulator Group.

Hardwick, R.C. (1986). Physiological consequences of modular growth in plants. *Philosophical Transactions of the Royal Society of London*, **B313**, 161–73.

Harper, J.L. (1977). *The Population Biology of Plants*. London: Academic Press.

Hunt, E.R., Weber, J.A. & Gates, D.M. (1985). Effects of nitrate application on *Amaranthus powelii* Wats. I. Changes in photosynthesis, growth rates, and leaf area. *Plant Physiology*, **79**, 609–13.

Hunt, R. (1982*a*). *Plant Growth Curves: The Functional Approach to Plant Growth Analysis*. London: Edward Arnold.

Hunt, R. (1982*b*). Plant growth analysis: second derivatives and compounded second derivatives of splined growth curves. *Annals of Botany*, **50**, 317–28.

Iwasa, Y. & Roughgarden, J. (1984). Shoot/root balance of plants: optimal growth of a system with many vegetative organs. *Theoretical Population Biology*, **25**, 78–105.

Janzen, D.H. (1981). Evolutionary physiology of personal defence. In *Physiological Ecology: An Evolutionary Approach to Resource Use*, eds. C.R. Townsend & P. Calow, pp. 145–64. Oxford: Blackwell Scientific Publications.

Jarvis, S.C. (1987). The effects of low, regulated supplies of nitrate and ammonium on the growth and composition of perennial ryegrass. *Plant and Soil*, **100**, 99–112.

Kimura, M. & Ohta, T. (1971). *Theoretical Aspects of Population Genetics*. Princeton: Princeton University Press.

Lacey, E.P. (1986). Onset of reproduction in plants: size- versus age-dependency. *Trends in Ecology and Evolution*, **1**, 72–5.

Lambers, H. & Dijkstra, P. (1987). A physiological analysis of genotypic variation in relative growth rate: can growth rate confer ecological advantage? In *Disturbance in Grasslands*, eds. J. Van Andel, J.P. Bakker & R.W. Snaydon, pp. 237–52. Dordrecht: Junk.

Levins, R. (1968). *Evolution in Changing Environments*. Princeton: Princeton University Press.

Mahmoud, A. & Grime, J.P. (1974). A comparison of negative relative growth rates in shaded seedlings. *New Phytologist*, **73**, 1215–19.

Maynard Smith, J. (1982). *Evolution and the Theory of Games*. Cambridge University Press.

McClintock, B. (1984). The significance of responses of the genome to challenge. *Science*, **226**, 792–801.

McGraw, J.B. & Wulff, R. (1983). The study of plant growth: a link between the physiological ecology and population biology of plants. *Journal of Theoretical Biology*, **103**, 21–8.

McNaughton, S.J. (1983). Physiological and ecological implications of herbivory. In *Encyclopedia of Plant Physiology, Vol. 12B, Physiological Plant Ecology III*, eds. O.L. Lange, P.S. Nobel, C.B. Osmond & H. Ziegler, pp. 657–77. Berlin: Springer-Verlag.

Niklas, K.J. (1976). The role of morphological biochemical reciprocity in early land plant evolution. *Annals of Botany*, **40**, 1239–54.

Niklas, K.J. (1988). Biophysical limitations on plant form and evolution. In *Evolutionary Plant Biology*, eds. L.D. Gottlieb & S.K. Jain, pp. 185–220. London: Chapman & Hall.

Partridge, L. & Harvey, P.H. (1988). The ecological context of life-history evolution. *Science*, **241**, 1449–55.

Robinson, D. (1986*a*). Compensatory changes in the partitioning of dry matter in relation to nitrogen uptake and optimal variations in growth. *Annals of Botany*, **58**, 841–8.

Robinson, D. (1986*b*). Limits to nutrient inflow rates in roots and root systems. *Physiologia Plantarum*, **68**, 551–9.

Robinson, D. (1989). Phenotypic plasticity in roots and root systems: constraints, compensations and compromises. *Roots and the Soil Environment, Aspects of Applied Biology*, **22**, 49–55.

Robinson, D. & Rorison, I.H. (1983). A comparison of the responses of *Lolium perenne* L., *Holcus lanatus* L. and *Deschampsia flexuosa* (L.) Trin. to a localized supply of nitrogen. *New Phytologist*, **94**, 263–73.

Robinson, D. & Rorison, I.H. (1988). Plasticity in grass species in relation to nitrogen supply. *Functional Ecology*, **2**, 249–57.

Rorison, I.H. (1968). The response to phosphorus of some ecologically distinct plant species. I. Growth and phosphorus absorption. *New Phytologist*, **67**, 913–23.

Rorison, I.H. (1987). Mineral nutrition in time and space. In *Frontiers of Comparative Plant Ecology*, eds. I.H. Rorison, J.P. Grime, R. Hunt, G.A.F. Hendry & D.H. Lewis, pp. 79–92. London: Academic Press.

Rorison, I.H. (1991). Ecophysiological aspects of nutrition. In *Plant Growth: Interactions with Nutrition and Environment*, eds. J.R. Porter & D.W. Lawlor, Society for Experimental Biology, Seminar Series 43, pp. 157–76. Cambridge University Press.

Roughgarden, J., May, R.M. & Levin, S.A. (eds.) (1989). *Perspectives in Ecological Theory*. Princeton: Princeton University Press.

Russell, R.S. (1977). *Plant Root Systems: Their Function and Interaction with the Soil*. Maidenhead: McGraw-Hill.

Schlichting, C.D. (1989). Phenotypic integration and environmental change. *BioScience*, **39**, 460–4.

Schultz, J.C. (1988). Plant responses induced by herbivores. *Trends in Ecology and Evolution*, **2**, 45–9.

Schulze, E.-D. & Chapin, F.S. (1987). Plant specialization to environments of different resource availability. In *Potentials and Limitations of Ecosystem Analysis*, eds. E.-D. Schulze & H. Zwölfer, pp. 120–48. Berlin: Springer-Verlag.

Shipley, B. & Keddy, P.A. (1988). The relationship between relative growth rate and sensitivity to stress in twenty-eight species of emergent macrophytes. *Journal of Ecology*, **76**, 1101–10.

Sibly, R. & Calow, P. (1983). An integrated approach to life-cycle evolution using selective landscapes. *Journal of Theoretical Biology*, **102**, 527–47.

Sibly, R. & Grime, J.P. (1986). Strategies of resource capture by plants: evidence for adversity selection. *Journal of Theoretical Biology*, **118**, 247–50.

Silvertown, J.W. (1982). *Introduction to Plant Population Ecology*. London: Longman.

Southwood, T.R.E. (1988). Tactics, strategies and templets. *Oikos*, **52**, 3–18.

Stephens, D.W. & Krebs, J.R. (1987). *Foraging Theory*. Princeton: Princeton University Press.

Taylor, D.R. & Aarssen, L.W. (1988). An interpretation of phenotypic plasticity in *Agropyron repens* (Graminae). *American Journal of Botany*, **75**, 401–13.

Tilman, D. (1982). *Resource Competition and Community Structure*. Princeton: Princeton University Press.

Tilman, D. (1988). *Plant Strategies and the Dynamics and Structure of Plant Communities*. Princeton: Princeton University Press.

Tilman, D. & Cowan, M.L. (1989). Growth of old field herbs on a nitrogen gradient. *Functional Ecology*, **3**, 425–38.

Tomlinson, P.B. (1982). Chance and design in the construction of plants. *Acta Biotheoretica*, **31A**, 162–83.

Trewavas, A.J. (1986*a*). Resource allocation under poor growth conditions. A major role for growth substances in developmental plasticity. In *Plasticity in Plants*, eds. D.H. Jennings & A.J. Trewavas, pp. 31–76. Cambridge: The Company of Biologists.

Trewavas, A.J. (1986*b*). Understanding the control of plant development and the role of growth substances. *Australian Journal of Plant Physiology*, **13**, 447–57.

Via, S. & Lande, R. (1985). Genotype-environment interaction and the evolution of phenotypic plasticity. *Evolution*, **39**, 505–22.

Wagner, G.P. (1988). The influence of variation and of developmental constraints on the rate of multivariate phenotypic evolution. *Journal of Evolutionary Biology*, **1**, 45–66.

Wareing, P.F. & Phillips, I.D.J. (1970). *The Control of Growth and Differentiation in Plants*. Oxford: Pergamon Press.

Weiner, J. (1988). Influence of competition on plant reproduction. In *Plant Reproductive Ecology*, eds. J. Lovett Doust & L. Lovett Doust, pp. 228–45. New York: Oxford University Press.

Wessler, S. (1988). Phenotypic diversity mediated by the maize transposable elements *Ac* and *Spm*. *Science*, **242**, 399–405.

Westoby, M. (1984). The self-thinning rule. *Advances in Ecological Research*, **14**, 167–225.

Westoby, M. (1988). Selective forces exerted by vertebrate herbivores on plants. *Trends in Ecology and Evolution*, **4**, 115–117.

Wild, A. (ed.) (1988). *Russell's Soil Conditions and Plant Growth*, 11th edn. London: Longman.

Williams, G.C. (1966). *Adaptation and Natural Selection*. Princeton: Princeton University Press.

R.F. HUETTL AND S. FINK

Pollution, nutrition and plant function

Introduction

Plants are exposed to many natural and unnatural environmental factors that can affect their vitality and growth. In forest ecosystems these influences may cause changes in the nutrient supply, eventually leading to forest damage. However, due to complex interactions between these processes it is generally difficult to determine specific stress factors responsible for the decline of growth in forest trees. With respect to air pollution, 'acid rain' has been named frequently as an important human-made environmental change causing 'new type' forest damage in central Europe and North America. In this context, a distinction is made between direct and indirect damage. Foliar tissue can be injured directly by increased amounts of various air pollutants and/or by acid precipitation (particularly acid fog) causing increased leaching of nutrients. Beside proton production by plants, proton input from acid deposition accelerates soil acidification. This causes loss of nutrients and the mobilisation of potentially toxic and/or antagonistic ions in the root zone, provoking nutrient deficiencies. High atmosphere N deposition can lead to imbalances in the nutrient supply.

Compensation for nutrient losses from the foliage is possible through increased nutrient uptake from the soil, depending on the nutrient availability in particular sites. However, forest soils are frequently deficient in one or more nutrients. As soil nutrients play an important role for both the vigour and the growth of trees, an improvement in the nutrient supply by application of fertilisers generally reduces or even removes forest damage associated with nutritional disturbances. Fertilisation based on the specific requirements of both sites and species can therefore be used as diagnostic as well as therapeutic tools. Histological analyses of conifer needles show how tissue may regenerate after fertilisation. Furthermore, needles of nutrient deficient trees reveal characteristic tissue damage when compared to needles exposed to high sulphur dioxide (SO_2) concentrations.

Thus, histological investigations can be used to differentiate between direct and indirect forest damage.

Interest in the effects of air pollutants on trees and forests has increased greatly over the last few years. However, apprehension about the effects of air pollution has been expressed for a long time. Previously, concern focused on local forest damage directly caused by emissions of pollutants at a short distance from the site of damage. The main pollutants at that time were sulphur dioxide and particulates, e.g. dust. With increasing industrialisation and associated increase in the combustion of fossil fuels together with the recent 'high-stake policy', i.e. replacing short stakes with higher ones the air pollution problem has become far more serious and complex (cf. Kozlowski & Constantinidou, 1986). Since the mid-1970s the so-called 'new type' forest damage has been observed in West Germany. The damage is considered 'new', particularly with respect to its wide spatial distribution and rapid temporal development. The dieback of silver fir (*Abies alba*) in the Bavarian Forest and in the Black Forest was soon followed by severe damage to stands of Norway spruce (*Picea abies*) and Scots pine (*Pinus sylvestris*). Since 1983, increased damage has also been found in deciduous trees such as European beech (*Fagus sylvatica*) and oak (*Quercus* spp.) (Forschungsbeirat Waldschaeden/Luftverunreinigungen der Bundesregierung und der Laender, 1986, Advisory Service for Forest Damage/Air Pollution of the Federal German Government and the Laender).

Extent of damage

To evaluate the area of damage and its development, annual terrestrial inventories have been performed since 1983. But only since 1984 have the inventory parameters been uniformly applied to all eleven German states. Beside discoloration symptoms, foliage losses are the main parameters to determine damage class (Forschungsbeirat Waldschaeden/Luftverunreinigungen der Bundesregierung und der Laender, 1986). Initially the results of the inventory suggested a rapid increase in damage. In 1983 34% and in 1984 50% of the stands were already marked by slight to heavy damage symptoms. Thereafter the rate of increase of damage was much reduced. The damage increased in both 1985 and 1986 by 2% annually. Thus, in the autumn of 1986 54% of the total forest area in West Germany was believed to be suffering, at least, some damage. In 1987 there was a 2% decrease in damage across the country. In 1988, the same amount of damage as in the previous year was found, i.e. 52%. However, the 1988 survey indicated that 37% of the damaged area was in the lowest category of damage, less than 14% was moderately damaged and only 1.3% showed severe damage symptoms. The different types of damage were and are often attributed to the adverse effects of acid rain and other air pollutants (BML, 1988). However, it seems worth noting that since this type of damage inventory has been used, a significant part of the scientific community involved in research on 'new type' forest decline has agreed that foliar losses can be caused by multiple factors of unspecific nature.

The amount of healthy foliage on trees of a particular species varies with site conditions, genetic variability and other factors, indicating that no 'standard' leaf area can be assumed for a specific tree or stand. However, a standard leaf area must be assumed in order to determine the amount of foliar losses, as the amount of foliage on a tree before the inventory is unknown in almost all cases. Therefore, it must be concluded that the parameter 'foliage loss' is not a reliable means for evaluating the decline of forest trees and stands when research is focused on determining specific causes (cf. Huettl & Wisniewski, 1987; Innes, 1987).

Types of decline and nutritional status

Recently, a wide variety of symptoms related to forest decline has been described and attributed to the 'new type' forest damage. This section briefly describes the common symptoms of damage for Norway spruce, the tree species most affected in West Germany. As the Norway spruce is also the most common and economically most important species of trees in central Europe, is has been studied in many research projects prior to its recently observed decline in vigour (Huettl, 1991).

According to the Forschungsbeirat Waldschaeden/Luftverunreinigungen der Bundesregierung und der Laender (1986) damage to spruce can be differentiated into five types:

- yellowing in stands at the higher elevations of the German mountain chains,
- crown thinning in stands at middle elevations of the German central mountain chains,
- needle necroses in older stands in southern Germany,
- yellowing in stands on calcareous sites at higher elevations of the Alps,
- crown thinning in coastal areas.

Except for the short-term occurrence of needle necrosis in older stands of southern West Germany, all damage in Norway spruce is, at least to some degree, associated with nutritional disturbances which depend on the site (Table 1). We take, as an example, the yellowing in stands at higher elevations of the German central mountain chains. This damage is new. It is marked by symptoms of Mg deficiency and is widespread (Zech & Popp, 1983; Bosch *et al.*, 1983; Zoettl & Mies, 1983; Hauhs, 1985). The yellowing starts in the older needles of the lower- and middle-crown and occurs only

Table 1. *Relation between nutrient deficiency and type of damage to Norway spruce (from Forschungsbeirat Waldschaeden/Luftverunreinigungen der Bundesregierung und der Laender, 1986).*

Type of damage	Nutrient deficiency	Reference
1, tip-yellowing	Mg, Ca; K, Zn	Bosch *et al.*, 1983; Hauhs, 1985; Prinz, Krause & Stratmann, 1982; Zech & Popp, 1983; Zoettl & Mies, 1983
2, crown thinning	K, Mg; Ca, Zn	Kenk *et al.*, 1985; Lutz & Breininger, 1986; Zech, Suttner & Kotschenreuther, 1983; Zech, Suttner & Popp, 1985; Huettl, 1989
3, reddish needle discolorations	(K)	Rehfuess & Rodenkirchen, 1984; Zoettl & Huettl, 1985
4, yellowing on calcareous sites	K, Mn	Huettl, 1985; Rehfuess, 1985
5, crown thinning in coastal areas	Mg, K, P	Reemtsma, 1988

on those needles on the side of twigs that are directly exposed to sunlight. After a time, which cannot be predicted, the needles may become necrotic and may finally fall off. Generally, the current year's shoots are not affected. In severe cases of deficiency the discoloration is 'transferred' to the needles that are currently green, when they move to the older age-classes as new shoots are developed in spring. The heaviest losses of needles are frequently found in the middle-crown area, provoking the 'sub-top-dying' symptom. Finally, the trees may die completely. This type of damage is found in younger as well as in older spruces. The same phenomenon has been observed at similar sites in many countries of Europe (Huettl, 1987).

In some cases when younger Norway spruce trees were intensively studied for several years the yellowing was found to remain constant or even to be reversed (Kandler, Miller & Ostner, 1987). Such natural regeneration processes are probably related to more favourable climatic conditions, e.g. greater and more evenly distributed precipitation during the vegetative period which allows absorption of more nutrients, particularly on nutrient poor soils. For this as for all other types of damage to Norway spruce, a strong correlation can be found with the chemical conditions of the soil linked to nutritional disorders in the trees (Zoettl & Huettl, 1986).

When comparing data of needle analysis for 1975 to values for 1983, using comparable Norway spruce stands at similar sites of the Black Forest on crystalline soils, a distinct decrease of foliar Mg content was found which coincided temporarily with the appearance of Mg deficiency symptoms. Similar observations were made for Norway spruce stands on moranic soils in south-west Germany. There, a dramatic reduction in foliar K content was observed from analyses made over the last two decades (Huettl, 1985).

The problem of an unbalanced nutrient supply is greater in areas with a large supply of N owing to atmospheric deposition where decreasing concentrations of Mg, K, Ca, and P and increasing N contents in foliar tissue have been demonstrated. For example, 20 to 30 years ago the mean N content in dry matter of one-year-old needles of Douglas fir (*Pseudotsuga menziesii*) on sandy soils in the south-east of the Netherlands was 14 mg g^{-1}. Recently, mean values of 24 mg g^{-1} have been reported (van den Burg, 1991). Similar values have been found for Norway spruce stands in north-west Germany (Reemtsma, 1988).

Also, in North America acute nutritional disturbances have been observed during forest decline. For example, Friedland *et al.* (1984) reported damage in red spruce (*Picea rubens*) from sites at high elevation in the north-east of the United States, e.g. Whiteface Mountain, Adirondacks, USA. The frequently observed necrosis phenomenon in younger shoots was interpreted, at first, as frost damage related to increased N deposition which reduced frost resistance. However, beside this phenomenon in red spruce, yellowing of older age-classes of needles can also be observed in Balsam fir (*Abies balsamea*). At higher elevations there are fir forests which are in the decline and regeneration phases. Whether this widespread, wave-like decline of old fir and spruce trees on acidic crystalline soils is a natural biological process or caused, at least to some extent, by anthropogenic causes (e.g. air pollution) is controversial (Bruck & Robarge, 1988; Cook & Johnson, 1988). In areas undergoing natural regeneration which are generally growing well, occasional firs with yellowing symptoms can be found next to healthy trees. From Table 2 it is evident that discoloured firs are marked by lower K content, probably caused by deficient K supply, and also poor Mg contents. For all other elements the nutrition is most probably near optimal. In contrast to the discoloured trees healthy firs have much larger K content. Only the Mg content seems small. Also year-to-year fluctuations in N and P contents are pronounced. Below about 1200 m in

Table 2. *Elemental composition of needles of different age-classes of 10 to 20-year-old spruces and firs on base-poor granite sites at Whiteface Mountain, Adirondacks, New York, USA ($n = 10$), sampling occured during the non-reproduction period (data from Zoettl et al., 1989).*

Elevation above sea level	Tree species	Whorl (counting from top)	Needle year class	Extent of discolouration sample trees	N	P	K	Ca	Mg	Mn	Zn	Al
					mg g^{-1} dry matter					µg g^{-1} dry matter		
1250–1300	*A. balsamea*	I	1985	light	19.5	2.8	3.0	5.4	1.0	280	45	355
(fir forests)		IV	1985		18.8	2.1	3.5	4.5	0.9	280	46	300
	IV	1984			17.5	1.9	3.3	6.3	0.8	320	49	310
	IV	1983			15.2	1.4	2.9	6.1	0.6	280	43	315
	IV	1982			13.3	1.2	2.5	7.0	0.5	260	36	360
A. balsamea	I	1986	mod	14.8	1.7	2.3	5.1	0.7	222	31	300	
	A. balsamea	I	1986	(green)	14.9	1.7	5.4	4.7	0.8	214	38	510
1150–1200*A.*	I	1985	(green)	13.9	1.4	4.8	6.3	0.9	450	49	300	
balsamea												
(mixed	*P. rubens*	I	1985	(green)	10.1	1.2	8.1	2.1	0.6	370	20	70
fir-spruce	*P. rubens*	I	1986	moderate	11.9	1.2	9.2	1.7	0.6	535	19	130
forests)												

the Adirondacks there are mixed forests of spruce and fir. There, young healthy firs have higher N, P, and K contents (Table 2). The Ca values are also high, whereas the Mg contents are rather small. Surprisingly, there are differences in the content of some of the elements of healthy spruces and firs which are close neighbours. The spruce may suffer from an insufficient N and Mg supply, have small P and Ca but high K contents. In contrast, K contents of fir are small and the Ca values are very large. In this unmanaged ecosystem the differences can probably be explained, to a large extent, by the different root systems. On these sites *Abies balsamea* generally develops a rather shallow root system whereas *Picea rubens* grows roots into deeper soil layers (E. White, State University of New York at Syracuse, personal communication). Therefore, the spruce is able to take up enough K from a greater depth of mineral soil. The fir collects sufficient amounts of N and P from the organic layer and top soil to maintain its nutrient balance. Whether *Abies balsamea* has a greater uptake potential for Ca, or whether content of Ca is related to competition between nutrients or to other mechanisms is unclear.

Even though the moderately discoloured spruce trees have somewhat higher N contents than trees without symptoms, the available data do not indicate that N is oversupplied. If the necrosis observed in younger needles were caused by nutrient disorders it seems more likely that the damage would be related to the low Mg content (Tomlinson, 1990). Comprehensive investigations into the nutritional status of red spruce in the north-east United States by Friedland, Hawley & Gregory (1988) could not verify their hypothesis that the supply of N was too large but they frequently found an inadequate supply of Mg. Unfortunately no K values were reported. It is not surprising that the N hypothesis could not be verified as N deposition is rather small in this area (Mohnen, 1988). Further nutritional disturbances in declining forests of North America have been discussed by Huettl (1988) and Tomlinson (1990).

Diagnostic fertiliser trials and Mg deficiency

Diagnostic fertiliser trials have been conducted in West Germany since the early 1980s in forest stands with the 'new type' of damage symptoms (Zech, 1983; Huettl, 1985; Zoettl & Huettl, 1986; Bosch & Rehfuess, 1986; Kaupenjohann *et al.*, 1987; Kandler *et al.*, 1987; Isermann, 1987; Liu, 1988; Zoettl *et al.*, 1989). The indications are that acute nutritional deficiencies of Mg and K can be removed or completely overcome by the appropriate application of rapidly soluble fertilisers. Mg deficiency can also be compensated by applying limestone containing Mg as, for example, dolomitic material.

In this context, the question remains – how long will the effect of the

Table 3. *Chemical analyses[a] of soil on the diagnostic fertilisation trial at Elzach, Black Forest, West Germany.*

Exchangeable cations (at soil pH); $\mu eq\ g^{-1}$ d m			
	Ca^{2+}	Mg^{2+}	Al^{3+}
	1.9	0.7	64.2
pH ($CaCl_2$)	Mg:Al (mol)	Ca:Al (mol)	Base saturation (%)
3.6	0.016	0.044	5.7

[a] NH_4 Cl extraction with a control soil at 20–30 cm soil depth.

applied fertilisers last under the given site and stand conditions? Furthermore, in view of the nitrogen oversupply-hypothesis, the effect of unbalanced nutrition needed to be investigated in field experiments.

Diagnostic fertilisation trial at Elzach

In autumn 1983, the 12-year-old spruce stand in the forest district of Elzach (Black Forest) exhibited symptoms of needle discoloration indicating Mg deficiency. The trees grow on an acidic brown-earth derived from solifluction debris over granite. The humus form is mull, the slope has a northwesterly exposure at an elevation of 900 m and the site receives about 1500 mm mean annual precipitation. The soil analysis data (Table 3) show a poor Ca and Mg supply. This can clearly be seen from the low base saturation value and the pH level, as well as the very small Mg:Al and Ca:Al ratios.

Chemical analysis of needles, indicating dramatic Mg deficiency (Table 4) are clearly correlated with the observed damage. In needles which were still green in the first whorl, extremely small Mg contents as well as relatively small Ca and Zn values were found. The supply of N, P, and K was optimal and the Mn contents are considered normal for this site.

Overcoming Mg deficiency

To counteract acute Mg deficiency part of this young stand was fertilised with Kieserite (150 g per tree, 160 kg MgO ha^{-1}) in spring 1984. Four to five months later, the fertilised plot was revitalised when compared to the control trees. To a large part of the yellow spruce trees had regreened, needle loss was reduced and the current shoots were well developed and green. The general, visual, improvement was verified by the needle analyses (Table 4).

Table 4. *Elemental composition of current year's needles (first whorl; n = 10; autumn sampling) in a diagnostic fertiliser trial on 12-year-old Norway spruce (Picea abies) at Elzach, Black Forest, West Germany.*[a]

Element[b]	Control					$MgSo_4 \cdot H_2O$				$(NH_4)_2SO_4$		
($mg\ g^{-1}$ d m)	1983	1984	1985	1986	1987	1984	1985	1986	1987	1985	1986	1987
N	16.0	16.2	17.8	14.6	14.2	15.1	16.9	14.2	12.8	21.0	15.2	14.3
P	2.6	2.5	2.3	2.4	1.2	2.6	2.5	2.3	1.9	2.5	1.9	2.0
K	7.5	8.6	7.9	6.4	5.6	10.1	8.9	6.2	5.9	9.1	6.6	5.8
Ca	2.4	2.0	3.2	1.7	3.2	2.4	4.0	1.6	2.7	3.2	1.5	1.7
Mg	**0.20**	**0.32**	**0.31**	**0.28**	**0.25**	**0.56**	**0.56**	**0.56**	**0.55**	**0.28**	**0.18**	**0.37**
Mn	0.640	0.690	0.830	0.750	0.870	0.740	0.530	0.430	0.470	0.990	0.520	0.560
Zn	0.018	0.018	0.022	0.022	0.022	0.025	0.030	0.020	0.028	0.025	0.015	0.019
Element ratio[c]												
N:P	6.2	6.5	7.7	6.1	12.0	5.8	6.8	6.2	6.7	8.4	8.0	7.2
N:K	2.1	1.8	2.3	2.3	2.5	1.5	1.9	2.3	2.2	2.3	2.3	2.5
N:Ca	6.7	5.4	5.6	8.6	4.4	6.3	4.2	8.9	4.7	6.5	10.1	8.4
N:Mg	**80.0**	**51.0**	**57.0**	**52.0**	**57.0**	**27.0**	**30.0**	**25.0**	**23.0**	**75.0**	**84.0**	**39.0**

[a] Soil application (150 g fertiliser, per tree): Kieserite ($MgSO_4 \cdot H_2O$; 27 % MgO) in spring 1984; ammonium sulphate ($(NH_4)_2SO_4$; 21 % N) in spring 1985.
[b] Deficiency range ($mg\ g^{-1}$ d m): N, 12–13; P, 1.1–1.2; K, 4.0–4.5; Ca, 1.0–2.0; Mg, 0.7–0.8; Mn, 0.02–0.08; Zn, < 0.013.
[c] Balanced nutrition: N:P = 6–12; N:K = 1–3; N:Ca = 2–7; N:Mg = 8–30.

Compared with the control (the analytical data of the 1984 samples varied within the normal range), the Mg content of the fertilised trees almost doubled. The supply of all other nutrient elements investigated was optimal, except for Ca. This positive effect of fertiliser application continued in all subsequent years, i.e. in autumn 1988, the plot to which Mg was added was completely green whereas control trees still showed severe deficiency symptoms. Within the five-year research period a pronounced difference in needle biomass resulted, i.e. the unfertilised trees showed approximately 35% higher needle losses than the regreened spruces (cf. Huettl, 1988).

Even though the Mg supply was improved remarkably, the Mg values were still below the threshold for deficiency (Baule & Fricker, 1967). Until the end of 1985, this finding correlated with the presence of weak discoloration in some of the trial trees. In this context, the Mg content in the needles of the fertilised trees does not vary from year to year. In contrast, the fluctuation of the N contents is large. For both plots, the N contents reached the highest value in autumn 1985 and then decreased. Also the P and K values tended to decrease. However, P and K were always in the range of sufficient to optimal supply. The Ca content was also marked by large fluctuations. The complete regreening of the fertilised trees coincided temporarily with more balanced N:Mg ratios (Table 4). However, because of the fluctuation of the Mg content in the unfertilised trees no trend in the N:Mg ratios could be found. The extremely large N:Mg value occurred in the very dry and warm summer of 1983. The temporary improvement in availability of N (1983–1985) at this elevated site might be related to increased mineralisation and nitrification due to more active microbial decomposition or more favourable climatic conditions (Ulrich, 1986; Feger, Zoettl & Brahmer, 1988). The declining N values were correlated with cooler and wetter vegetative periods.

Increasing Mg deficiency

Another plot of the Elzach trial received ammonium sulphate (126 kg N ha^{-1}) in spring 1985, to test the hypothesis that increased anthropogenic N input, particularly HN_4-N, would induce nutrient imbalances and thus provoke deficiencies. In autumn of the same year, the trees of the plot had worse yellowing symptoms. Compared with the control trees, most of these spruces had tip-yellowing in the younger needles. The data from needle analyses underlined this observation (Table 4). The increased N uptake resulted in a very unbalanced N:Mg ratio. This situation was even more extreme in 1986. In 1987, however, the N effect was not observed by needle analysis, and the Mg supply was somewhat improved. It is interesting that in comparison with the control trees the Ca content of the fertilised trees did not recover from its low value of 1986. However, the K content of both plots stayed at about the same level. In contrast to both the Mg and the control plots the P contents of the N plot remained stable over the last two years.

It may be concluded from the Elzach trial that Mg fertilisation is an appropriate tool to reduce forest damage associated with Mg deficiency. However, the application of N, particularly in form of NH_4-N, may impede Mg uptake. Similar results were obtained from K fertiliser trials at sites where coniferous trees showed K deficiency (Huettl, 1988). Furthermore, Norway spruce stands suffering from K and Mg deficiency could be revitalised by applying K and Mg fertilisers.

Experiences from earlier liming and fertilisation trials

The findings of the above trials raised the question of whether the application of Mg and K fertilisers at times prior to the occurrence of the decline led to an improvement in the vigour of trees and stands in areas nowadays affected by the 'new type' of forest damage.

Generally, former fertiliser trials were aimed at increasing wood production of forest stands. Therefore Mg fertilisation was not of major concern. Furthermore, because Mg deficiency in high altitude forests as well as K deficiency on certain moranic sites, have developed only recently, the number of relevant trials is limited (Huettl, 1989).

The evaluation of older liming trials has produced varying results. Aldinger & Kremer (1985) did not find significantly increased growth or improved vigour in older Norway spruce stands on acidic sandstone sites in the Black Forest that were limed between 1964 and 1975. However, the application of lime containing Mg in an area that has been marked by Mg deficiency symptoms since the late 1970s did not show the appearance of tip-yellowing. Furthermore, growth of these stands was improved when compared with the control trees. In addition, Bauch *et al.* (1985) demonstrated better growth and less damage in Norway spruce stands that had been treated between 1953 and 1980 with lime containing Mg. Kreutzer's (1984) investigations of older liming trials (to which a certain amount of Mg limestone had been applied and in some cases combined with P, K, and Mg fertilisation) indicated for spruce and pine stands that fertilisation and liming significantly reduced the effects of the 'new type' forest damage in Lower Frankonia. These stands were limed and/or fertilised between 1973 and 1979. Similar conclusions were also arrived at by Andersson & Persson (1988) when they reviewed the literature on forest liming. Evers (1984) evaluated a fertiliser trial on an older Norway spruce stand that had been established at a sandstone site in the Oden Forest of southwest Germany at the end of the 1950s. More than 20 years after the application of 'Kalimagnesia' (sulphate of potash and magnesium) needle

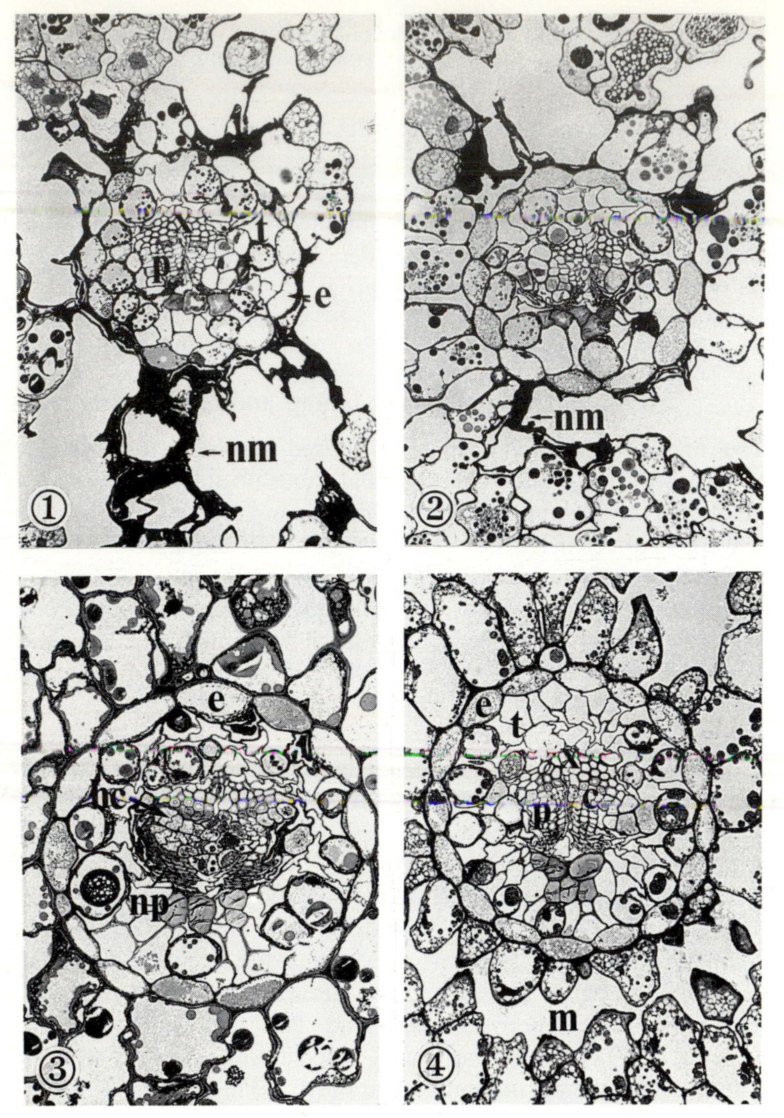

Fig. 1. Cross-section through a two-year-old needle of Norway spruce, exposed for two years to fluctuating concentrations (30–200 μg m^{-3}) of SO_2. Note the intact vascular bundle in the centre, encircled by the endodermis and transfusion tissue, containing xylem tracheids in the upper centre and functional phloem with open sieve cells below. Outside the endodermis there are extensively collapsed necrotic mesophyll cells (cf. with healthy, intact mesophyll in Fig. 4). Abbreviations: c = cambium, e = endodermis, hc = hypertrophied cambium, m = mesophyll, nm =

analyses still showed a positive response to fertiliser. In contrast to the control spruce trees the tip-yellowing observed in this area since about 1980 did not occur in the K + Mg fertiliser plot. In addition, growth of wood volume was increased due to Mg application. At comparable sites Evers also found similar results in beech stands.

In a fertiliser trial on Norway spruce that was established on a sandstone site in Lower Saxony in 1964, Baule (1984) demonstrated a remarkable improvement in tree vigour. An inventory of the crown biomass showed large loss of needles for the control trees as well as for the N + P plot. The NP + KMg plot was almost undamaged. In this case, the visible differences in vigour of the spruce trees also correlated with analyses of element content of their needles.

Already in the late 1920s, K and Mg fertiliser trials were established for a number of different tree species at Hammelspring-Werderhof in Neubrandenburg, in the former German Democratic Republic. For almost 60 years these trials have been maintained and intensively investigated (Bruening, 1959; Trillmich & Uebel, 1972; Uebel & Fiedler, 1977). The soils are derived from acidic glacial sands degraded by former agricultural practices. Due to the extremely low K and Mg supply of the soil, aforestation was difficult. Various tree species could only be established by means of appropriate fertiliser applications. Recent re-investigations of these trial stands illustrated a sustained positive fertiliser effect, even though this area is affected by relatively high air pollution (D. Heinsdorf, Institute of Forest Science, Eberswalde, FRG, personal communication). Similar, long-term effects were reported by Nowak *et al.* (1988) when they re-investigated the Pack Forest fertiliser trials in the Adirondacks of northeastern USA where soil conditions are comparable to those described above.

---

necrotic mesophyll, np = necrotic phloem, p = phloem, rp = regenerating phloem, t = transfusion tissue, x = xylem.

Fig. 2. Cross-section through a one-year-old needle of Norway spruce exposed for seven weeks to 300 $\mu g\ m^{-3}$ of ozone. Similar to Fig. 1, the vascular bundle is intact but several mesophyll cells outside the endodermis are necrotic and collapsed.

Fig. 3. Cross-section through a two-year-old yellow needle of a Norway spruce from a declining stand in the Black Forest. The mesophyll is intact, but in the vascular bundle the phloem is necrotic and collapsed, separated from the xylem by hypertrophic and hyperplastic cambium.

Fig. 4. Cross-section through a two-year-old green needle of a healthy Norway spruce. Note intact mesophyll and vascular bundle with open, functional sieve cells in the phloem.

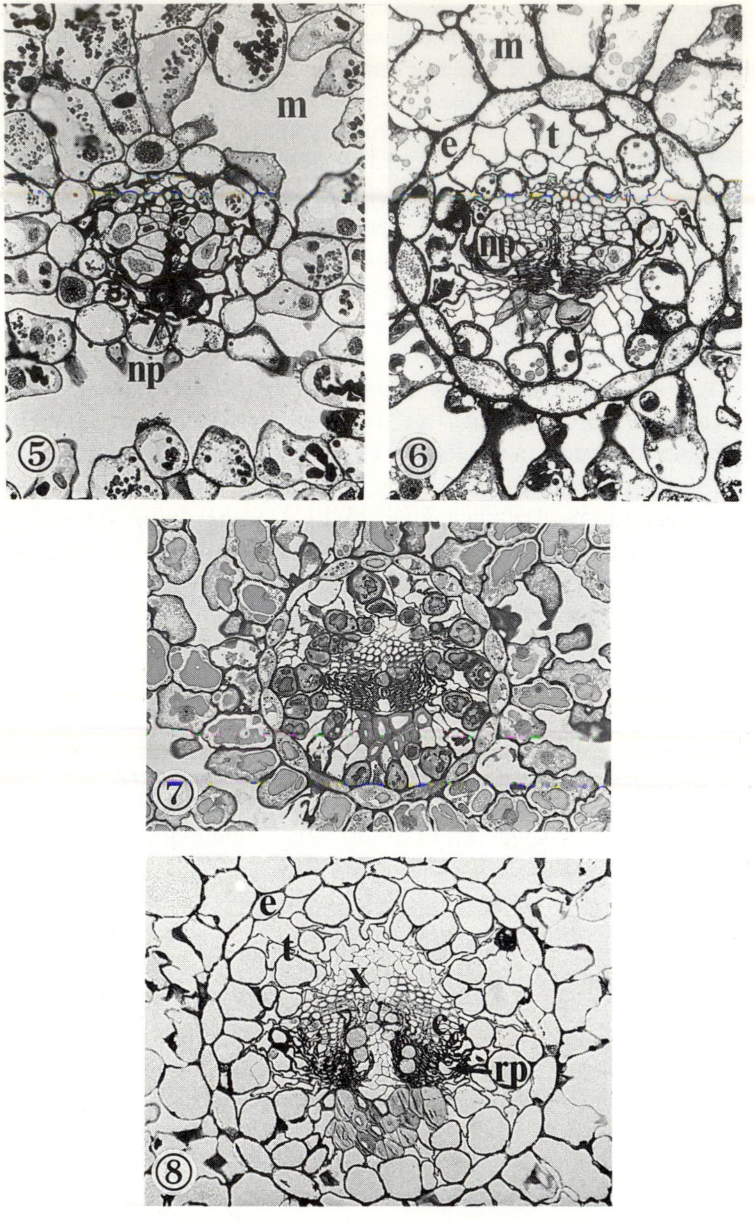

Fig. 5. Cross-section through a one-year-old yellow-tip Norway spruce needle from a seedling grown in Mg-free nutrient solution. The Mg-deficiency has caused severe necrosis of phloem with hypertrophy and hyperplasia of cambial cell; in principle same reaction as in the needles

Histological findings

Fink (1983) and Parameswaran, Fink & Liese (1985) found changes in the microscopic structure of the needle tissue of Norway spruce trees suffering from damage. This raised the question of whether the tissue damage was caused directly by air pollutants or indirectly by other factors.

Necroses of the mesophyll tissue, close to the stomata, were detected in needles of Norway spruce trees fumigated with SO_2 (Fig. 1) as well as with O_3 (Fig. 2) Only in very heavily damaged needles was the tissue of the central bundle affected. In contrast, Mg-deficient, tip-yellowed needles of Norway spruce suffering from the decline syndrome revealed damage within the vascular bundle; Fig. 3 indicates that mainly the phloem cells had collapsed. For comparison, Fig. 4 shows the appearance of a healthy Norway spruce needle. To verify these observations young Norway spruce plants were grown in a Mg-free nutrient solution to induce Mg deficiency. At the end of the first growing season the trees already showed the typical tip-yellowing and the needles had the same anatomical damage as those found in the tip-yellowed needles of trees suffering from decline at field sites (Fig. 5). Comparable findings were obtained by analysing Norway spruce needles affected by K deficiency (Fig. 6). The same histological phenomenon is shown (Fig. 7) in needles of a K deficient white spruce (*Picea pungens glauca*) from the Adirondacks in the USA. Furthermore, Huettl & Fink (1988) demonstrated the revitalisation of Norway spruce at an histological level; after regreening due to Mg fertilisation new phloem cells adjacent to the necrotic phloem tissue were regenerated by the cambium (Fig. 8).

Histological investigations of green needles from the fertiliser trial at Lutterberg (Lower Saxony; Baule, 1984) on acidic, base-poor sandstone soil, indicated that two-year-old needles of unfertilised 80-year-old spruces were marked by phloem damages typical of Mg and/or K deficiency. Initial damage to the phloem was already observable in current needles at the end

---

from declining trees (Fig. 3).

Fig. 6. Cross-section through a two-year-old needle of K-deficient Norway spruce from Saulgau, FRG. Note the partially collapsed phloem.

Fig. 7. Cross-section through a two-year-old needle of K-deficient white spruce needles from a trial at Lake Cushman (USA). Note also partially collapsed phloem.

Fig. 8. Cross-section through a three-year-old needle of Norway spruce, which originally was yellow but regreened following Mg-fertilisation. Adjacent to the necrotic phloem new sieve cells were regenerated by renewed divisions of the still living cambium. Hypertrophic parenchyma and ray cells ('Strasburger cells') are still visible in the damaged older phloem.

of the vegetation period. However, no damage was found in the needles of trees that had been fertilised with K and Mg 22 years before. Indeed, Mg deficiency and insufficient K supply were indicated by chemical analysis of needles for the unfertilised trees whereas the fertilised trees were marked by adequate nutrient concentrations (Huettl, 1988).

Summary

The 'new type' of forest damages in the FRG, central Europe, and in North America are frequently associated with acute nutritional disturbances. These disorders are mainly related to the elements Mg and K, which are mobile in the phloem. By means of appropriate fertilisation to improve the soil in which trees are grown the specific symptoms of damage can be reduced or completely removed. This holds also true for associated loss of foliage. These positive effects have now been observed for five years. By the application of N fertilisers, particularly in the form of NH_4-N, Mg deficiency can be induced on Mg-poor soils. Adequate Mg and K supply due to earlier Mg and K fertilisation of forest sites impeded the appearance of visible symptoms of Mg and K deficiency in forests where damage such as tip-yellowing have been observed in the unfertilised stands for some time. Histological investigations can be used as a diagnostic tool to differentiate between damages caused to trees by direct (air pollutants) and indirect (nutrient deficiencies) causes.

Acknowledgements

The authors wish to gratefully acknowledge the guidance and support of Professor Dr H.W. Zoettl, Director of the Institute of Soil Science and Forest Nutrition at the Albert Ludwigs University in Freiburg, FRG, and to thank the Federal Ministry of Science and Technology in Bonn, FRG, for funding most of the work presented. The excellent clerical effort was provided by Ms I. Mantel.

References

Aldinger, E. & Kremer, W.L. (1985). Zuwachsuntersuchungen an gesunden und geschaedigten Fichten und Tannen auf Praxiskalkungsflaechen. *Forstwissenschaftliches Zentralblatt*, **104**, 360–73.

Andersson, F. & Persson, T. (eds.) (1988). *Liming as a measure to improve soil and tree condition in areas affected by air pollution*. National Swedish Environmental Protection Board, Report 3518, 131 pp.

Bauch, J., Stienen, H., Ulrich, B. & Matzner, E. (1985). Einfluß einer Kalkung bzw. Duengung auf den Elementgehalt in Feinwurzeln und das Dickenwachstum von Fichten aus Waldschadensgebieten. *Allgemeine Forstzeitschrift*, **40**, 1148–50.

Baule, H. (1984). Zusammenhaenge zwischen Naehrstoffversorgung und Walderkrankungen. *Allgemeine Forstzeitschrift*, **39**, 775–8.

Baule, H. & Fricker, C. (1967). *Die Duengung der Waldbaeume*. München: Landwirtschaftsverlag. 259 pp.

Bosch, C., Pfannkuch, E., Baum, U. & Rehfuess, K.E. (1983). Ueber die Erkrankung der Fichte (*Picea abies* Karst.) in den Hochlagen des Bayerischen Waldes. *Forstwissenschaftliches Zentralblatt*, **102**, 167–81.

Bosch, C. & Rehfuess, K.E. (1986). Die Erkrankung der Fichte in den Hochlagen des Bayerischen Waldes – ein diagnostischer Duengungsversuch. In Proceedings, IMA-Querschnitt-Seminar *Restabilisierungsmaßnahmen – Duengung*, Karlsruhe, West Germany, 15–16 April, 1986, 15–20.

Bruck, R.I. & Robarge, W.P. (1988). Forest decline in boreal montane ecosystems of the southern Appalachian Mountains. In International Symposium, *Acidic deposition and forest decline*, Rochester, New York, USA, 20–21 October, 1988.

Bruening, D. (1959). *Forstduengung*. Neumann-Verlag, 210 pp.

BML (*Bundesministerium fuer Ernaehrung, Landwirtschaft und Forsten*) (1988). Waldschadenserhebung 1988. 83 pp.

Burg, J. van den (1991). N-Deposition, Naehrstoffversorgung und Duengungsversuche in den Niederlanden. *Forst und Holz* (in press).

Cook, E. & Johnson, A.H. (1988). Climate and forest decline – a review of the red spruce case. In International Symposium *Acidic deposition and forest decline*, Rochester, New York, USA, 20–21 October, 1988.

Evers, F.H. (1984). Welche Erfahrungen liegen bei K- und Mg-Duengungsversuchen auf verschiedenen Standorten in Baden–Wuerttemberg vor? *Allgemeine Forstzeitschrift*, **39**, 767–8.

Feger, K.-H., Zoettl, H.W. & Brahmer, G. (1988). Project ARINUS: II. Einrichtung der Meß-Stellen und Vorlaufphase. *Proceedings Kernforschungsanlage Karlsruhe–Projekt Europäischens Forschungszentrum für Maßnahmen zur Luftreinhaltung*, **35**, 27–38.

Fink, S. (1983). Histologische und histochemische Untersuchungen an Nadeln erkrankter Tannen und Fichten im Suedschwarzwald. *Allgemeine Forstzeitschrift*, **38**, 660–3.

Forschungsbeirat Waldschaeden/Luftverunreinigungen der Bundesregierung und der Laender (1986), 2 Bericht, 229 pp.

Friedland, A.J., Gregory, R.A., Karenlampi, L. & Johnson, A.H. (1984). Winter damage to foliage as a factor in red spruce decline. *Canadian Journal of Forest Research*, **14**, 963–5.

Friedland, A.J., Hawley, G.J. & Gregory, R.A. (1988). Red spruce (*Picea rubens* Sarg.) foliar chemistry in northern Vermont and New York, USA. *Plant and Soil*, **105**, 189–93.

Hauhs, M. (1985). Wasser- und Stoffhaushalt im Einzugsgebiet der Langen Bramke (Harz). Berichte des Forschungszentrums Waldoekosysteme/Waldsterben der Universitaet Goettingen, **17**, 206 pp.

Huettl, R.F. (1985). 'Neuartige' Waldschaeden und Naehrelementversorgung von Fichtenbestaenden (*Picea abies* Karst.) in Suedwestdeutschland. *Freiburger* Bodenkundliche Abhandlung, **16**, 195 pp.

Huettl, R.F. (1987). 'Neuartige' Waldschaeden, Ernaehrungsstoerungen und Deungung. *Allgemeine Forstzeitschrift*, **42**, 289–99.

Huettl, R.F. (1988). Vergleichende Analyse von Duengungsversuchen in der Bundesrepublik Deutschland und in den USA. IMA-Querschnitt-Seminar, Bayreuth, W. Germany, 20–21 November, 1988. *Proceedings Kernforschungsan lage Karlsruhe–Projekt Europäisches Forschungszentrum für Maßnahmen zur Luftreinhaltung*, **55**, 221–47.

Huettl, R.F. (1989). 'New types' of forest damages in central Europe. In *Air Pollutions Toll on Forests and Crops*, eds. J.J. McKenzie & M.T. El-Ashry, ch. 2, pp. 22–74. New Haven and London: Yale University Press.

Huettl, R.F. (1991). Nutrient supply and fertiliser experiments in view of N saturation. FERN Conference, Aberdeen, UK, 21–23 September, 1988. *Plant and Soil* (in press).

Huettl, R.F. & Fink, S. (1988). Diagnostische Duengungsversuche zur Revitalisierung geschaedigter Fichtenbestaende (*Picea abies* Karst.) in Suedwestdeutschland. *Forstwissenschaftliches Zentralblatt*, **107**, 173–83.

Huettl, R.F. & Wisniewski, J. (1987). A critique of forest fertilisation efforts in 'new type' decline forests associated with nutrient deficiencies: the West German and United States experience. (Manuscript.)

Innes, J.L. (1987). The interpretation of international forest health data.

In *Acid Rain: Scientific and Technical Advances*, eds. R. Perry, R.M. Harrison, J.N.B. Bell & J.N. Lester, pp. 633–40.

Isermann, K. (1987). Revitalisierung geschaedigter Fichten-Altbestaende durch Mineralduengung. *Allgemeine Forstzeitschrift*, **42**, 997–1000.

Kandler, O., Miller, W. & Ostner, R. (1987). Dynamik der 'akuten Vergilbung' der Fichte. *Allgemeine Forstzeitschrift*, **42**, 715–23.

Kaupenjohann, M., Zech, W., Hantschel, R. & Horn, R. (1987). Ergebnisse von Duengungsversuchen mit Magnesium an vermutlich immissionsgeschaedigten Fichten (*Picea abies* (L) Karst.) im Fichtelgebirge. *Forstwissenschaftliches Zentralblatt*, **106**, 78–84.

Kenk, R., Unfried, P., Evers, F.H. & Hildebrand, E. (1985). Zur langfristigen Duengung auf Zuwachs, Ernaehrung und Gesundheitszustand in einem Fichtenbestand des Buntsandstein–Odenwaldes. *Mitteilung der Forstl. Versuchs- und Forschungsanstalt Baden–Wuerttemberg, H.*, 114.

Kozlowski, T.T. & Constantinidou, H.A. (1986). Responses of woody plants to environmental pollution. *Forestry Abstracts*, **47**, 5–51.

Kreutzer, K. (1984). Mindern Duengungsmaßnahmen die Waldschaeden? *Allgemeine Forstzeitschrift*, **39**, 771–3.

Liu, J. (1988). Ernaehrungskundliche Auswertungen von diagnostischen Duengungsversuchen in Fichtenbestaenden (*Picea abies* Karst.) Suedwestdeutschlands. *Freiburger Bodenkundliche Abhandlung*, **21**, 193.

Lutz, H.J. & Breininger, M.Th. (1986). Erste Ergebnisse von Auswaschungsversuchen mit Fichten (*Picea abies*) in Wasserkultur. *Kali-Briefe*, **18**, 1–7.

Mohnen, V.A. (1988). Air pollutant distribution pattern: Elevational gradients, local chemistry. In *Air pollution and forest decline*, Proceedings of 15th international meeting for specialists in air pollution on forest ecosystems, Interlaken, Switzerland, 2–8 October, 1988, pp. 79–82.

Nowak, C.A., Shepard, U.P., Downard, R.W., White, E.H., Raynal, D.J. & Mitchell, M.J. (1988). Nutrient cycling in declining Adirondack conifer plantations: is acidic deposition an influencing factor? In *Acidic deposition and forest decline*, International Symposium, Rochester, New York, USA, 20–21 October, 1988.

Parameswaran, N., Fink, S. & Liese, W. (1985). Feinstrukturelle Untersuchungen an Nadeln geschaedigter Tannen und Fichten aus Waldschadensgebieten im Schwarzwald. *European Journal of Forest Pathology*, **15**, 168–82.

Prinz, B., Krause, G.H.M. & Stratmann, H. (1982). Waldschaeden in der Bundesrepublik Deutschland. *LIS-Berichte*, **28**, 1–154.

Reemtsma, J.B. (1988). *Ernaehrungsverhaeltnisse der Fichte im niedersaechsischen Kuestenraum.* Tagung der Sektion Waldernaehrung im DVFF, Wingst, W. Germany, 27–28 September, 1988.

Rehfuess, K.E. (1985). Vielfältige Formen der Fichtenerkrankung in Suedwestdeutschland. In *Was wir ueber das Waldsterben wissen*, eds. E. Niesslein & G. Voss, pp. 124–30.

Rehfuess, K.E. & Rodenkirchen, H. (1984). Ueber die Nadelroete-Erkrankung der Fichte (*Picea abies* Karst.) in Sueddeutschland. *Forstwissenschaftliches Zentralblatt*, **103**, 245–62.

Tomlinson, G.H. (1990). *Effects of Acid Deposition on the Forests of Europe and North America.* Boca Raton, Ann Arbor, Boston: CRC-Press.

Trillmich, H.-D. & Uebel, E. (1972). Ein Duengungstest zu Fichte. *Beiträge für die Forstwirtschaft*, **3**, 31–9.

Uebel, E. & Fiedler, H.J. (1977). Nachweis der Wirkung einer KMg-Duengung auf den Stoffkreislauf von Kiefernbestaenden unter besonderer Beruecksichtigung des Zellulosetests. *Zentralblatt für Bakteriologie, Mikrobiologie und Hygiene*, **132**, 515–31.

Ulrich, B. (1986). Die Rolle der Bodenversauerung beim Waldsterben: langfristige Konsequenzen und forstliche Moeglichkeiten. *Forstwissenschaftliches Zentralblatt*, **105**, 421–35.

Zech, W. (1983). Kann Magnesium immissionsgeschaedigte Tannen retten? *Allgemeine Forstzeitschrift*, **38**, 237.

Zech, W. & Popp, E. (1983). Magnesiummangel – einer der Gruende fuer das Fichten- und Tannensterben in NO-Bayern. *Forstwissenschaftliches Zentralblatt*, **102**, 50–5.

Zech, W., Suttner, Th. and Kotschenreuther, R. (1983). Mineralstoffversorgung vermutlich immissionsgeschaedigter Baeume in NO-Bayern. *Kali-Briefe*, **16**, 565–71.

Zech, W., Suttner, Th. & Popp, E. (1985). Elemental analysis and physiological responses of forest trees in SO_2-polluted areas of NE-Bavaria. *Water, Air, and Soil Pollution*, **25**, 175–83.

Zoettl, H.W. & Huettl, R.F. (1985). Schadsymptome und Ernaehrungszustand von Fichtenbestaenden im suedwestdeutschen Alpenvorland. *Allgemeine Forstzeitschrift*, **40**, 197–9.

Zoettl, H.W. & Huettl, R.F. (1986). Nutrient supply and forest decline in southwest Germany. *Water, Air, and Soil Pollution*, **31**, 449–62.

Zoettl, H.W., Huettl, R.F., Fink, S., Tomlinson, G.H. & Wisniewski, J. (1989). Nutritional disturbances and histological changes in declining forests. *Water, Air, and Soil Pollution*, **48**, 87–109.

Zoettl, H.W. & Mies, E. (1983). Naehrelementversorgung und Schadstoffbelastung von Fichten-Oekosystemen im Suedschwarzwald unter Immissionseinfluß. *Mitteilungen Deutsche Bodenkundliche Gesellschaft*, **38**, 429–34.

J.J.R. GROOT and J.H.J. SPIERTZ

The role of nitrogen in yield formation and achievement of quality standards in cereals

Introduction

Grain yield, nitrogen content of the grain and nitrogen yield of wheat may vary widely from site to site and from year to year under suboptimal growing conditions, as was shown by Benzian & Lane (1979) in their analysis of wheat production between 1954 and 1973 at Rothamsted. Growth and yield of a crop are the result of the interactive response of the plant to dynamic changes in weather and soil conditions and to the interference of pests, diseases and weeds. Recent crop physiological research has resulted in a better understanding of the dependence of crop growth and yield on weather conditions and availability of water and nutrients.

During the last two decades considerable effort has been exerted to obtain maximum grain yields under the growing conditions of northwestern Europe. It was shown by various authors (De Vos & Sinke, 1981; De Wit, Van Laar & Van Keulen, 1979; Evans, 1987) that grain yields of wheat have increased considerably. Part of this yield increase can be attributed to genetic improvement of the dry matter distribution as was found when ancient and modern wheat and barley varieties were compared (Austin *et al.*, 1980; Riggs *et al.*, 1981). The major part of the yield increase, however, is the result of improved management, especially time and rate of nitrogen application and pest and disease control (Spiertz & Ellen, 1978). One of the most important features of management responsible for the dramatic increase in crop yields has been the great increase in use of nitrogen fertiliser (Prins, Dilz & Neeteson, 1988).

Comparing ancient and modern wheat varieties under the same growing conditions shows that modern varieties allocate more carbohydrates to the grain, but the partitioning of nitrogen is relatively unaffected (Table 1) (De Vos & Sinke, 1981; Spiertz & De Vos, 1983; Van Dobben, 1962), leading to a smaller grain nitrogen content of higher-yielding varieties. In this regard the observations of Pommer (1983) should also be mentioned: in

Table 1. *Yield characteristics of five winter wheat cultivars.*

Variety (year of introduction)	Total above-ground biomass (t ha^{-1})	Grain yield (15% moist.) (t ha^{-1})	Harvest index dry matter (g g^{-1})	Nitrogen yield (kg ha^{-1})	Harvest index nitrogen (g g^{-1})
Staring (1941)	18.1	7.90	0.37	212	0.73
Felix (1958)	17.4	8.78	0.43	195	0.78
Manella (1964)	16.2	8.94	0.47	197	0.79
Arminda (1977)	16.8	9.69	0.49	207	0.82
Hobbit (1975)	17.7	10.18	0.49	213	0.79

Based on data of De Vos & Sinke, 1981.

spite of the changed assimilate distribution pattern, both the rate of root growth and the final root density did not differ between old and modern cereal varieties. So it is not surprising that genetic improvement of grain yield tends to be associated with a lower grain nitrogen content (Austin *et al.*, 1980; Kramer, 1979). However, in modern cereal production the grain nitrogen content has been maintained at a level of about 20 g N per kg dry weight, in spite of pronounced yield increases. The greater grain nitrogen requirement has been met by a larger input of nitrogen on varieties with considerably improved lodging resistance, use of growth regulators and by the introduction of split nitrogen dressings (Darwinkel, 1987; Spiertz & De Vos, 1983). The current situation in the EC, with decreasing prices for cereals, and the forthcoming stricter regulations to minimise nitrogen losses to the environment might lead to reduced use of nitrogen in cereal production systems, although high nitrogen input in cereals does not necessarily lead to increased losses to the environment (Addiscott, 1988; Dilz, 1988). However, lower nitrogen inputs may decrease grain nitrogen content in high yielding varieties.

In the normal range of protein contents from 90 to 180 g per kg dry weight, baking quality of commercial wheat flours depends on cultivar and directly on protein content. Generally, baking quality of wheat is defined as the suitability of a flour for baking a loaf with a large volume and with a light, fine and regular crumb. When loaf volume is plotted against protein content for samples of wheat cultivars, the rate of increase of loaf volume with protein content varies among cultivars (Bushuk, 1987). These differences among varieties are attributed to gluten content and gluten characteristics, amino acid composition, molecular weight distribution and protein fraction.

Several properties of wheat storage proteins have been identified as possible causes of intercultivar differences in baking quality. The key protein in this regard appears to be glutenin. Thus baking quality is strongly dependent on variety traits. Therefore, it is necessary to choose varieties which combine properties for good baking quality with agronomic characteristics that are most suitable under the specific growing conditions.

Quality standards can be defined for cereals other than wheat (e.g. barley) and different species (e.g. legumes) according to their use as food, feed or industrial purposes (e.g. starch production). In this paper the main emphasis will be on the agronomical and physiological aspects of the role of nitrogen in yield formation of cereals, within the limits set for protein yield and with regard to the specific quality standards.

Quality standards in cereal production

The agronomic value of a variety is still mainly determined by such traits as yield level and stability, disease-, lodging- and sprouting resistance. More recently, due to overproduction of poor-quality cereals and substantial imports of bread wheat, much research is focussed on improvement of quality traits without sacrificing yields (Bhatia & Rabson, 1987). Some specific quality traits are a high protein content and specific biochemical properties of the chemical compounds in the grain of bread-making wheat, and a low protein content and a large grain size for malting barley.

Depending on industrial technology a wide range of additional criteria has to be met. Generally, cereal producers are facing higher quality standards and a demand for less variability in quality traits between varieties and sites. At the same time there is a need for modification of current farming systems, because of the side-effects on nature and environment. A high priority will be given to prevent nitrate leaching and reduction of the number of applications and the amount of fungicides, insecticides and herbicides. This undoubtedly will affect cereal breeding and selection of new varieties (Fischbeck, 1988). A conceptual model of an ideotype of cereals for low-input agriculture in an environment with a high yield potential has to be developed. It may be hypothesised that the following features will be required: a high resistance to a wide range of pests and diseases, a high ability to compete with weeds, a prolonged root activity after anthesis, and earliness in spring and at anthesis.

Constraints in yield formation and quality achievement

Austin (1988) distinguished three types of limitations, which prevent a crop from reaching its potential yield, namely limitations set by climatic constraints, soil fertility and water availability, and physiological processes.

For each crop, the rate and duration of plant physiological processes are determined by the interactive response of plants to the growing conditions.

Plant response depends on genetic traits which can only be improved through breeding. The question of how yield and quality goals can be realised within a given set of environmental and nutritional constraints cannot be answered without a sound knowledge of the functioning of the various physiological processes involved in crop growth and root development. Modelling studies show that increased photosynthetic capacity of cereal leaves is a beneficial trait to attain higher yields, but it would increase the crop's requirement for nitrogen, and will therefore only be manifest when nitrogen is not limiting (Austin, 1982; Evans, 1989). Nitrogen shortage normally will be overcome by fertilisation, which, however, will be affected by environmental conditions. Drought not only causes moisture stress, but also hampers nitrogen mineralisation and nitrogen transport towards the root surface. Low soil temperatures reduce mineralisation, and also root growth; thus both the level of available soil nitrogen and consequently the nitrogen uptake capacity of the crop are reduced. Finally, weather conditions may favour the occurrence of diseases, thus impairing the translocation of carbon assimilates and nitrogen compounds from vegetative plant parts to the ears.

Carbon and nitrogen sources for grain growth

A schematic presentation of the interactions between carbohydrate and nitrogen metabolism, as well as the targets of action both of environmental and nutritional constraints during crop growth, is given in Fig. 1. C- and N-assimilation processes are strongly linked with external growing conditions, such as solar radiation, water and nutrient availability. However, grain growth depends on the internal supply of assimilates. For carbohydrates, the demand of the grains is mainly met by current photosynthesis, about 80%, and only to a limited extent by relocation of reserves, about 20% (Gent & Kiyanota, 1989; Spiertz, 1982). For proteins the grains rely mainly on the relocation of reserves, about 80%, and less on current uptake. Grain growth is strongly influenced by temperature, with a higher Q_{10}-value for N-accumulation than for C-accumulation (Spiertz & Ellen, 1978; Vos, 1981). Carbohydrate and especially protein reserves are required to maintain maximum rates of grain growth independent of short periods (ranging from hours to a week) of a reduced rate of C- and/or N-assimilation.

Grains are the major sink both for carbohydrates and nitrogen during grain growth. The sink size depends on the number and size of grains set. Growth rate of grains, both in terms of C and N, will never exceed a maximum value (Sofield *et al.*, 1977), even when assimilates and nitrogen are sufficiently available. This maximum rate is cultivar-specific, is related

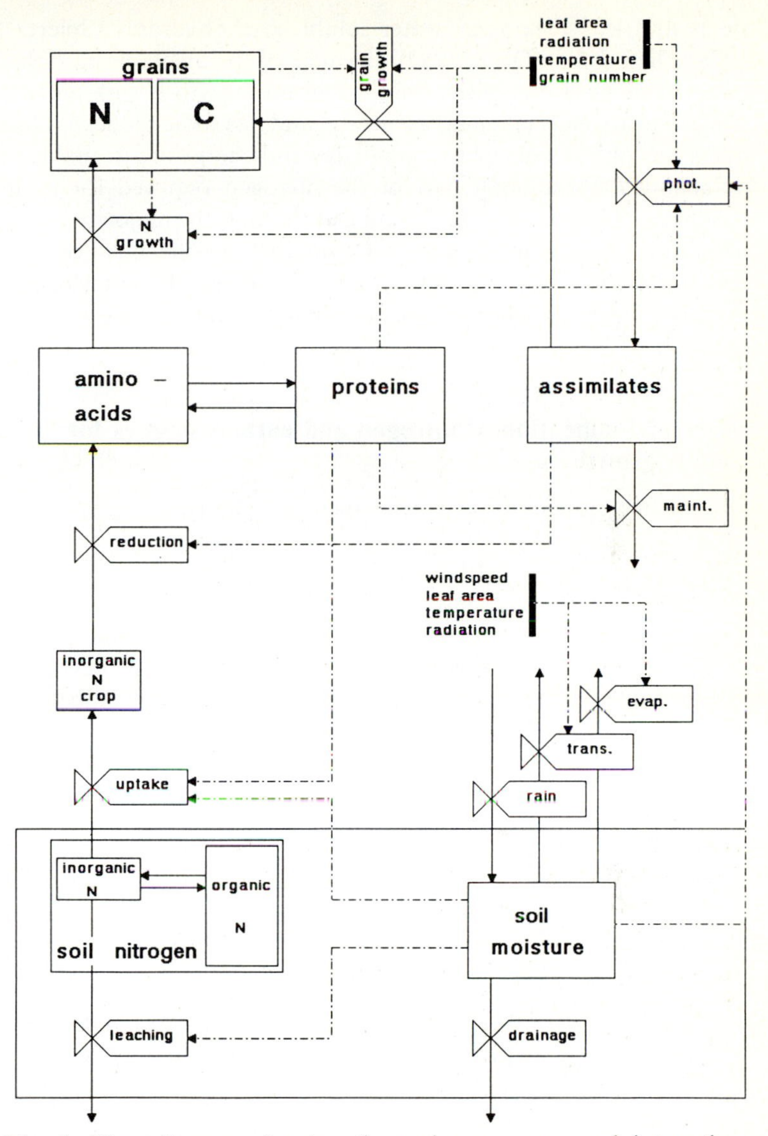

Fig. 1. Flow diagram showing the major processes and interactions between them during grain filling of cereals. Solid lines with arrows are material flows, those with broken lines are interactions.

to ambient temperature and can be described by empirical functions (Van Keulen & Seligman, 1987).

Shortly before and during anthesis, vegetative growth ceases, and assimilates are not yet used for grain growth but are stored temporarily in stems and leaf sheaths as water-soluble carbohydrates (Spiertz & Ellen, 1978; Ellen & Van Oene, 1989). Because of the available carbohydrates in this reserve pool, the initial phase of grain growth is sink-limited.

Most nitrogen is taken up as nitrate, and has to be reduced. During grain filling this only occurs when assimilates and energy are available for nitrate reduction. Only a small part of the nitrogen required for grain growth originates from root uptake during grain filling; the majority is translocated from vegetative plant material. Depending on the balance between the demand of the grains and the amount of reserves, this translocation results in a reduction in photosynthetic capacity and an acceleration of leaf senescence.

Modification of nitrogen and carbon sources for grain growth

The greatest potential for improvement of quality standards in cereals lies in the manipulation of the uptake pattern of nitrogen and in the proportion of nitrogen translocated from vegetative plant parts to developing grains. The uptake pattern of nitrogen consists of two distinct phases. Before anthesis, nitrogen uptake by the crop mainly increases the photosynthetic surface, the number of tillers and the number of grains per ear, while the photosynthetic capacity of the individual leaves is barely affected (Gregory, Marshall & Biscoe, 1981; Morgan, 1988). When nitrogen is supplied in excess of the crop's demand, a slight increase in maximum leaf nitrogen content is observed (Groot, 1988). During the second phase, after anthesis and, depending on the nitrogen supply, nitrogen uptake ceases and most of the nitrogen for grain growth is translocated from vegetative parts. Leaf proteins are hydrolysed and nitrogen is supplied in amino-acid or amide form to grains (Dalling, Boland & Wilson, 1976; Makino, Mae & Ohira, 1984).

In cereals, 40–60 % of leaf proteins consists of ribulose-1,5-bisphosphate carboxylase (Rubisco) (Mae, Makino & Ohira, 1983; Wittenbach, 1979), the key enzyme in photosynthesis. During grain filling, so-called self-destruction may occur (Sinclair & De Wit, 1976), as relocation of nitrogen decreases the amount and activity of the photosynthetic machinery. This self-destruction can be offset in part by prolonged nitrogen uptake, but as the root system senescences, and often nitrogen transport towards roots is hampered by drought, sooner or later this self-destruction occurs. A slower decrease of crop photosynthesis during grain filling can also result from a high nitrogen uptake before anthesis. It has been observed that the relationship between leaf nitrogen content and the rate of carbon exchange of single leaves (CER) is curvilinear (Evans, 1983; Hirose & Werger, 1987). Groot & Spiertz (1989) showed that the relationship between CER

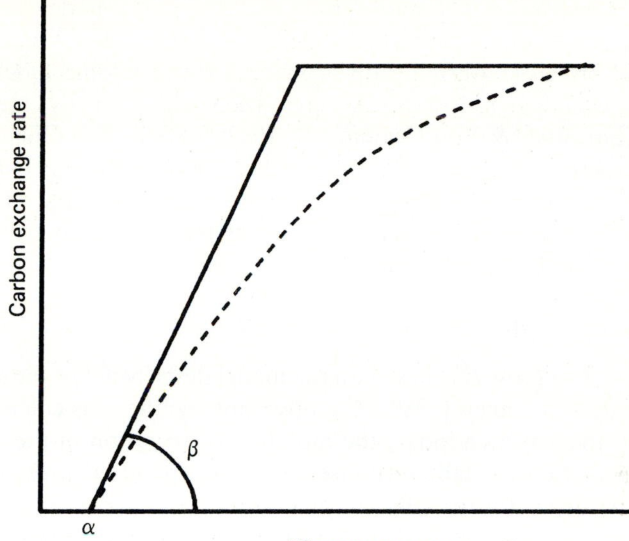

Fig. 2. Solid lines: schematic representation of the relation between leaf nitrogen content and rate of carbon exchange of single leaves, CER. α represents the residual nitrogen content and β represents the specific activity of Rubisco. Broken line: relation between leaf nitrogen content as it is normally found in experiments (after Groot & Spiertz, 1989).

and leaf nitrogen content consists of two parts (Fig. 2). During the first phase, the rate of CO_2 diffusion limits CER and leaf nitrogen content declines without affecting CER. During the second phase, nitrogen limits CER, and CER and leaf nitrogen content decline proportionally. The authors propose that the observed curvilinearity in the relationship is caused by a gradual transition between phase one and two. In Fig. 2, the intercept α represents residual nitrogen which is not available for translocation, as it is incorporated into cell-wall material (Charles-Edwards *et al.*, 1987; Van Keulen & Seligman, 1987).

Modelling carbon and nitrogen economy of cereal crops

Due to the multiple interactions between physiological processes and the fluctuations in environmental and nutritional limitations (Fig. 1), the extent to which fertiliser affects yield and quality cannot be predicted easily. The same applies to identification of crop characteristics that can be improved to increase yield and quality: the degree of improvement depends on the conditions experienced during the growing season.

Dynamic simulation of crop growth improves the assessment of the effects of fertiliser or the effects of changes in crop parameters within a given set of constraints. The model used in this study was developed for agronomic purposes, and not for detailed study of biochemical or plant physiological processes during crop growth. Although some biochemical and plant physiological processes have been analysed by means of simulation (Farquhar & Von Caemmerer, 1982; Spek & Van Oijen, 1988), the time constants in these models make simulation possible only over short time intervals, while crop physiologists are interested in the effects of agronomic measures or changes in crop characteristics on crop growth and yield during a whole growing season.

Crop growth

Simulation of crop growth is based on the model described by Spitters, Van Keulen & Van Kraalingen (1989). Canopy photosynthesis is calculated as a function of the leaf area index, the radiation distribution in the canopy, and the photosynthesis–light response curve of individual leaves. Maintenance requirements for the different plant organs, calculated as a function of their weight and chemical composition according to Penning de Vries (1975), are subtracted from daily gross assimilation and the remaining assimilates are allocated to leaves, stems and roots, depending on the development stage of the crop, according to empirical functions. In the model both the rate of photosynthesis and the maintenance respiration rate increase with a rise in nitrogen content of vegetative plant parts. Assimilates allocated to the various plant parts are converted into structural plant material. The energy required for the conversion (growth respiration), depends on the chemical composition of the growing material (Penning de Vries, 1974). In the present model, chemical composition is considered only in terms of proteins and carbohydrates.

In order to calculate water and nitrogen uptake, rooting depth and root distribution have to be known. The rate of root extension is related to soil moisture content and to the temperature of the soil compartment in which root growth occurs. It is assumed that root length density decreases exponentially down the profile.

Nitrogen uptake by the crop

Crop nitrogen demand before anthesis is based on the concept of nitrogen deficiency of leaves, stems and roots. As long as the nitrogen content of a given plant part is below its maximum possible value corresponding with the current development stage, there will be a sink for nitrogen. The values used for the maximum nitrogen content are given by Groot (1987), and are in the range of 6.7% for leaves at emergence to 0.75% at maturity of the crop. The actual nitrogen uptake proceeds according to crop demand if nitrogen transport through the soil is not limiting, otherwise the uptake has its maximum value under the prevailing conditions.

Crop development

Several events during the growth cycle of a crop are related to the development stage of the crop. The rate of crop development depends on the ambient air temperature, and is modified to account for the effects of vernalisation and photoperiod (Porter, 1984; Reinink, Jorritsma & Darwinkel, 1986).

Soil moisture

The soil is treated as a multilayer system. Changes in soil water content are treated as a combined effect of infiltration, extraction of soil moisture as a result of soil surface evaporation and by the crop (transpiration), and downward movement from one soil compartment to the next when the moisture storage capacity of a compartment is exceeded. Potential soil surface evaporation and potential crop transpiration are calculated according to a modified Penman method. When actual transpiration is smaller than potential transpiration, gross canopy assimilation is reduced accordingly.

Soil nitrogen

Soil nitrogen supply depends on the application of fertiliser nitrogen, decomposition of old organic matter (humus) and fresh organic matter (crop residues), crop nitrogen uptake, and downward movement of nitrogen by leaching. Denitrification is not taken into account. Decomposition of both types of organic matter is treated as a process with first-order kinetics, and the specific rate is modified to account for the effects of soil temperature and soil moisture content (Johnsson *et al.*, 1987). Decomposition results in either mineralisation or immobilisation of nitrogen, depending on the C:N ratio of the substrate.

According to De Willigen & Van Noordwijk (1987), nitrogen uptake proceeds according to crop demand if transport through the soil is not limiting. Otherwise the roots act like a 'zero-sink', i.e. uptake has a maximum under the prevailing conditions. Diffusion towards the root surface is considered to be the major limiting process in nitrogen uptake as soil water content considerably affects the diffusion coefficient (Barraclough & Tinker, 1981).

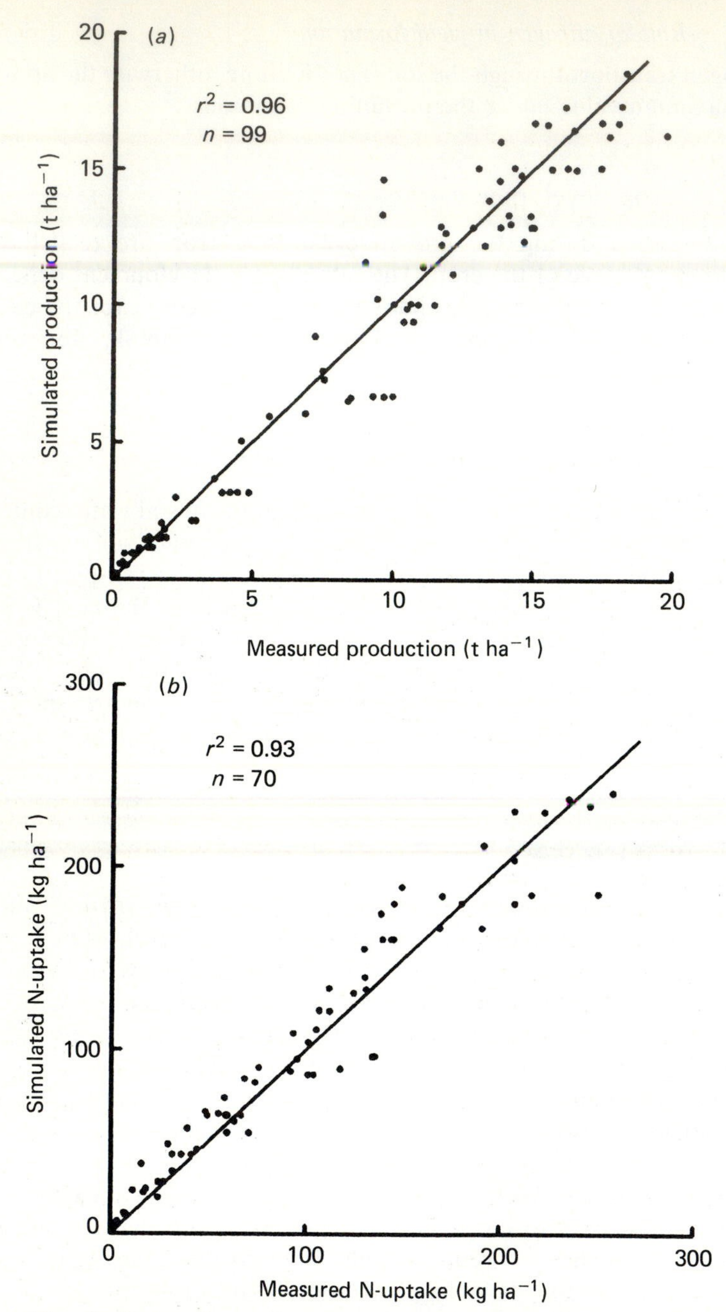

Fig. 3. Comparison of measured total dry matter production with simulated total dry matter production (*a*) and comparison of measured crop nitrogen uptake with simulated nitrogen uptake (*b*) for a series of experiments with nitrogen applications ranging from 0 to 200 kg ha^{-1}.

Validation

For a series of winter wheat experiments crop growth and crop nitrogen uptake were simulated for a range of fertiliser applications (Groot, 1989). The parameter values used in the model were derived independently of the studied experiments. The total dry matter production and nitrogen uptake (Fig. 3) were simulated within approximately 10 % of the measured values, which is considered a satisfying simulation.

Case studies by means of simulation

The greatest potential for improvement of grain quality lies in the manipulation of the uptake pattern of nitrogen and the proportion of nitrogen translocated from vegetative plant parts to developing grains as argued earlier. Model studies by Austin (1982) suggested that increased photosynthetic capacity was beneficial for higher yields, but that this would increase the crop's demand for nitrogen. In Fig. 2, the relationship between the rate of carbon exchange of single leaves and the leaf nitrogen content is given. The slope, β, of this relationship is assumed to be fixed for different cereal varieties, as it is a measure of the specific activity of RuBPCase. However, the amount of nitrogen which is part of the structural material (mainly cell walls) and which is not actively involved in carbon exchange, shows some genetic variability. Therefore, the effects of changes in the amount of structural nitrogen (α in Fig. 2) were predicted with the model; when α changes from 0.5 % to 1.5 %, grain yield decreases, but the nitrogen content of the grains increases considerably (Fig. 4*a*). In Fig. 4*b* the effects of changes in α are given in relative terms; the decrease of yield is greater than the positive effect on grain nitrogen content.

Increased nitrogen content of leaves up to a certain level (Fig. 2) increases the rate of carbon exchange, and with it the production capacity of the canopy. In the model it is assumed that the nitrogen uptake capacity is reflected in the maximum nitrogen content the crop can attain. In Fig. 5, the simulated effects of an increase in the maximum nitrogen content are given. An increase of 5 % in the nitrogen content of all plant organs increases the grain yield more or less proportionally. However, the nitrogen uptake by the crop will increase more dramatically. These results agree with Austin's (1982) findings that an increased photosynthetic capacity will increase the nitrogen requirement. The effect of a rise of 5 % in maximum nitrogen levels of the canopy on the protein content of grains was small but positive; grain protein content increased by 4 %.

In the introduction, the effects of nitrogen on yield and quality were discussed. The model was used to investigate these effects for growing conditions representative of those for cereal production in the Netherlands.

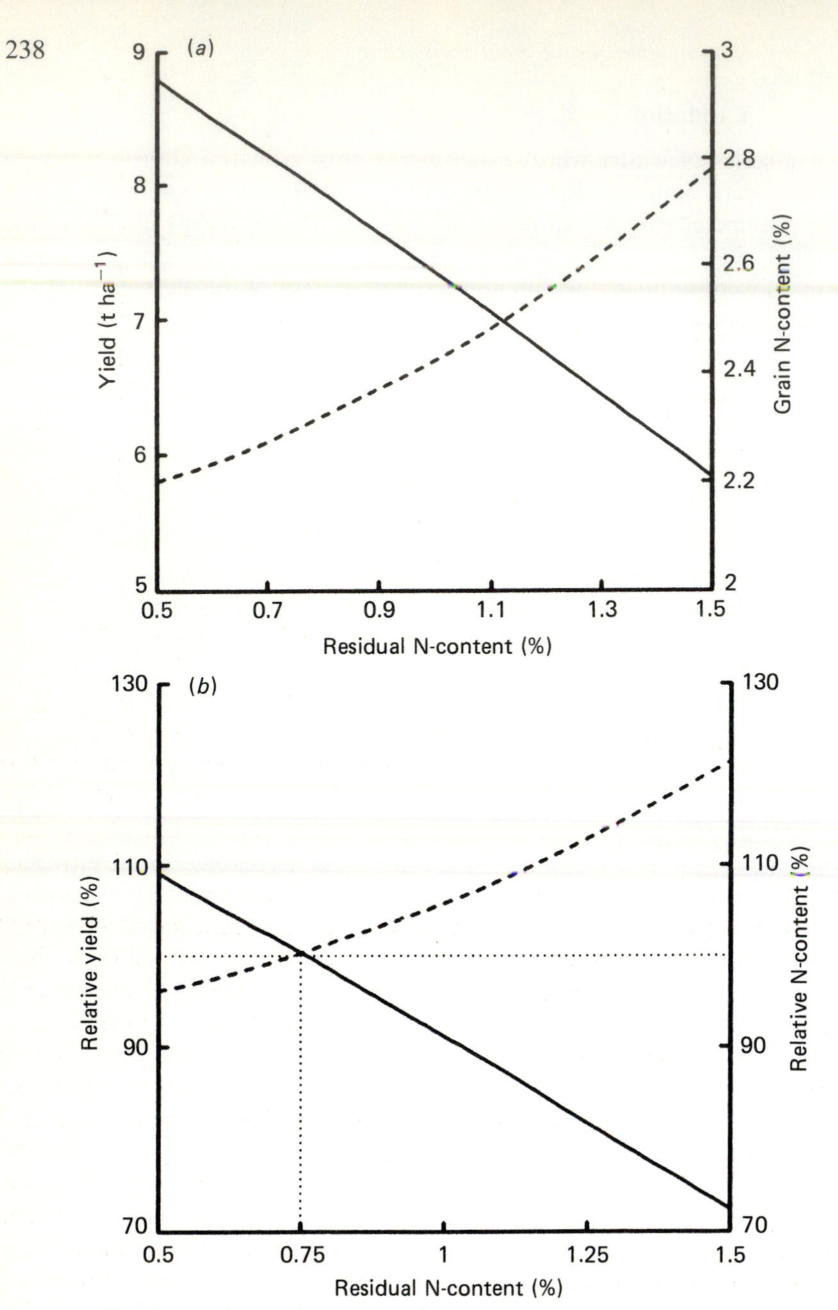

Fig. 4. Simulated grain yield (——) and simulated grain nitrogen content (---) as a function of the amount of residual nitrogen (nitrogen in structural material) of leaves, in absolute values of yield and nitrogen content (*a*) and in relative values of yield and nitrogen content (*b*).

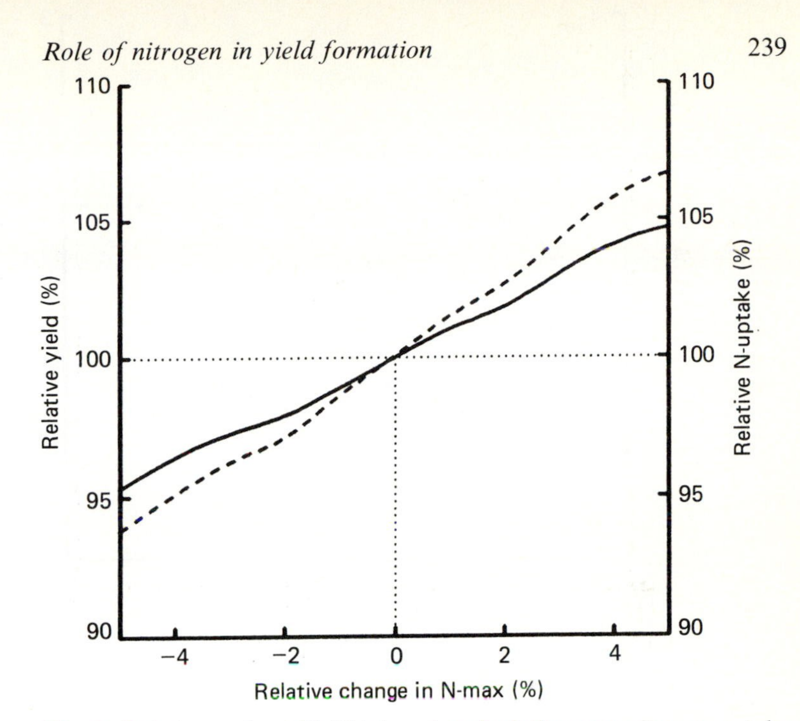

Fig. 5. Relative grain yield (%) (——) and relative crop nitrogen uptake (%) (---) for simulations in which the maximum nitrogen content of all plant organs varied within the range of 95% to 105% of the standard value.

For a given year, and for a Dutch polder soil (silty loam), the effect of the amount of N-fertiliser on final grain yield and on grain nitrogen content was simulated. Fig. 6*a* shows that without fertiliser a yield of 5300 kg ha^{-1} was calculated. The maximum grain yield of 8200 kg dry matter ha^{-1} (which is slightly above the average production in the Netherlands) is reached with a nitrogen application of 120 kg ha^{-1} and a nitrogen yield of 250 kg ha^{-1}. Higher application rates only seem to raise the grain nitrogen content. This simulated pattern is in agreement with experimental results of Darwinkel (1987), who showed that increased rate of N application above a certain threshold increased protein contents, without effect on grain yield.

Simulation of the nitrogen yield of the crop, and the amount of mineral nitrogen present in the upper 100 cm of the soil profile at harvest (Fig. 6*b*) show that for the chosen conditions, nitrogen applications up to 150 kg are almost completely taken up by the crop. Dilz (1988) and Addiscott (1988) reported that nitrogen applications for optimum grain growth may not increase soil nitrogen at the time of harvest. The simulations presented in Fig. 6*b* confirm their findings. They emphasise, however, that attempts to obtain high grain protein contents by applying more fertiliser are not

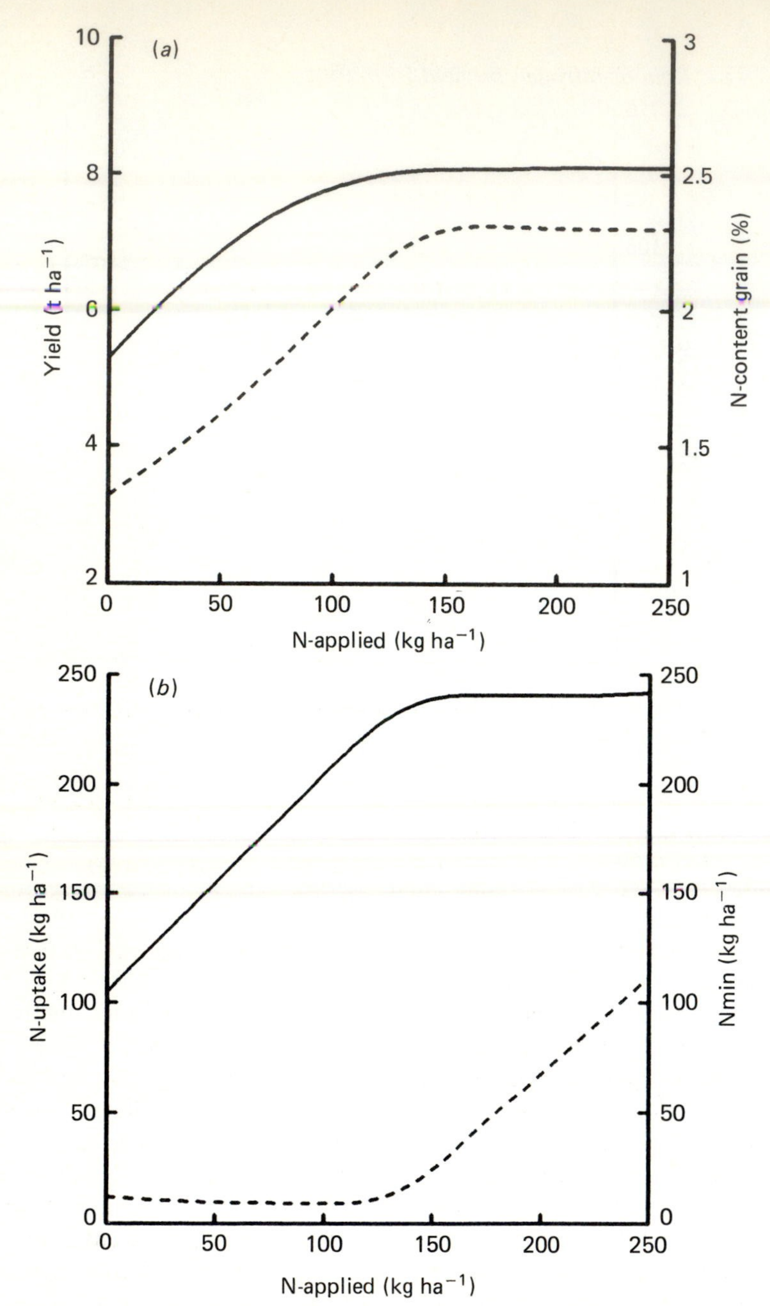

Fig. 6. (*a*) Simulated grain yield (——) and simulated grain nitrogen content (---) as a function of the amount of N-fertiliser applied. (*b*) Simulated nitrogen uptake by the crop (——) and simulated mineral nitrogen in the upper 100 cm of the profile (---) as a function of the amount of N-fertiliser applied.

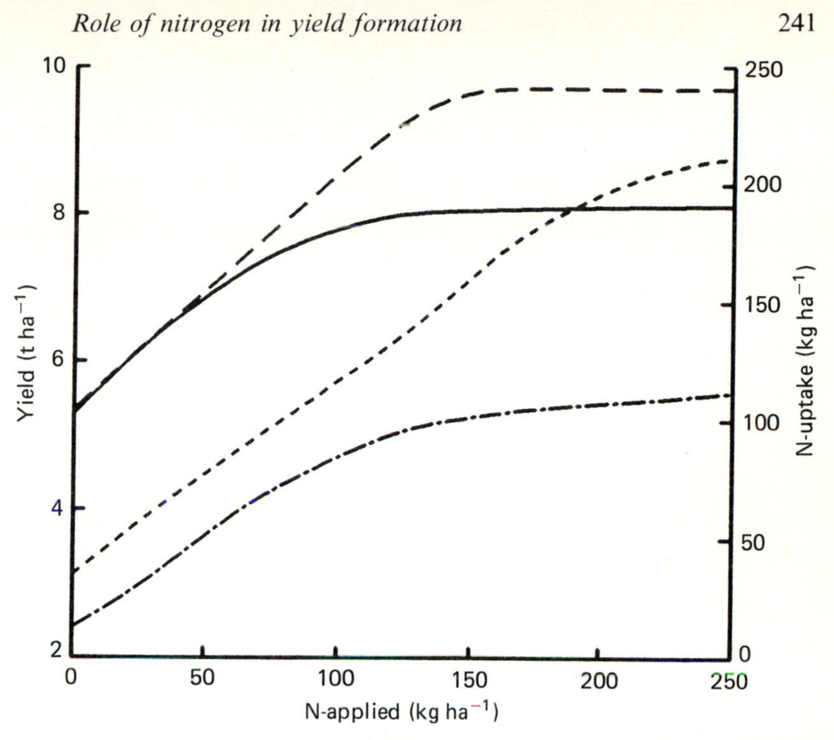

Fig. 7. Simulated grain yield for a clay soil (——) and for a sandy soil (– · –), and simulated nitrogen uptake by the crop for a clay soil (– –) and for a sandy soil (- - -).

without risk, and can increase nitrate in the soil at harvest. Nitrate present in autumn is easily leached during winter.

At low nitrogen applications the relation between grain yield and nitrogen uptake is linear; at higher rates the yield response curve deviates from the straight line, reflecting an increase in the nitrogen content of the grain. The level of the plateau where increased nitrogen uptake does not increase grain yields is determined by genetic traits (nitrogen partitioning efficiency and nitrogen content of the grains) and by environmental constraints (radiation, temperature, water shortage, diseases). Usually, grain yield of wheat grown on sandy soils is restricted by environmental constraints. As a result, nitrogen use efficiency is reduced, but nitrogen concentration of grains is larger.

The grain yield response and nitrogen uptake for soils differing in water holding capacity, but with comparable amounts of organic material and mineral nitrogen in the spring, are shown in Fig. 7. The smaller grain yield simulated for sandy soil can be partly attributed to a lower nitrogen uptake caused by a lower rate of nitrogen diffusion towards roots when the soil dries. The effect of reduced transpiration on photosynthesis of the canopy is the second major cause of yield reduction.

Discussion

Strategies for optimisation of plant characteristics

It is apparent from the extensive literature that, so far, breeding efforts for the improvement of grain quality have partly been successful. Difficulties encountered in breeding programmes aiming at high yields of superior quality cereals have to be solved through a better understanding of the biochemical basis of quality and of the interaction between the carbon and nitrogen economies in cereals. Often a negative correlation between grain yield and grain protein concentration has been observed among varieties (Kramer, 1979). In a three-year experiment with seven winter wheat varieties, Darwinkel (1987) showed clearly that the inverse relationship between protein content and grain yield holds for different levels of nitrogen supply. However, for each individual variety, grain yield as well as protein content are raised by increased nitrogen dressings. In general, bread wheat cultivars should be able to accumulate large amounts of nitrogen in vegetative material before anthesis, and to translocate nitrogen efficiently during grain filling. In addition, roots should maintain their uptake capacity throughout grain filling (Groot, 1988).

Grain protein concentrations and grain protein yield in response to nitrogen fertilisation were described by Fowler, Brydon & Baker (1989) with mathematical equations. They found a sigmoidal response of grain protein concentrations to nitrogen fertilisation, with concentrations ranging from 98.8 to 231.0 g protein per kg dry grain for wheat. It confirms the response found in our simulation study. A better understanding is needed of the genetic traits regulating protein accumulation in wheat grains under conditions of low and high nitrogen availability to plants.

In the simulation study the nitrogen content of the grains increased considerably with higher residual nitrogen contents in vegetative plant parts. This positive correlation may be the result of sacrificing yield at the expense of larger protein concentrations. To prevent possible artefacts in experimentation and modelling, the correlations between grain yield and protein concentrations have to be studied at optimum levels of nitrogen fertilisation and comparable grain protein yields.

For malting barley a different strategy has to be developed. Here, nitrogen is mainly needed for prolonged photosynthesis during grain filling, and a restricted relocation of nitrogen from the vegetative parts to the grains would be desirable in order to prevent high protein contents. However, it is postulated that starch and prolamin synthesis in developing barley grains are interdependent, and normal prolamin synthesis is essential for starch synthesis (Bhatia & Rabson, 1987).

Strategies for optimising grain yield and quality through crop management

Recommendations for nitrogen application are based on the amount of available soil mineral nitrogen at the beginning of the growing season and on the demand of the crop in a given environment. Both components may vary considerably due to environmental constraints. Despite inaccuracies in the assessment of soil mineral nitrogen content and the predicted nitrogen requirement of the crop, the recommendations for splitting nitrogen application in winter wheat have considerably increased the efficiency of nitrogen fertiliser use and also grain quality (Darwinkel, 1987; Spiertz & De Vos, 1983). However, for malting barley, nitrogen uptake after flowering should be minimised in order to avoid high protein contents.

It is apparent that availability of nitrogen as well as soil water is required for maximum uptake of nitrogen. Without adequate available soil water, roots cannot obtain sufficient nitrogen even though it may be present in the soil. This is often the case for cereals after anthesis, when soil water reserves are being depleted. The importance of soil water for root penetration and distribution and the importance of soil moisture and nitrogen being available simultaneously, support the concept that a certain amount of nitrogen should be available in deeper soil layers for better nitrogen uptake during grain filling of winter wheat, especially during dry years (Smika & Greb, 1973). Relatively high nitrogen uptake after anthesis in an extremely dry season was found by Spiertz & Ellen (1978).

Up to now nitrogen recommendations have been aimed mainly at obtaining high grain yields. At present, there is a need to include also quality and environmental objectives. To prevent pollution due to nitrate leaching it is important to determine mineral soil nitrogen; also threshold values should be established for the amount of available nitrogen required for achieving given quality standards. These threshold values will be site-specific with only marginal effects of varietal differences.

Acknowledgments

The authors are grateful to Dr K. Dilz for his comments on the manuscript.

References

Addiscott, T. (1988). Farmers, fertilizers and the nitrate flood. *New Scientist*, 8 October 1988, pp. 50–4.

Austin, R.B., Bingham, J., Blackwell, R.D., Evans, L.T., Ford, M.A., Morgan, Ch. & Tailor, M. (1980). Genetic improvements in winter wheat yields since 1900 and associated physiological changes. *Journal of Agricultural Science, Cambridge*, **94**, 675–89.

Austin, R.B. (1982). Crop characteristics and the potential yield of wheat. *Journal of Agricultural Science, Cambridge*, **98**, 447–53.

Austin, R.B. (1988). A different idiotype for each environment? In *Cereal Breeding Related to Integrated Cereal Production*, eds. M.L. Jorna & L.A.J. Slootmaker, pp. 47–61. Proceedings Cereal Section Conference of EUCARPIA. Wageningen: PUDOC.

Barraclough, P.B. & Tinker, P.B. (1981). The determination of ionic diffusion coefficients in field soils. I. Diffusion coefficients in sieved soils in relation to water content and bulk density. *Journal of Soil Science*, **32**, 225–36.

Benzian, B. & Lane, P. (1979). Some relationships between grain yield and grain protein of wheat experiments in South-East England and comparisons with such relationships elsewhere. *Journal of the Science of Food and Agriculture*, **30**, 59–70.

Bhatia, C.R. & Rabson, R. (1987). Relationship of grain yield and nutritional quality. In *Nutritional Quality of Cereal Grains; Genetic and Agronomic Improvement*, eds. R.A. Olson & K.J. Frey, pp. 11–43. Agronomy Monograph No. 28. Madison, Wisconsin, USA: ASA.

Bushuk, W. (1987). Aspects of chemical and physical structure of wheat proteins that determine breadmaking quality. In *Agriculture. Hard Wheat: Agronomic, Technological, Biochemical and Genetic Aspects*, ed. Basilio Borgho, pp. 7–18. CEC Report EUR 11172en, Brussels.

Charles-Edwards, D.A., Stutzel, H., Ferraris, R. & Beech, D.F. (1987). An analysis of spatial variation in the nitrogen content of leaves from different horizons within a canopy. *Annals of Botany*, **60**, 421–6.

Dalling, M.J., Boland, G. & Wilson, J.H. (1976). Relation between acid proteinase activity and redistribution of nitrogen during grain development in wheat. *Australian Journal of Plant Physiology*, **3**, 721–30.

Darwinkel, A. (1987). Improvement of grain quality of winter wheat in arable farming in the Netherlands. In *Agriculture. Hard Wheat: Agronomic, Technological, Biochemical and Genetic Aspects*, ed. Basilio Borgho, pp. 73–84. CEC Report EUR 11172en, Brussels.

De Vos, N.M. & Sinke, J. (1981). Progress in yield of winter wheat during the period 1940–1980. (Original in Dutch.) *Bedrijfsontwikkeling*, **12(6)**, 615–18.

De Willigen, P. & Van Noordwijk, M. (1987). Roots, plant production and nutrient use efficiency. Ph.D. Thesis, Agricultural University Wageningen, the Netherlands, 282 pp.

De Wit, C.T., Van Laar, H.H. & Van Keulen, H. (1979). Physiological potential of crop production. In *Plant Breeding Perspectives*, eds. J. Sneep & A.J.T. Hendriksen, pp. 47–82. Wageningen: PUDOC.

Dilz, K. (1988). Efficiency of uptake and utilization of fertilizer nitrogen by plants. In *Nitrogen Efficiency in Agricultural Soils*, eds. D.S. Jenkinson & K. Smith, pp. 1–26. London, New York: Elsevier Applied Science.

Ellen, J. & Van Oene, H. (1989). Effects of light intensity on yield components, carbohydrate economy and cell-wall constituents in spring barley (*Hordeum distichum* L.). *Netherlands Journal of Agricultural Science*, **37**, 83–95.

Evans, J.R. (1983). Nitrogen and photosynthesis in the flag leaf of wheat (*Triticum aestivum* L.). *Plant Physiology*, **72**, 297–302.

Evans, L.T. (1987). Opportunities for increasing the yield potential of wheat. In The future development of maize and wheat in the Third World, pp. 79–93. CIMMYT, Mexico.

Evans, L.T. (1989). Assimilation, allocation, explanation, extrapolation. In *Theoretical Production Ecology: Reflections and Prospects*. Simulation Monographs, pp. 77–87. Wageningen: PUDOC.

Farquhar, G.D. & Caemmerer, S. von (1982). Modelling of photosynthetic response to environmental conditions. In *Encyclopedia of Plant Physiology, Vol. 12B, Physiological Plant Ecology*, eds. O.L. Lange, P.S. Nobel, C.B. Osmond & H. Siegler, pp. 547–587. Berlin: Springer-Verlag.

Fischbeck, G. (1988). Cereal breeding and input reductions in cultivation of cereals. In *Cereal Breeding Related to Integrated Cereal Production*, eds. M.L. Jorna & L.A.J. Slootmaker, pp. 9–27. Proceedings Cereal Section Conference of EUCARPIA. Wageningen: PUDOC.

Fowler, D.B., Brydon, J. & Baker, R.J. (1989). Nitrogen fertilization of no-till winter wheat and rye. II. Influence on grain protein. *Agronomy Journal*, **81**, 72–7.

Gent, M.P.N. & Kiyanota, R.K. (1989). Assimilation and distribution of photosynthate in winter wheat cultivars differing in harvest index. *Crop Science*, **29**, 120–5.

Gregory, P.J., Marshall, B. & Biscoe, P.V. (1981). Nutrient relations in winter wheat. 3. Nitrogen uptake, photosynthesis of flag leaves and translocation of nitrogen to grain. *Journal of Agricultural Science, Cambridge*, **96**, 539–47.

Groot, J.J.R. (1987). *Simulation of nitrogen balance in a system of winter wheat and soil.* Simulation Report CABO-TT No. 13. Centre for Agrobiological Research & Department of Theoretical Production Ecology, Agricultural University, Wageningen, pp. 195.

Groot, J.J.R. (1988). Post floral nitrogen behaviour in grain crops: Measurements and simulation. In *Cereal Breeding Related to Integrated Cereal Production*, eds. M.L. Jorna & L.A.J. Slootmaker, pp. 93–7. Proceedings Cereal Section Conference of EUCARPIA. Wageningen: PUDOC.

Groot, J.J.R. (1989). Possibilities for improvement of N-fertilizer recommendation in winter wheat; a simulation study. (Original in Dutch.) *Meststoffen*, 1989–1, 1989–2/3, 31–5.

Groot, J.J.R. & Spiertz, J.H.J. (1989). Photosynthesis and nitrogen translocation in cereals during grain filling and implications for crop yield. In *Proceedings International Congress Plant Physiology*, New Delhi, India, 15–20 February, 1988 (in press).

Hirose, T. & Werger, M.J.A. (1987). Nitrogen use efficiency in instantaneous and daily photosynthesis of leaves in the canopy of a Solidago altissima stand. *Physiologia Plantarum*, **70**, 215–22.

Johnsson, H., Bergström, L., Jansson, P.-E. & Paustian, K. (1987). Simulated nitrogen dynamics and losses in a layered agricultural soil. *Agriculture, Ecosystems and Environment*, **18**, 333–56.

Kramer, T. (1979). Environmental and genetic variation for protein content in winter wheat (*Triticum aestivum* L.). *Euphytica*, **28**, 209–18.

Mae, T., Makino, A. & Ohira, K. (1983). Changes in the amount of Ribulose biphosphate Carboxylase synthesized and degraded during the life span of rice leaf (*Oryza sativa* L.). *Plant and Cell Physiology*, **24**, 1079–86.

Makino, A., Mae, T. & Ohira, K. (1984). Relation between nitrogen and Ribulose-1,5-biphosphate Carboxylase in rice leaves from emergence through senescence. *Plant and Cell Physiology*, **25**, 429–37.

Morgan, J.A. (1988). Growth and canopy carbon dioxide exchange rate of spring wheat as affected by nitrogen status. *Crop Science*, **28**, 95–100.

Penning de Vries, F.W.T. (1974). Substrate utilization and respiration in relation to growth and maintenance in higher plants. *Netherlands Journal of Agricultural Science*, **22**, 40–4.

Penning de Vries, F.W.T. (1975). The cost of maintenance processes in plant cells. *Annals of Botany*, **39**, 77–92.

Pommer, G. (1983). Root growth of old and new cereal cultivars. In *Root Ecology and its Practical Application*. International Symposium Gumpestein, 1982, Bundesanstalt Gumpenstein, A-8952 Irdning, pp. 483–90.

Porter, J.R. (1984). A model of canopy development in winter wheat. *Journal of Agricultural Science, Cambridge*, **102**, 383–92.

Prins, W.H., Dilz, K. & Neeteson, J.J. (1988). Current recommendations for nitrogen fertilisation within the EEC in relation to nitrate leaching. In *Proceedings of the Fertiliser Society*, no. 276, 27 pp. London.

Riggs, T.J., Hanson, P.R., Start, N.D., Miles, D.M., Morgan, C. & Ford, M.A. (1981). Comparison of spring barley varieties grown in England and Wales between 1880 and 1980. *Journal of Agricultural Science, Cambridge*, **97**, 599–610.

Reinink, K., Jorritsma, I. & Dawinkel, A. (1986). Adaptation of the AFRC wheat phenology model for Dutch conditions. *Netherlands Journal of Agricultural Science*, **34**, 1–13.

Sinclair, T.R. & De Wit, C.T. (1976). Analysis of the carbon and nitrogen limitations to soybean yield. *Agronomy Journal*, **68**, 319–24.

Smika, D.E. & Greb, B.W. (1973). Protein content of winter wheat grain as related to soil and climatic factors in the semi-arid central Great Plains. *Agronomy Journal*, **65**, 433–6.

Sofield, I., Wardlaw, I.F., Evans, L.T. & Zee, S.Y. (1977). Nitrogen, phosphorus and water contents during grain development and maturation in wheat. *Australian Journal of Plant Physiology*, **4**, 799–810.

Spek, L. & Van Oijen, M. (1988). A simulation model of root and shoot growth at different levels of nitrogen availability. *Plant and Soil*, **111**, 191–7.

Spiertz, J.H.J. (1982). Physiological and environmental determinants of potential crop productivity. In *Optimizing Yields – The Role of Fertilizers*, pp. 27–46. Proc. 12th Congr. Int. Potash Institute, Bern.

Spiertz, J.H.J. & De Vos, N.M. (1983). Agronomical and physiological aspects of the role of nitrogen in yield formation of cereals. *Plant and Soil*, **75**, 379–91.

Spiertz, J.H.J. & Ellen, J. (1978). Effects of nitrogen on crop development and grain growth of winter wheat in relation to assimilation and utilization of assimilates and nutrients. *Netherlands Journal of Aricultural Science*, **26**, 210–31.

Spitters, C.J.T., Van Keulen, H. & Van Kraalingen, D.W.G. (1989). A simple and universal crop growth simulator: SUCROS87. In *Simulation and Systems Management in Crop Protection*, eds. R. Rabbinge, S.A. Ward & H.H. Van Laar, pp. 147–81. Simulation Monographs. Wageningen: PUDOC.

Van Dobben, W.H. (1962). Influence of temperature and light conditions on dry matter distribution, development rate and yield in arable crops. *Netherlands Journal of Agricultural Science*, **10**, 337–89.

Van Keulen, H. & Seligman, N.G. (1987). *Simulation of water use, nitrogen nutrition and growth of a spring wheat crop*. Simulation Monographs, pp. 310. Wageningen: PUDOC.

Vos, J. (1981). *Effects of temperature and nitrogen supply on post-floral growth of wheat; measurements and simulations*. Agricultural Research Reports, 911, pp. 164. Wageningen: PUDOC.

Wittenbach, V.A. (1979). Ribulose biphosphate carboxylase and proteolytic activity in wheat leaves from anthesis through senescence. *Plant Physiology*, **64**, 884–7.

J.P. GRIME

Nutrition, environment and plant ecology: an overview

Introduction

The formative years of both agricultural research and plant ecology were characterised by a strong preoccupation with deducing the correlations which could be observed between soil characteristics and the mineral nutrition, growth and survival of native and cultivated plant species (Liebig, 1840; Tansley, 1917; Hoagland, 1919; Pearsall, 1922; Lundegardh, 1931; Hewitt, 1951; Balme, 1953; Steele, 1955; Gauch, 1957). During the middle third of the present century, excitement mounted as technical developments (controlled environment cabinets, nutrient solution culture and improved chemical analytical procedures) brought this field of research within the reach of refined experimental methods. Perusal of *Ecological Aspects of the Mineral Nutrition of Plants* (Rorison, 1969) conveys an accurate impression of the extent to which the growing competence and scale of such research was perceived as a major step towards a general explanation of plant distribution and crop performance.

Twenty years later it is instructive to re-examine those endeavours and to ask how far the high ambitions so evident in Rorison (1969) have been realised. Although more time must be allowed to pass before a complete perspective can emerge, it is already evident that research on mineral nutrition did not follow the predicted path and quite clearly the subject no longer enjoys such a high priority in agricultural and ecological research. This is because some of its former objectives are now revealed as founded upon false premises. New sub-disciplines have captured the attention of many agriculturalists and ecologists and the research goals of those who remain interested in the mineral nutrition of plants have often tended to become more specialised with respect to the range of species considered and experimental variables applied.

The initial purpose of this review is to identify briefly some of the main reasons why, after such promising beginnings, mineral nutritional studies have declined in number and coherence. Secondly, it is to suggest that economic and environmental concerns, allied to a more dynamic understanding of mineral nutrient capture and utilisation by plants of contrasting ecology, are leading to a renewed interest in broadly-based, integrated and comparative studies of the mineral nutrition of vascular plants.

Reasons for the recent decline of research into mineral nutrition

The reasons for the recent decline in the importance attached to mineral nutritional studies fall into two broad categories. First, agricultural technology and philosophy changed radically so that many of the applied research objectives of the preceding era became invalid. Second, developments within the science of plant ecology led to a re-evaluation of research priorities; this included both the promotion of studies unrelated to mineral nutrition and a realisation that the controlling effects of particular mineral nutrients on plant distribution and abundance were more indirect than was popularly supposed by many earlier ecologists.

The green revolution

Early agriculture relied heavily upon the idea of matching the demands of each crop species against the nutrient-supplying ability of particular soils. This gave rise to considerable curiosity and consequent research activity on the role of particular nutrient elements and metal toxicities as determinants of survival and yield. The need for this type of research has declined sharply as the modern revolution in agriculture first introduced the notion of 'changing the soil to match the crop', and then, plant-breeding and biotechnology began to manipulate substantially the crop itself. Removal or reduction of the limiting effect of mineral nutrients on crop productivity through the use of heavy dressings of mineral fertilisers and other ameliorative measures have allowed other research priorities (notably those related to pest and weed control) to move centre-stage.

Developments within plant ecology

Since 1969 there has been a rapid erosion of confidence in research approaches founded on the assumption that vegetation patterns arise from the direct effects of mineral nutrients on the distributions of plant species. This change in thinking is the result of several developments, all of which have their roots in an earlier literature.

Plant demography

The changed perspective, alluded to above, owes much to a widening of interest in the dynamic properties of both plant populations and plant communities. Building upon the pioneer studies of Salisbury (1942), Watt (1947) and Tamm (1956), it has been accepted that plant distributions have proximate, as opposed to ultimate, causal explanations in terms of the 'births, deaths and immigrations' (Harper, 1977) of individuals and that much can be learned by direct observation of these phenomena in the field. The incursion of demographic approaches has exerted both positive and negative effects upon our understanding of the ecological importance of mineral nutrition. At its best, demography has focused attention upon critical events in the life histories of individuals where mineral nutrition *may be suggested* as a predisposing factor in success or failure. At its worst, demography has merely produced a description of events through time with little penetration as to the possible role of mineral nutrients as an underlying cause. The difficulty in attempting to recognize nutritional effects by direct field observation was foreseen:

> Localized studies of this type are most effective where the vegetation pattern is imposed by rapid and catastrophic effects of environment. More typically however, fatalities are due to a complex of factors and in particular, the contribution of mineral nutritional factors remains obscure. Seedlings may persist for an indefinite period in a state of chronic nutrient deficiency and whilst it is often possible to recognize terminal phenomena, it is difficult to measure the extent to which plants may be predisposed to killing factors, by nutritional disorders.
>
> The direct observational approach is often extremely difficult to apply in older perennial communities where a plant distribution is the result of a long history of events arising from interactions between a plant population and a changing environment. In vegetation of some antiquity therefore it may be impossible to recognize mechanisms critical in the past, by examination of the contemporary situation.
>
> Grime & Hodgson (1969)

Therefore, we may conclude that demography has been useful to ecology in describing the effects of critical events but it has revealed little of the nutritional mechanisms which, acting over long periods, often predispose plant populations to such events.

Plant population genetics

The foundations of plant ecology demanded a system in which field data could be organised by reference to identifiable botanic units. This led to an early reliance upon taxonomy and, most particularly, upon descriptions of the field distributions of *species*. It was quite logical therefore to suppose that ecological understanding would follow from a closer study of species. This is quite explicit in A.R. Clapham's Presidential Address to the British Ecological Society:

> Our main concern as plant ecologists is to know why a plant of this species and not of that, is growing in a given spot; that whatever views we might entertain about the community as organism(s) or as a mere assemblage of individuals we should show ourselves to be interested primarily in autecological problems, that is enlarging our knowledge and understanding of the biology of individual species of plants.
>
> Clapham (1956)

Since Clapham's address, a large number of studies, many of them concerned with mineral nutrition, have established the principle that local populations of most plant species differ in genetic constitution according to local habitat and history (Bradshaw, 1959; Snaydon & Bradshaw, 1961). This discovery has made the study of micro-evolutionary processes within populations interesting in its own right and some have argued that studies of contemporary evolution below the level of the species provides the most promising basis for the elucidation of ecological mechanisms.

> Ultimate ecological explanation has to be focused on the nature of present evolutionary processes in action, which must imply the study of genetic individuals and their descendants.
>
> Harper (1982)

This view, which will be contested in the later parts of this paper, has been highly influential in shaping the activities of the present generation of research workers and, more than any other single factor, it has deterred plant ecologists from pursuing the broad comparative studies of species advocated by Clapham (1956).

The proliferation of non-equilibrium systems

In common with many other early ecological theories, those which emerged from the correlative phase of mineral nutritional study were primarily associated with vegetation of low productivity and considerable antiquity and the popularity of terms such as 'calcicole' and 'calcifuge' reflected the widely-held belief that, in these circumstances at least, vegetation resulted from an equilibration between the nutritional potential of the soil and the demands or tolerances of the species present. In modern landscapes, such as those of lowland Britain, it has become increasingly necessary to widen this perspective to include circumstances where disruptive impacts, of urbanisation, intensive agriculture, mechanised forestry, and other forms of human intervention, prevent such an equilibrium from forming. This transition is a general feature of the 'developed' world and non-equilibrium systems now occupy a large and rapidly-expanding sector of the land surface. The vegetation and mineral nutritional modes associated with non-equilibrium systems fall into two broad categories which differ according to the frequency and severity of habitat disturbance.

1. *Ephemeral vegetation.* Here the impact of climatic or mechanical disturbance of the habitat is frequent, severe and unpredictable and a selective advantage will accrue to plants which germinate and develop rapidly, absorb mineral nutrients at high rates and allocate them swiftly into offspring.
2. *Perennial vegetation.* When the intervals between major disturbance are longer, a dense, rapidly-expanding cover of fast-growing perennial vegetation may be expected. The sustained pulse of mineral nutrient release which normally follows habitat disturbance, will be gradually replaced by a dynamic spatial mosaic (Bhat & Nye, 1973) resulting from localised depletion of nutrients by the roots. In these circumstances a selective advantage will be enjoyed by those individuals which most rapidly and continuously project new, short-lived roots into the undepleted parts of the soil (Drew, 1975; Drew & Saker, 1975; Crick & Grime, 1987).

For rather different reasons, therefore, both types of non-equilibrium systems encourage plant species with high potential growth rates. This suggests an important nutritional feed-back in that the photosynthetic mechanisms of fast-growing plants require larger amounts of enzymes (Field & Mooney, 1986) and foliar concentrations of mineral nutrients especially phosphorus and nitrogen higher than those occurring in potentially slow-growing species (Fig. 1). Under the impact of increasing habitat disturbance and eutrophication, the composition of vegetation is therefore becoming more dependent upon the rates at which nutrients are captured and less upon the capacity to tolerate particular nutrient limitations.

The future

A renewed interest in mineral nutrition?

For two main reasons, we may suspect that a revival of research activity in mineral nutrition is imminent. Firstly, there is widespread recognition that

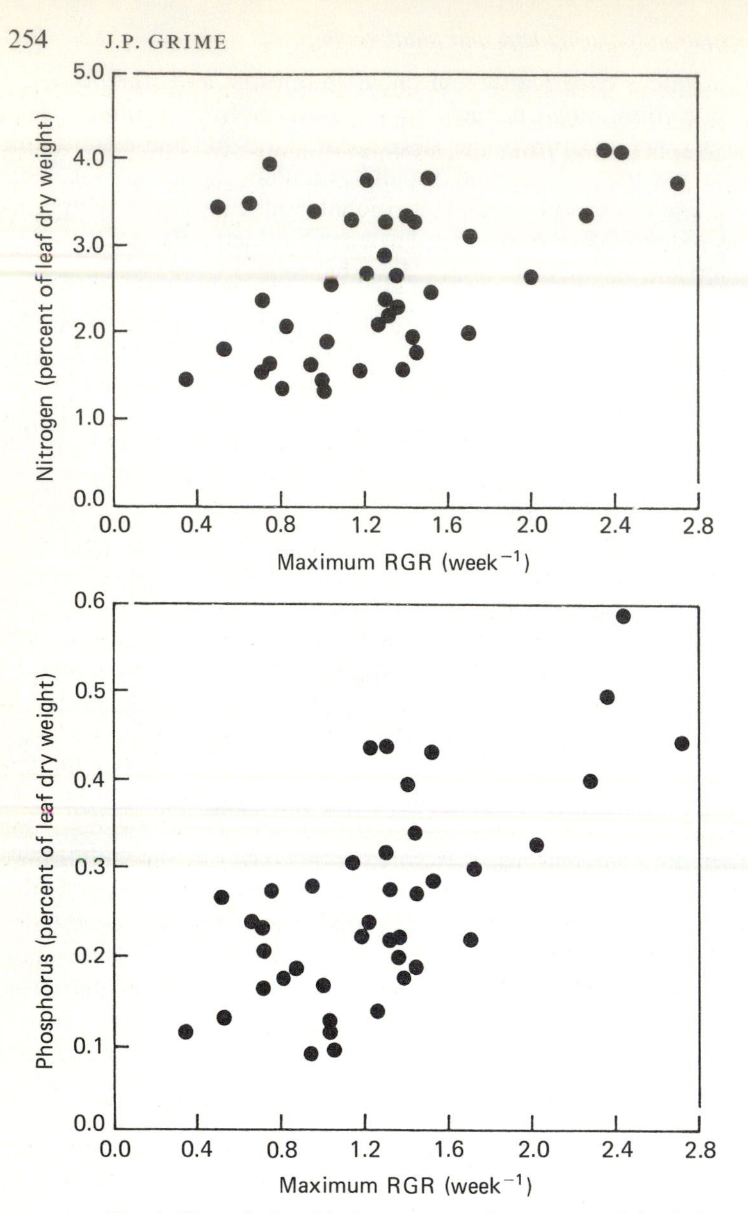

Fig. 1. The relationship between maximum potential relative growth rate (RGR) in the seedling phase and the average concentration of nitrogen and phosphorus in the leaf. Relative growth rates are from Grime & Hunt (1975). Each determination of nutrient concentration refers to mature, non-senescent leaves and is the mean of populations sampled from a wide range of natural habitats distributed within an area of 2400 km^2 in northern England (Band & Grime 1981).

the intensive use of fertilisers in agriculture is costly and wasteful and can cause the pollution of water courses and aquifers; this is likely to lead to renewed attempts to find plant species and techniques of cultivation which can sustain dry matter production at modest levels of soil fertility. Secondly, among ecologists there is a growing awareness that the recent tendency to neglect the study of physiological mechanisms and to rely upon demographic approaches to analyse population and vegetation processes is generating uncertainty with regard to the contribution of mineral nutrition to the mechanisms controlling plant distribution and community structure. Although the impacts of mineral nutrition upon vegetation patterns and plant population dynamics are less direct and dramatic than those of killing factors (e.g. predation, frost and drought), there is an abundance of circumstantial evidence pointing to mineral nutrients as a more subtle yet pervasive controller of local plant community distributions and rates of vegetation succession. Indeed, given the large, non-limiting supply of photosynthetic energy worldwide and the tendency in both temperate and tropical regions for temperature and in some instances water also to control the timing rather than the magnitude of plant growth, there is a strong case for the view that mineral nutrients, particularly phosphorus and nitrogen, provide the most important currency in ecosystem function. Mineral nutrients not only exert a dominant effect upon the quality and quantity of phytomass; they also control the rates of both cyclical and successional vegetation change. This broad generalisation is consistent with another hypothesis (Loveless, 1961; Grime, 1977; Coley, Bryant & Chapin, 1985) which identifies a widely recurring dichotomy between plants which capture mineral nutrients at rapid rates from productive habitats (and, inevitably, also release them at high rates through rapid tissue turnover) and those which capture mineral nutrients at much slower rates from infertile soils but, through slow tissue turnover and antiherbivore defence, retain them more effectively.

Is the species a legitimate subject for nutritional study?

If major effort is again devoted to ecological studies of mineral nutrition, the existence of intra-specific variation in nutritional characteristics could be viewed as a potential obstacle to the building of a predictive database. However, this presents a serious problem only where the form and amplitude of variation are unpredictable and relatively independent of the species. Analyses of extensive sets of field data (Hundt, 1966; Grime & Lloyd, 1973; Hansen & Jensen, 1974; Zarzycki, 1976; Grime, Hodgson & Hunt, 1988) reveal that despite genetic variation most common plant species have distinctive ecologies with different yet consistently-defined edaphic limits. Experimental evidence suggests that differences in ecological amplitude between species are frequently associated with constitutive features detectable from the responses of individuals. This is illustrated in Figs. 2 and 3 in which a mixed sowing of a large number of grassland species, each derived from a single local seed population, was introduced to experimental plots which provided controlled gradients of soil fertility and vegetation disturbance. After two years, the distribution of each species over the experimental matrix was measured. The comparison between *Rumex acetosa* and *Danthonia decumbens* (Fig. 2), reveals consistent and inter-specific differences in ecological amplitude (the ranges of nutritional stress and vegetation disturbance tolerated). In Fig. 3, measures of the evenness of persistence over the experimental matrix, for all species included in the experiment, are plotted against field estimates of the frequency of occurrence of the same species calculated from vegetation surveys over a 2400 km^2 area of varied landscape at a boundary between Upland and Lowland Britain (Grime *et al.* 1988). Therefore, for each species the test compares persistence of individuals from a small sample of a single population over the experimental matrix against a field distribution which results from the behaviour of numerous populations; the consistent relationship evident in Fig. 3 strongly suggests that, in the species examined, the capacity to exploit a wide range of soil conditions and other habitat variables is related to characteristics which are constitutive in the species and are manifested in each component population.

Another source of evidence which suggests the need to consider evolutionary specialisation at the level of the species is heavy metal tolerance. Although initial studies (Jowett, 1958; Jain & Bradshaw, 1966; Antonovics, 1968; Bradshaw, McNeilly & Gregory, 1965) emphasised the specificity and rapidity with which some local populations had developed metal tolerance on contaminated sites, there is growing evidence that the potential to evolve extreme tolerance is often species-specific and may coincide with a lower degree of constitutive, genetically-based tolerance of one or more heavy metals (McNaughton *et al.*, 1974; Bradshaw, 1975; Cox & Hutchinson, 1980; Reeves & Baker, 1984). Here, as in the nutritional examples presented in Fig. 3, it now appears that a balanced perspective of metal tolerance must include consideration of ancestral ecological specialisations. Whilst some of these are lodged at the level of the species, others are associated with families (Grime, 1984; Hodgson, 1986) and some may even coincide with larger and more ancient taxa, e.g. *Pteridophyta* (Grime, 1985).

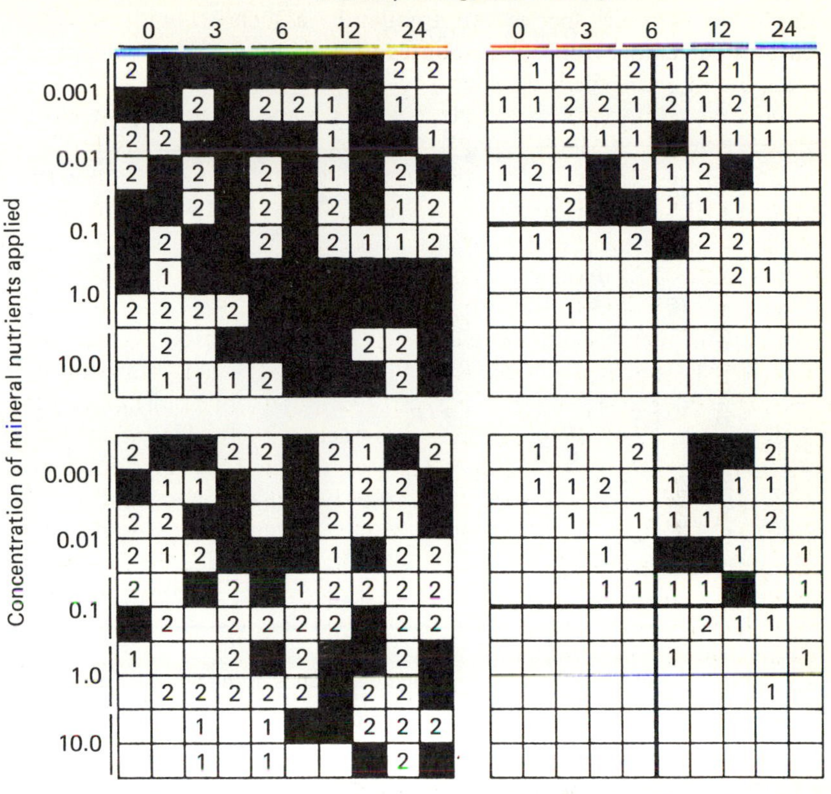

Fig. 2. The distribution of *Rumex acetosa* (left) and *Danthonia decumbens* (right) within plots of vegetation synthesised from a mixed sowing of a large number of grassland species. Each plot contains orthoganally-crossed gradients of soil fertility and vegetation damage (simulated grazing and trampling). Species occurrences were recorded in 100 mm × 100 mm subsections of three replicate plots. Symbols: ■, present in all three replicates; 2, present in two replicates; 1, present in one replicate; □, absent from all replicates. Upper diagrams: mixture without legumes. Lower diagrams: the same mixture with legumes present. Unpublished data of J.P. Grime and B.D. Campbell.

Patch and pulse: the dynamics of nutrient capture

With the notable exception of the models of Tilman (1982, 1987), contemporary theories of mineral nutrition have abandoned the notion that soils can be characterised by reference to the *concentrations* of particular nutrient ions. Most authors recognise spatial and temporal heterogeneity in

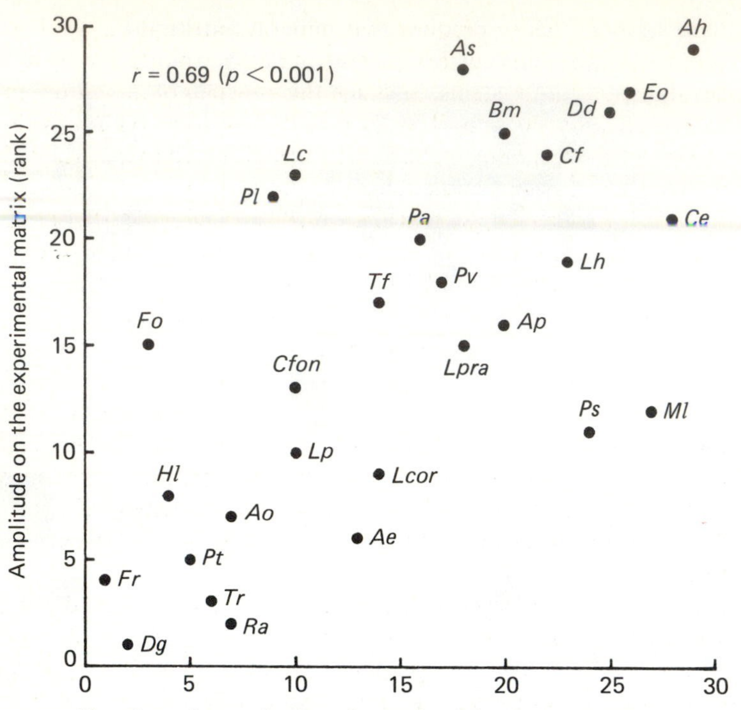

Fig. 3. Rank correlation between a field estimate of ecological amplitude (frequency in a stratified survey of grasslands in a 2400 km² area of Central England) and an index of the evenness of persistence of the same species in vegetation synthesised from a mixed sowing, including legumes, on a matrix of controlled gradients in soil fertility and vegetation damage. Key to species: *Ae, Arrhenatherum elatius; Ah, Arabis hirsuta; Ao, Anthoxanthum odoratum; Ap, Avenula pratensis; As, Anthriscus sylvestris; Bm, Briza media; Ce, Centaurium erythraea; Cf, Carex flacca; Cfon, Cerastium fontanum; Dd, Danthonia decumbens; Dg, Dactylis glomerata; Eo, Euphrasia officinalis; Fo, Festuca ovina; Fr, Festuca rubra; Hl, Holcus lanatus; Lc, Luzula campestris; Lcor, Lotus corniculatus; Lh, Leontodon hispidus; Lp, Lolium perenne; Lpra, Lathyrus pratensis; Ml, Medicago lupulina; Pa, Poa annua; Pl, Plantago lanceolata; Ps, Pimpinella saxifraga; Pt, Poa trivialis; Pv, Prunella vulgaris; Ra, Rumex acetosa; Tf, Trisetum flavescens; Tr, Trifolium repens.*

nutrient availability within the soil. In plant strategy theory (Grime, 1974) 'foraging' concepts analogous to those used to analyse food capture by animals have been applied to the very different mechanisms of mineral nutrient capture associated with fertile and infertile soils.

Plant strategy theory predicts that mineral nutrition on fertile soils will involve 'active foraging' which consists of patch exploitation by morphologically dynamic root systems, in which the life-span of individual fine roots is short. On infertile soils it is predicted that root systems will be less dynamic but by remaining functional throughout the year will be capable of intercepting the brief mineralisation pulses which are characteristic of many infertile soils (Gupta & Rorison, 1975; Taylor, De-Felice & Havill, 1982). Experimental tests of these predictions have been conducted (Crick & Grime, 1987; Grime *et al.*, 1989). These involve partitioned containers allowing presentation of mineral nutrients in various spatial and temporal patterns and a new technique which maintains patchiness without partitions (Campbell & Grime, 1989*a*, *b*). The results confirm the superior ability of species associated with fertile soils to capture nitrogen from the undepleted sectors of a nutritionally-heterogeneous rooting volume. This capacity is related to high rates of dry matter production and high specific N-absorption rates and is not the result of greater plasticity in the partitioning of dry matter between parts of the root system located in rich and poor sectors.

In a recent experiment providing nutrient pulses of varying duration (Fig. 4), evidence has been obtained of the differential ability of the roots of *Festuca ovina*, a slow-growing species of infertile soils, to exploit brief episodes of mineral nutrient enrichment; this appears to be associated with low rates of tissue turnover and the capacity of the roots to remain viable under chronic mineral nutrient stress.

In Fig. 5 an inverse relationship is apparent between the proportion of the root dry matter allocated to the undepleted sectors of a patchy rooting volume and the status achieved by the same species in an experimental plant community synthesised in productive conditions. This relationship has been interpreted (Grime, 1987) as the result of a difference in the scale and precision with which dominant and subordinate plants exploit the soil mosaic. 'Coarse-grained' foraging, in potentially dominant species such as *Arrhenatherum elatius*, involves the production of a robust and extensive system exploring a large rooting volume. In contrast, the smaller and predominantly fine-rooted systems of subordinates such as *Campanula rotundifolia* exhibits a more precise or 'fine-grained' adjustment of root development to local patchiness in mineral nutrient supply.

Mycorrhizas

Over the last 20 years one of the most conspicuous developments in ecological studies of mineral nutrition has been an increasing emphasis upon the role of mycorrhizal infections. Investigation has now progressed

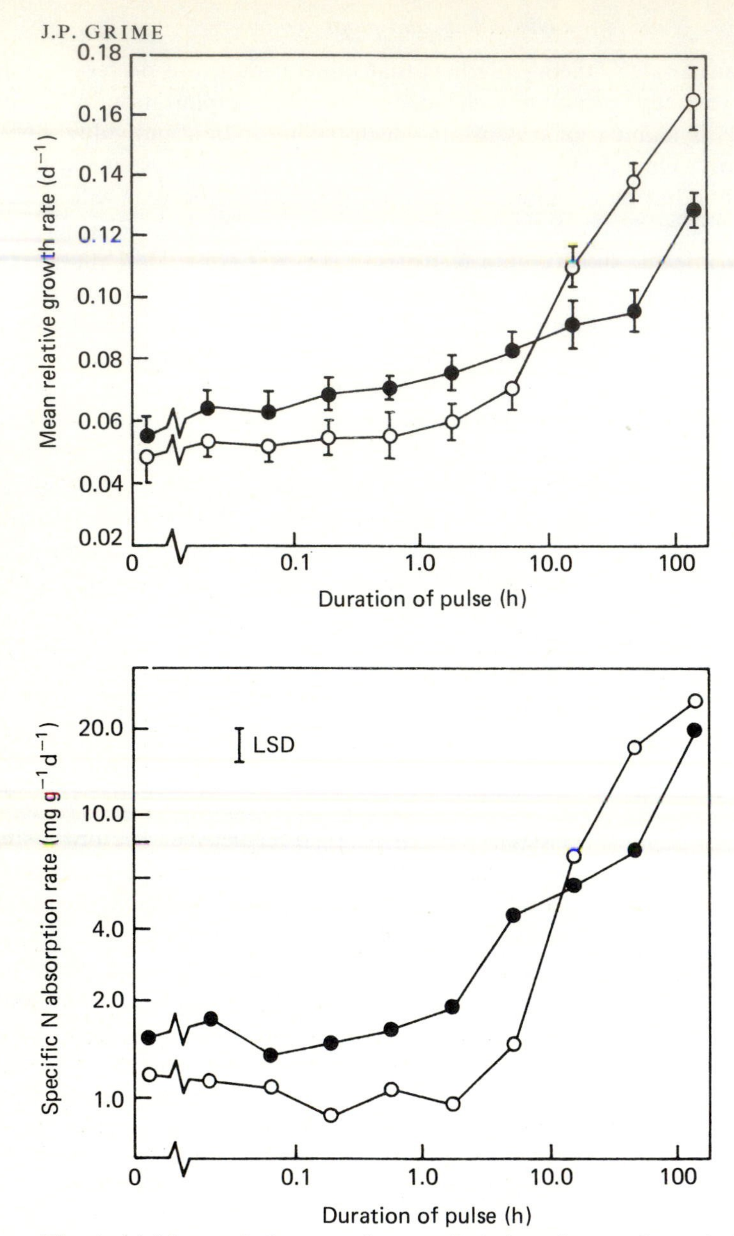

Fig. 4. (*a*) Mean relative growth rate of *Arrhenatherum elatius* (○) and *Festuca ovina* (●) plants exposed once every six days to nutrient pulse treatments of differing duration. Means are shown ±95% confidence limits. (*b*) Mean specific N-absorption rate of *Arrhenatherum elatius* (○) and *Festuca ovina* (●) plants exposed once every six days to nutrient pulse treatments of differing duration. Vertical bar is LSD ($p < 0.05$) for comparing means on logarithmic scale.

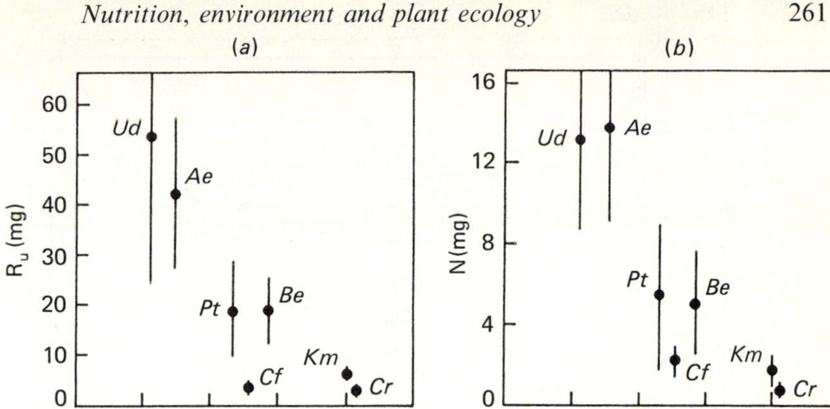

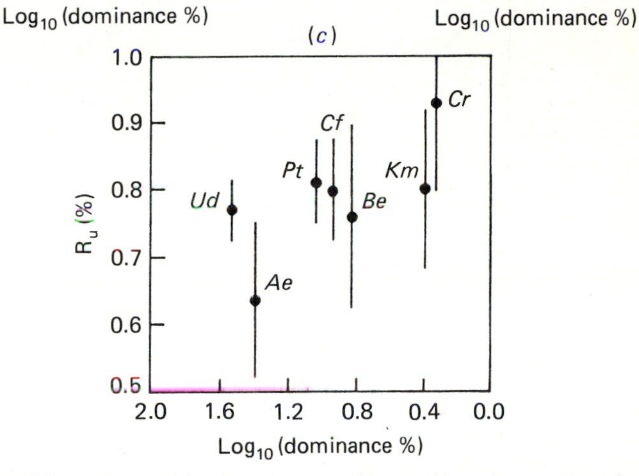

Fig. 5. The relationship between species ranking in an experimental community and the ability to modify root distribution and capture nitrogen in a nutritionally-heterogeneous rooting volume. Species were ranked according to the shoot biomass attained in the mixed experimental community after 16 weeks' growth in a productive greenhouse environment. 95% confidence limits are indicated by the vertical lines. Root distribution (Fig. 5*a*) is expressed as R_u, the increment of root dry matter to the undepleted sectors of a cylindrical mass of sand during a period of two weeks in which half the sand volume is deprived of nutrients using the standardised technique of Campbell & Grime (1989*a*, *b*). Fig. 5*b* describes the total nitrogen (N) increment over the two week depletion period. Fig. 5*c* expresses the increment of root dry matter to the undepleted sectors as a proportion of total root production during the depletion period. Key to species: *Ae, Arrhenatherum elatius; Be, Bromus erectus; Cr, Campanula rotundifolia; Cf, Cerastium fontanum; Km, Koeleria macrantha; Pt, Poa trivialis; Ud, Urtica dioica.*

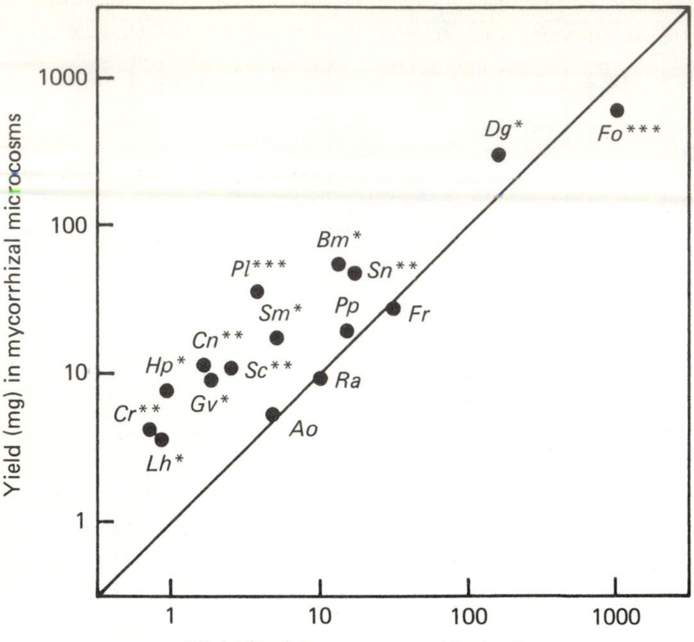

Fig. 6. Effects of mycorrhizal infection on the average shoot weight of individuals of various grassland species grown together for one year in laboratory microcosms. The statistical significance of changes in weight associated with infection are indicated as follows: *, $p < 0.05$; **, $p < 0.01$; ***, $p < 0.001$. Key to species: *Ao, Anthoxanthum odoratum; Bm, Briza media; Cr, Campanula rotundifolia; Cn, Centaurea nigra; Dg, Dactylis glomerata; Fo, Festuca ovina; Fr, Festuca rubra; Gv, Galium verum; Hp, Hieracium pilosella; Lh, Leontodon hispidus; Pl, Plantago lanceolata; Pp, Poa pratensis; Ra, Rumex acetosa; Sm, Sanguisorba minor; Sc, Scabiosa columbaria; Sn, Silene nutans.*

beyond the study of nutrient uptake by infected and control plants grown in isolation and has begun to inform theories of plant community structure and ecosystem function. An experiment in which plant communities were synthesised in the laboratory in the presence and absence of vesicular–arbuscular mycorrhizal fungi (Grime *et al.*, 1987) revealed (Fig. 6) that mycorrhizas can increase diversity markedly by raising the biomass of the subordinate species relative to that of the canopy dominant (in this case, *Festuca ovina*). It is established that VA mycorrhizal fungi usually develop an extensive network through which transport of mineral nutrients and assimilate between connected individuals has been demonstrated (Francis, Finlay & Read, 1986; Grime *et al.*, 1987). This experimental evidence suggests that transfer of mineral nutrients and assimilate through mycorrhizas may be one of the factors which reduce the intensity of competition and encourage species co-existence on infertile soils.

Conclusions

During the last 20 years, mineral nutritional studies have receded in importance in both agricultural research and in plant ecology. A revival of interest now seems likely because: (1) leakage of nutrients from intensively-fertilised farmland is prompting the research for more conservative technologies; and (2) mineral nutrients, especially phosphorus and nitrogen, are emerging as indirect but potent controllers of global and local variation in the quantity and quality of phytomass. Mineral nutrients frequently control the rates of succession and cyclical change in vegetation and they appear to provide the most influential currency of ecosystem function.

Future research into mineral nutrition will benefit from an evolutionary perspective which recognises the existence of continuing edaphic specialisation within species and populations. However, a coherent, predictive data base will depend also upon recognition of constitutive, ancestral specialisations manifested at the level of the species or even higher taxonomic units.

Ecological studies of mineral nutrition are increasingly convergent in design with those used to investigate foot capture by animals. These include 'patch' exploitation by dynamic root systems in fertile soils and 'pulse' interception by long-lived roots in infertile soils.

A relationship has been detected between root behaviour and the status achieved by species in an experimental plant community; this suggests a trade-off between the *scale* of root foraging (total volume of nutrient-rich soil explored) and its *precision* (proportion of root located in nutrient-rich soil).

Mycorrhizas are an increasingly indispensable component of ecological studies of mineral nutrition. Experiments involving the synthesis of species-rich turf in microcosms suggest that transfer of resources between species through underground hyphal connections may exert a modifying effect on plant community structure.

References

Antonovics, J. (1968). Evolution in closely adjacent plant populations. V. Evolution of self-fertility. *Heredity*, **23**, 219–38.

Balme, O.E. (1953). Edaphic and vegetational zoning on the carboniferous limestone of the Derbyshire Dales. *Journal of Ecology*, **41**, 331–44.

Band, S.R. & Grime, J.P. (1981). *Chemical composition of leaves*. Annual Report 1981, pp. 6–8. University of Sheffield: Unit of Comparative Plant Ecology (NERC).

Bhat, K. & Nye, P.H. (1973). Diffusion of phosphate to plant roots in soil. I. Quantitative autoradiography of the depletion zone. *Plant and Soil*, **38**, 161–75.

Bradshaw, A.D. (1959). Population differentiation in *Agrostis tenuis*. 3. Populations in varied environments. *New Phytologist*, **59**, 92–103.

Bradshaw, A.D. (1975). The evolution of heavy metal tolerance and its significance for vegetation establishment on metal contaminated sites. In *Heavy Metals in the Environment*, ed. T.C. Hutchinson, pp. 599–622. Toronto: Toronto University Press.

Bradshaw, A.D., McNeilly, T.S. & Gregory, R.P.G. (1965). Industrialization, evolution and the development of heavy metal tolerance in plants. In *Ecology and the Industrial Society*, eds. G.T. Goodman, R.W. Edwards & J.M. Lambert, pp. 327–44. New York: Wiley.

Campbell, B.D. & Grime, J.P. (1989*a*). A new method of exposing developing root systems to controlled patchiness in mineral nutrient supply. *Annals of Botany*, **63**, 395–400.

Campbell, B.D. & Grime, J.P. (1989*b*). A comparative study of plant responsiveness to the duration of episodes of mineral nutrient enrichment. *New Phytologist*, **112**, 261–7.

Clapham, A.R. (1956). Autecological studies and the Biological Flora of the British isles. *Journal of Ecology*, **44**, 1–11.

Coley, P.D., Bryant, J.P. & Chapin, F.S. (1985). Resource availability and plant antiherbivore defence. *Science*, **230**, 895–9.

Cox, R.M. & Hutchinson, T.C. (1980). Multiple metal tolerances in the grass *Deschampsia cespitosa* (L.) from the Sudbury smelting area. *New Phytologist*, **84**, 631–47.

Crick, J.C. & Grime, J.P. (1987). Morphological plasticity and mineral nutrient capture in two herbaceous species of contrasted ecology. *New Phytologist*, **107**, 403–14.

Drew, M.C. (1975). Comparison of the effects of a localized supply of phosphate, nitrate, ammonium and potassium on the growth of the seminal root system, and the shoot, in barley. *New Phytologist*, **75**, 479–90.

Drew, M.C. & Saker, L.R. (1975). Nutrient supply and the growth of the seminal root system in barley. II. Localized, compensatory increases in lateral root growth and rates of nitrate uptake when nitrate supply is restricted to only part of the root system. *Journal of Experimental Botany*, **26**, 79–90.

Field, C. & Mooney, H.A. (1986). The photosynthesis-nitrogen relationship in wild plants. In *On the Economy of Plant Form and Function*, ed. T.J. Givnish, pp. 25–55. Cambridge University Press.

Francis, R., Finlay, R.D. & Read, D.W. (1986). Vesicular–arbuscular mycorrhiza in natural vegetation systems IV Transfer of nutrients in inter- and intra-specific combinations of host plants. *New Phytologist*, **102**, 103–11.

Gauch, H.L. (1957). Mineral nutrition of plants. *Annual Review of Plant Physiology*, **8**, 31–63.

Grime, J.P. (1974). Vegetation classification by reference to strategies. *Nature*, **250**, 25–31.

Grime, J.P. (1977). Evidence for the existence of thre primary strategies in plants and its relevance to ecological and evolutionary theory. *American Naturalist*, **111**, 1169–94.

Grime, J.P. (1979). *Plant Strategies and Vegetation Processes. Chichester: John Wiley.*

Grime, J.P. (1984). The ecology of species, families and communities of the contemporary British flora. *New Phytologist*, **98**, 15–33.

Grime, J.P. (1985). Factors limiting the contribution of pteridophytes to a local flora. In *The Biology of Pteridophytes*, eds. A.F. Dyer & C.N. Page. Proceedings of the Royal Society of Edinburgh, **86B**, 403–21.

Grime, J.P. (1987). Dominant and subordinate components of plant communities: implications for succession, stability and diversity. In *Colonization, Succession and Stability*, eds. A.J. Gray, M.J. Crawley & P.J. Edwards, pp. 413–428. Oxford: Blackwell Scientific Publications.

Grime, J.P., Campbell, B.D., Mackey, J.M.L. & Crick, J.C. (1989). Root plasticity, nitrogen capture and competitive ability. In *Plant Root Systems: their Effect on Ecosystem Composition and Structure*, ed. D. Atkinson, BES Symposium. Oxford: Blackwell Scientific Publications (in press).

Grime, J.P. & Hodgson, J.G. (1969). An investigation of the ecological significance of lime-chlorosis by means of large-scale comparative experiments. In *Ecological Aspects of the Mineral Nutrition of Plants*, ed. I.H. Rorison, BES Symposium No. 9, pp. 67–99. Oxford: Blackwell Scientific Publications.

Grime, J.P. & Hunt, R. (1975). Relative growth rate: its range and adaptive significance in a local flora. *Journal of Ecology*, **63**, 393–422.

Grime, J.P. & Lloyd, P. (1973). *An Ecological Atlas of Grassland Plants*. London: Edward Arnold.

Grime, J.P., Hodgson, J.G. & Hunt, R. (1988). *Comparative Plant Ecology. A Functional Approach to Common British Species*. London: Unwin Hyman.

Grime, J.P., Mackey, J.M.L., Hillier S.H. & Read, D.J. (1987). Floristic diversity in a model system using experimental microcosms. *Nature*, **328**, 420–2.

Gupta, P.J. & Rorison, I.H. (1975). Seasonal differences in the availability of nutrients down a podzolic profile. *Journal of Ecology*, **63**, 521–34.

Hansen, K. & Jensen, J. (1974). Edaphic conditions and plant–soil relationships on roadsides in Denmark. *Danskk Botanisk Arkiv*, **28(3)**, 1–143.

Harper, J.L. (1977). *Population Biology of Plants*. London: Academic Press.

Harper, J.L. (1982). After description. In *The Plant Community as a Working Mechanism*, ed. E.I. Newman, Special Publication No. 1 BES, pp. 11–25. Oxford: Blackwell Scientific Publications.

Hewitt, E.J. (1951). The role of the mineral elements in plant nutrition. *Annual Review of Plant Physiology*, **2**, 25.

Hoagland, D.R. (1919). Relation of the concentration and reaction of the nutrient medium to the growth and absorption of the plant. *Journal of Agricultural Research*, **18**, 73–117.

Hodgson, J.P. (1986). Commonness and rarity in plants with special reference to the Sheffield flora. II. The relative importance of climate, soils and land use. *Biological Conservation*, **36**, 253–74.

Hundt, R. (1966). *Okologisch-geobotanische Untersuchungen an Pflanzender Mitteleuropaischen Wiesenvegetation*. Jena: Fisher.

Jain, S.K. & Bradshaw, A.D. (1966). Evolutionary divergence among adjacent plant populations. I. The evidence and its theoretical analysis. *Heredity*, **23**, 407–41.

Jowett, D. (1958). Populations of *Agrostis* spp. tolerant of heavy metals. *Nature*, **182**, 816–17.

Liebig, J. (1840). *Chemistry and its Application to Agriculture and Physiology*. London: Taylor and Walton.

Loveless, A.R. (1961). A nutritional interpretation of sclerophylly based on differences in chemical composition of sclerophyllous and mesophytic leaves. *Annals of Botany of Nova Scotia*, **25**, 168–76.

Lundegardh, H. (1931). *Environment and Plant Development*. London: Edward Arnold.

McNaughton, S.J., Folsom, T.C., Lee, T., Park, F., Price, C., Roeder, D., Schmitz, J. & Stockwell, C. (1974). Heavy metal tolerance in *Typha latifolia* without the evolution of tolerant races. *Ecology*, **55**, 1163–5.

Pearsall, W.H. (1922). Plant distribution and basic ratios. *Naturalist* (*London*), p. 269.

Reeves, R.D. & Baker, A.J.M. (1984). Studies on metal uptake by plants from serpentine and non-serpentine populations of *Thlaspi geosingense* Halacsy (Cruciferae). *New Phytologist*, **98**, 191–204.

Rorison, I.H. (ed.) (1969). *Ecological Aspects of the Mineral Nutrition of Plants*, BES Symposium No. 9. Oxford: Blackwell Scientific Publications.

Salisbury, E.J. (1942). *The Reproductive Capacity of Plants*. London: Bell.

Snaydon, R.W. & Bradshaw, A.D. (1961). Differential response to calcium within the species *Festuca ovina*. *New Phytologist*, **60**, 219–34.

Steele, D. (1955). Soil pH and base status as factors in the distribution of calcicoles. *Journal of Ecology*, **43**, 120–32.

Tamm, C.O. (1956). Further observations on the survival and flowering of some perennial herbs I. *Oikos*, **7**, 274–92.

Taylor, A.A., De-Felice, J. & Havill, D.C. (1982). Seasonal variation in nitrogen availability and utilization in an acidic and calcareous soil. *New Phytologist*, **92**, 141–52.

Tansley, A.G. (1917). On the competition between *Galium saxatile* and *Galium sylvestre* (*G. asperum*) on different types of soil. *Journal of Ecology*, **5**, 173.

Tilman, D. (1982). *Resource Competition and Community Structure*. Princeton: Princeton University Press.

Tilman, D. (1987). On the meaning of competition and the mechanisms of competitive superiority. *Functional Ecology*, **1**, 304–15.

Watt, A.S. (1947). Pattern and process in the plant community. *Journal of Ecology*, **35**, 1–22.

Zarzycki, K. (1976). Ecodiagrams of common vascular plants in the Pieniny mountains (the Polish West Carpathians). Part 1. Ecodiagrams of selected grassland species. *Fragmenta Floristica et Geobotanica*, **22**, 500–28.

INDEX

Page references in *italics* refer to figures or tables.